HYDROLOGY & WATER RESOURCES ENGINEERING

AUTHOR (S)

Dr. M. L. KANSAL
Professor (HAG)
Department of Water Resources Development & Management,
Indian Institute of Technology Roorkee,
ROORKEE 247 667

P. K. Agarwal
Former Scientist B
National Institute of Hydrology Roorkee,
ROORKEE 247 667

REVIEWER

Dr. Manish Kumar Goyal
Professor
Indian Institute of Technology Indore, INDORE 453552

KHANNA BOOK PUBLISHING CO. (P) Ltd.
4C/4344, Ansari Road, Darya Ganj, New Delhi, India

All India Council for Technical Education
Nelson Mandela Marg, Vasant Kunj, New Delhi, India

ISBN: 978-93-5538-555-0

Book Code: UG480EN

Hydrology & Water Resources Engineering

by Dr. M. L. Kansal and P. K. Agarwal

First Edition: 2026

Originally Published by:
All India Council for Technical Education, Ministry of Education, GOI

Exclusively Published & Distributed by:
Khanna Book Publishing Co. (P) Ltd.
4C/4344, Ansari Road, Daryaganj,
New Delhi (INDIA)

Visit us at: **www.khannabooks.com**
Write us at: **contact@khannabooks.com**

To view complete list of books, Please scan the QR Code:

Printed in Bharat.

प्रो. टी. जी. सीताराम
अध्यक्ष
Prof. T. G. Sitharam
Chairman

सत्यमेव जयते

अखिल भारतीय तकनीकी शिक्षा परिषद्
(भारत सरकार का एक सांविधिक निकाय)
(शिक्षा मंत्रालय, भारत सरकार)
नेल्सन मंडेला मार्ग, वसंत कुंज, नई दिल्ली–110070
दूरभाष : 011–26131498
ई–मेल : chairman@aicte-india.org

ALL INDIA COUNCIL FOR TECHNICAL EDUCATION
(A STATUTORY BODY OF THE GOVT. OF INDIA)
(Ministry of Education, Govt. of India)
Nelson Mandela Marg, Vasant Kunj, New Delhi-110070
Phone : 011-26131498
E-mail : chairman@aicte-india.org

FOREWORD

Engineers are the backbone of any modern society. They are the ones responsible for the marvels as well as the improved quality of life across the world. Engineers have driven humanity towards greater heights in a more evolved and unprecedented manner.

The All India Council for Technical Education (AICTE), have spared no efforts towards the strengthening of the technical education in the country. AICTE is always committed towards promoting quality Technical Education to make India a modern developed nation emphasizing on the overall welfare of mankind.

An array of initiatives has been taken by AICTE in last decade which have been accelerated now by the National Education Policy (NEP) 2020. The implementation of NEP under the visionary leadership of Hon'ble Prime Minister of India envisages the provision for education in regional languages to all, thereby ensuring that every graduate becomes competent enough and is in a position to contribute towards the national growth and development through innovation & entrepreneurship.

One of the spheres where AICTE had been relentlessly working since past couple of years is providing high quality original technical contents at Under Graduate & Diploma level prepared and translated by eminent educators in various Indian languages to its aspirants. For students pursuing 2nd year of their Engineering education, AICTE has identified 87 books, which shall be translated into 12 Indian languages - Hindi, Tamil, Gujarati, Odia, Bengali, Kannada, Urdu, Punjabi, Telugu, Marathi, Assamese & Malayalam. In addition to the English medium, books in different Indian Languages are going to support the students to understand the concepts in their respective mother tongue.

On behalf of AICTE, I express sincere gratitude to all distinguished authors, reviewers and translators from the renowned institutions of high repute for their admirable contribution in a record span of time.

AICTE is confident that these outcomes based original contents shall help aspirants to master the subject with comprehension and greater ease.

(Prof. T. G. Sitharam)

ACKNOWLEDGEMENT

We are grateful to the authorities of AICTE, particularly Prof. T. G. Sitharam, Chairman; Dr. Abhay Jere, Vice-Chairman; Prof. Rajiv Kumar, Member-Secretary, Dr. Sunil Luthra, Director, and Reena Sharma, Hindi Officer Training and Learning Bureau for their planning to publish the books on ***Hydrology & Water Resources Engineering***.

We sincerely acknowledge the valuable contributions of the reviewer Prof. (Dr.) Manish Kumar Goyal, Indian Institute of Technology, Indore.

We would also like to acknowledge the reference books that have been instrumental in the preparation of this book: "Engineering Hydrology" by K Subramanya; "Hydrology & Water Resources Engineering" by S K Garg; "Hydrology and Water Resources of India" by S K Jain, P K Agarwal, and V P Singh; "Irrigation and Water Resources Engineering " by G L Asawa; "Irrigation and Water Power Engineering" by B C Punmia and Pande B Lal.

We are also thankful to our family, as without their support, this work would not have come to fruition. Additionally, we extend our gratitude to our students and colleagues, over the years, who have made significant contributions that have enriched our experience and expertise in this subject. We are immensely grateful for their valuable input for making this book students' friendly and giving a better shape in an artistic manner. The authors are thankful to the Ph.D. students Ms. Neenu, Ms. Alka, Dr. Sacchidanand Singh, and Mr. Lingaraj Dhal at WRDM and to Ms. Pallavi and Saumya Agarwal for helping in preparing the manuscript of the book. The authors are also thankful to Dr. Sudhasil Bose, Research Associate for carrying out the proof reading.

This book is an outcome of various suggestions of AICTE members, experts and authors who shared their opinion and thought to further develop engineering education in our country. Acknowledgements are due to the contributors and different workers in this field whose published books, review articles, papers, photographs, footnotes, references, and other valuable information enriched us at the time of writing the book.

Dr. M L Kansal

P K Agarwal

PREFACE

Water is the necessity for the survival of life. It plays an important role in meeting with the several sustainable development goals. The basic concepts of hydrology and its input towards the water resources development and management are key to understand its role towards the societal sustainable development. In the civil engineering curriculum, hydrology and water resources engineering occupies an important position.

This book includes the topics recommended by AICTE, in a very systematic and orderly manner and serves as a comprehensive guide to understand the hydrological principles, concepts, and applications in the water resources engineering. Whether you are a student embarking on a journey of learning or a professional seeking to refresh your knowledge, this book is designed to provide you with a solid foundation in this fascinating field.

In this book, we will embark on a journey that will take us through the complexities of surface and sub-surface water resources engineering. Throughout the modules, we will emphasize a hands-on approach to learning. We will provide practical examples, step-by-step explanations, and opportunities for you to apply your knowledge through exercises and projects. We will not only focus on the theoretical aspects of hydrology and water resources engineering but also explore its practical applications. As you progress through this book, we encourage you to actively engage with the material, ask questions, and seek deeper understanding. Sometimes, it may be challenging, but with perseverance and practice, you will gain the necessary skills to analyse, design, and troubleshoot with confidence.

We would like to express our gratitude to the countless researchers, educators, and engineers who have contributed to the field of Hydrology and Water Resources Engineering over the years. Their collective efforts have paved the way for the advancements we witness today. We hope that this book will serve as a tribute to their contributions and inspire you to further explore the world of water resources engineering.

We sincerely hope that this book will be a valuable resource in your journey of learning and discovery.

Happy reading! Happy Learning!

Dr. M L Kansal

P K Agarwal

OUTCOME BASED EDUCATION

For the implementation of an outcome-based education the first requirement is to develop an outcome-based curriculum and incorporate an outcome-based assessment in the education system. By going through outcome-based assessments, evaluators will be able to evaluate whether the students have achieved the outlined standard, as well as specific and measurable outcomes. With the proper incorporation of outcome-based education, there will be a definite commitment to achieve a minimum standard for all learners without giving up at any level. At the end of the program running with the aid of outcome-based education, a student will be able to arrive at the following outcomes:

PO1. Engineering knowledge: Apply the knowledge of mathematics, science, engineering fundamentals, and an engineering specialization to solve complex engineering problems.

PO2. Problem analysis: Identify, formulate, review research literature, and analyze complex engineering problems, reaching substantiated conclusions using the first principles of mathematics, natural sciences, and engineering sciences.

PO3. Design / development of solutions: Design solutions for complex engineering problems and design system components or processes that meet the specified needs with appropriate public health and safety and cultural, societal, and environmental considerations.

PO4. Conduct investigations of complex problems: Use research-based knowledge and research methods, including design of experiments, analysis and interpretation of data, and synthesis of the information to provide valid conclusions.

PO5. Modern tool usage: Create, select, and apply appropriate techniques, resources, and modern engineering and IT tools, including prediction and modelling to complex engineering activities with an understanding of the limitations.

PO6. The engineer and society: Apply reasoning informed by the contextual knowledge to assess societal, health, safety, legal and cultural issues, and the consequent responsibilities relevant to the professional engineering practice.

PO7. Environment and sustainability: Understand the impact of professional engineering solutions in societal and environmental contexts and demonstrate the knowledge of and need for sustainable development.

PO8. Ethics: Apply ethical principles and commit to professional ethics, responsibilities, and norms of engineering practice.

PO9. Individual and team work: Function effectively as an individual and as a member or leader in diverse teams, as well as in multidisciplinary settings.

PO10. Communication: Communicate effectively on complex engineering activities with the engineering community and with society at large, such as being able to comprehend and write effective reports and design documentation, make effective presentations, and give and receive clear instructions.

PO11. Project management and finance: Demonstrate knowledge and understanding of engineering and management principles and apply these to one's own work as a member and leader in a team, as well as to manage projects and multidisciplinary environments.

PO12. Life-long learning: Recognize the need for and have the preparation and ability to engage in independent and life-long learning in the broadest context of technological change.

COURSE OUTCOMES

After completion of the course, the students will be able to:

CO-1: Understand the fundamentals of the hydrology cycle, water-budget and water balance, applications in engineering and socially relevant problems.

CO-2: Understand the different forms of precipitation, mean precipitation over an area, Depth-area-duration relationship, maximum intensity/ depth-duration-frequency relationship, and Probable Maximum Precipitation to design various hydrological problems.

CO-3: Learn about the abstractions from precipitation through various processes like evaporation, evapotranspiration, interception, depression storages, and infiltration. Evaporation from reservoirs, potential and actual evapotranspiration, measurement of infiltration, and infiltration indices.

CO-4: Learn about the Rainfall-runoff process and modelling, flow duration, flow-mass curves, hydrograph, base flow separation, unit hydrograph, factors affecting unit hydrograph, flood hydrograph, and environmental flow.

CO-5: Understand the groundwater and well hydrology, aquifer properties, equilibrium equations for confined and unconfined aquifers, and various aquifer tests.

CO-6: Learn about various water usage methods, such as water for energy production, agriculture, flood control, water supply, etc. Water requirements for crops, quality of irrigation water, soil-water relationship, methods of irrigation, etc.

CO-7: Learn about the canals as a system for water distribution, alignment of canals, canal losses, design of canals, Theories of regime channels, canal outlets, water logging and remedial measures like canal lining drainage of irrigation areas.

CO-8: Learn about the dams and spillways, types of dams like gravity, embankment, and arch dams, control of seepage, components of the spillway, types of reservoirs, reservoir regulations, sedimentation, selection of suitable sites for dams.

Course Outcomes	Expected Mapping with Programme Outcomes *(1- Weak Correlation; 2- Medium correlation; 3- Strong Correlation)*											
	PO-1	**PO-2**	**PO-3**	**PO- 4**	**PO-5**	**PO-6**	**PO-7**	**PO-8**	**PO-9**	**PO-10**	**PO-11**	**PO-12**
CO-1	3	3	-	-	-	-	-	-	-	-	-	1
CO-2	3	3	3	3	3	2	2	-	-	2	-	-
CO-3	3	3	3	3	-	2	2	-	-	-	-	1
CO-4	3	3	3	3	3	2	2	-	-	2	-	-
CO-5	3	3	3	3	3	3	2	-	-	-	-	1
CO-6	3	3	3	3	3	3	2	-	-	2	1	3
CO-7	3	3	3	3	3	3	2	-	-	2	1	3
CO-8	3	3	3	3	3	3	2	-	-	2	1	3

GUIDELINES FOR TEACHERS

To implement outcome-based education (OBE), the knowledge level and skill set of the students should be enhanced. Teachers should take a major responsibility for the proper implementation of OBE. Some of the responsibilities (not limited to) for the teachers in the OBE system may be as follows:

- Within reasonable constraints, they should manoeuvre time to the best advantage of all students.
- They should assess the students only upon certain defined criteria without considering any other potential ineligibility to discriminate against them.
- They should try to grow the learning abilities of the students to a certain level before they leave the institute.
- They should try to ensure that all the students are equipped with quality knowledge as well as competence after they finish their education.
- They should always encourage the students to develop their ultimate performance capabilities.
- They should facilitate and encourage group work and team work to consolidate newer approaches.
- They should follow Bloom's taxonomy in every part of the assessment.

Bloom's Taxonomy

Level	Teacher should Check	Students should be able to	Possible Mode of Assessment
Create	Students' ability to create	Design or create	Mini project
Evaluate	Students' ability to justify	Argue or defend	Assignment
Analyze	Students' ability to distinguish	Differentiate or distinguish	Project/ Lab. methodology
Apply	Students' ability to use information	Operate or demonstrate	Technical presentation/ Demonstration
Understand	Students' ability to explain the ideas	Explain or classify	Presentation/ Seminar
Remember	Students' ability to recall (or remember)	Define or recall	Quiz

GUIDELINES FOR STUDENTS

Students should take equal responsibility for implementing the OBE. Some of the responsibilities (not limited to) for the students in OBE system are as follows:

- Students should be aware of each UO before the start of a unit in each and every course.
- Students should be aware of each CO before the start of the course.
- Students should be aware of each PO before the start of the programme.
- Students should think critically and reasonably with proper reflection and action.
- Learning of the students should be connected and integrated with practical and real-life consequences.
- Students should be aware of their competency at every level of OBE.

ABBREVIATIONS AND SYMBOLS

Abbreviation / Symbol	Description
A	Average surface area of the lake/reservoir
A_b	Abstraction for irrigation, water supply and industrial use, inclusive of evaporation losses
AASW	Actual available soil-water at time t over the root depth
AD	Anno Domini
AET	Actual evapotranspiration
AM	Before midday
AMC	Antecedent moisture condition
B	Drainage Factor
BC	Before Christ
BWR	Basic water requirement
c_1	Pressure coefficient
°C	Celsius
C	Constant
C	Hydraulic Resistance
C	Unit cohesion
C_p	Pan coefficient
C_t	A regional constant representing watershed slope and storage
C_v	Coefficient of variation of the rainfall values (in %)
CCA	Cultivable Command Area
CM	Centimeter
CN	Curve Number
CSP	Concentrating solar power
CVR	Critical velocity ratio
d	Root zone depth

Abbreviation / Symbol	Description
D	Duty of water
D	Hydraulic diffusivity
D	Initial artesian pressure at the bottom of the aquifer
D_a	Volume of surface retention storage
e_a	Actual vapour pressure in the air
e_w	Saturation vapour pressure at the water temperature
E	Net evaporation losses
E_a	Parameter comprising wind velocity and saturation deficit
E_{act}	Actual evapotranspiration
E_{pm}	Pan evaporation loss
E_L	Rate of evaporation
E_x	Net export of water from the basin
EFs	Environmental flows
ELOHA	Ecological limits of hydrological abstractions
ENSO	El Niño-Southern Oscillation
ET	evapotranspiration
ET_0	Reference Crop Evapotranspiration
f	Actual rate of infiltration
°F	Fahrenheit
F	A sum of monthly consumption uses factors
FC	Field capacity
FDCs	Flow-duration curves
f(u)	Correction function for wind speed
f_p	Infiltration capacity
g	Acceleration due to gravity
G	Groundwater flow
G_o	Sub-surface outflow

Abbreviation / Symbol	Description
GCA	Gross Command Area
GIS	Geographic Information System
GW	Groundwater
h	Hydraulic head causing the flow/ Depth of reservoir.
h_w	Height of the wave
h_w	Artesian pressure in the well
H	Total height of confined aquifer
H_a	Sensible heat transfers from water's surface to the atmosphere
H_b	Long-wave back radiation from a water body.
H_e	Heat energy lost during evaporation.
H_g	Heat flux into the earth
H_i	Advected energy
H_L	Difference of water levels in the two observation wells
H_n	Net thermal energy derived from the surface of the water
H_s	heat retained in a water body
ΣH	Driving shear forces/ resultant horizontal forces
HEC-HMS	Hydrologic Engineering Centre-Hydrologic Modelling System
HSG	Hydrologic Soil Group
i	Intensity of rainfall
I	Inflow
$\bar{I}$	Mean inflow into the lake/reservoir for the time interval Δt
I_a	Initial losses
I_i	Interception loss
in	inches
IOD	Indian Ocean Dipole
IDF	Intensity-duration-frequency curves

Abbreviation / Symbol	Description
IMD	India Meteorological Department
IS	Indian standards
k	Coefficient of permeability/ Hydraulic conductivity
K	Coefficient for a given region
K_h	Horton's decay coefficient
K_i	Ratio of vegetal surface area to its projected area
K_M	Meyer's coefficient (0.36 for deep and 0.50 for shallow water)
K_r	Recession constant dependent upon the time and less than unity
K_x	Coefficients of permeability in the x direction
K_y	Coefficients of permeability in the y direction
km^3	Cubic kilometer
L	Abstraction losses
L	Latent heat of evaporation
L	Leakage Factor of a leaky aquifer
L_{ca}	Distance from the study point to the basin centroid
LU/LC	Land Use and Land Cover
m	meter
m	rank order for the probability of exceedance
m^2	Square meter
m^3/s	Cumec
M_a	Original slope of the mass curve
M_c	Corrected slope of the double-mass curve
MASW	Total available soil water content above the root depth
M km^2	Million square kilometers
mm	Millimeter
M m^2	Million square meters
n	Actual duration of bright sunshine hours

Abbreviation / Symbol	Description
n	Rugosity coefficient
n	Total number of flow events
N	Optimal number of Rain gauge stations
N	Time in days from the hydrograph peak
NRCS	Natural Resources Conservation Service
O	Outflow
$\overline{O}$	Mean outflow from the lake/reservoir for the time interval Δt
p_a	Average barometric pressure
P	Exceedance probability
P	Precipitation
$\overline{P}$	Mean precipitation over the catchment area
P_{cx}	Corrected precipitation at station X for any time period t
P_e	Hydrodynamic earthquake pressure normal to the face
P_o	Maximum value of rainfall in cm at the center of storm
$\sum P_{av}$	Accumulated precipitation of the average of the base station
$\sum P_x$	Cumulative precipitation
PET	Potential evapotranspiration
PWP	Permanent wilting point
Q	Constant at steady conditions
Q_o	Discharge at start of time period
Q	Design discharge/ Volume of Surface Runoff/Pumping rate of groundwater
$\bar{Q}$	Mean outflow rate from the lake/reservoir
Q_{ps}	Peak discharge of a unit hydrograph
Q_s	Annual siltation rate
Q_t	Discharge at end of time t
r	Reflection coefficient

Abbreviation / Symbol	**Description**
r	Distance from the pumping well to the observation well
r_w	Radius of the well
R	Runoff
R_n	Net Radiance flux
R_N	Natural runoff volume in time Δt
R_o	Observed runoff volume in time Δt at the terminal site
R_s	Surface runoff
RS	Remote sensing
S	Channel's slope
S	Coefficient of Storage or Storability
S	Distance between two observation wells
S	Potential maximum retention
S	Sedimentation rate
ΔS	Change in storage
s_a	drawdown in the aquifer at the effective radius of the pumping well
S_i	Interception storage
S_g	Water storage as Groundwater
S_r	Specific retention
S_s	Surface water storage/Specific storage
S_w	Storage coefficient for a water table aquifer/well loss
S_y	Specific yield of the aquifer
SCS	Soil conservation service
SCS-CN	Soil conservation service-curve number
SW	Surface water
SWAT	Soil and Water Assessment Tool
t	Base of natural logarithm
T	Transpiration

Abbreviation / Symbol	Description
T	Transmissivity
t_0	Time at the point where the straight line intersects the zero-drawdown line
T_a	Air temperature in degrees Celsius
t_c	Time at which steady-shape conditions develop
t_e	Rainfall excess duration
t_p	Duration of peak hydrograph from midpoint of hourly rainfall
T_w	Water surface temperature in degrees Celsius
T_L	Transpiration loss
Δt	Time interval
$\bar{X}$	Average precipitation
X_i	Magnitude of precipitation in the ith station
u	Velocity component in the x direction
u_9	Monthly average velocity at about 9 meter above the earth
U	Total Consumptive use
UH	Unit Hydrograph method
USGS	United States Geological Survey
v	Velocity component in the y direction
V	Mean velocity of flow
V_0	total volume of aquifer
V_d	Volume diverted flow of the stream for irrigation, domestic water supply and industrial use
V_D	The demanded volume of water
V_E	Water lost in evaporation
V_o	Critical velocity
V_r	Volume of return flow from irrigation, domestic water supply and industrial use
V_S	Supplied volume of water
V_{ig}	Groundwater inflow

Abbreviation / Symbol	Description
V_{is}	Surface inflow into the lake
V_{og}	Seepage outflow
V_{os}	Surface outflow from the lake
V_v	Volume of voids
W	Weight of the structure
W	Mean infiltration rate
W_m	Soil moisture difference content at the root zone at the beginning and end of plant growth.
W_t	Irrigation Water Supply
WG	Working Group
ΣW	Total vertical forces acting on the plane
y	Vertical distance from the reservoir surface to the elevation under consideration,
α_h	Horizontal acceleration factor
δe	Vapor pressure deficit
σ	Stefan-Boltzmann coefficient = 2.01×10^9 mm/day
σ_{m-1}	Standard deviation
ρ	Water density
ρ_d	Bulk density
ρ_m	Density of mineral particles
μ	Internal friction coefficient
μ	Co-efficient of viscosity
Σ	Degree of error to determine the estimate of mean rainfall
ϵ_{ex}	Predicted error (%)) in the estimation of average precipitation
γ	psychrometric constant = 0.49 mm of mercury/°C
γ	Soil density
γ	Specific weight of the fluid

Abbreviation / Symbol	Description
Δ	Delta
λ	Constant ranging from 0.1 to 0.4

LIST OF FIGURES

Fig. 1.1 Catchment area of a river 5
Fig. 1.2: Hydrological Cycle 6
Fig. 1.3: Horton's depiction of the water cycle 7
Fig. 2.1: Dates on which the nationwide monsoon begins 44
Fig. 2.2: Normal dates of withdrawal of monsoon 45
Fig. 2.3: Average annual rainfall over India 47
Fig. 2.4: Non-recording rain gauge (syphon's rain gauge) 49
Fig. 2.5: Estimation of missing rainfall data 56
Fig. 2.6: Double-mass curve 58
Fig. 2.7: Double mass curve of yearly rainfall at station M 61
Fig. 2.8: Mass Curve of rainfall 62
Fig. 2.9: Hyetograph of a storm 62
Fig. 2.10: Thiessen polygon 64
Fig. 2.11: Thiessen polygons 65
Fig. 2.12: Isohyets of a storm 67
Fig. 2.13: Typical DAD curve 70
Fig. 2.14: Streamflow distribution 71
Fig. 2.15: Depth-Area-Duration frequency curves 72
Fig. 2.16: Maximum Depth-Duration Curve 74
Fig. 2.17: Maximum Depth-Duration-Frequency Curves 75
Fig. 3.1:U.S. Class A Evaporation Pan 96
Fig. 3.2: ISI Evaporation Pan 96
Fig. 3.3: Colorado Sunken Pan Evaporimeter 97
Fig. 3.4: Energy Balance in a Water Body 104
Fig. 3.5: Variation of AET 113
Fig. 3.6 (a): Annual PET (cm) over India, (b): Monthly Variation of PET (mm) 122
Fig. 3.7: Typical Interception Loss Curve 127
Fig. 3.8: Infiltration Capacity Variation 131

Fig. 3.9: Section view of double-ring infiltrometer. 134
Fig. 3.10: Rainfall simulator (Source: I. S. Mcqueen, 1963) 135
Fig. 3.11: Infiltration Capacity and Cumulative Infiltration Capacity Curves 136
Fig. 3.12: Φ-index. (Source: Subramanya, 2006) 140
Fig. 4.1: Different routes of runoff 155
Fig. 4.2: Perennial Stream 159
Fig. 4.3 (a). Fan-shaped catchment (b) Leaf-shaped catchment 162
Fig. 4.4: Influent stream 166
Fig. 4.5: Hydrograph of an ephemeral stream 166
Fig. 4.6: Effluent stream 167
Fig. 4.7: Effect of catchment shape on the hydrograph 168
Fig. 4.8: Role of drainage density on the hydrograph 169
Fig. 4.9: Base flow separation based upon area 173
Fig. 4.10: Three-component hydrograph separation 175
Fig. 4.11: Determination of effective rainfall from rainfall-runoff 176
Fig. 4.12: SCS-Runoff Curve Number Method 191
Fig. 4.13: A typical Flow Duration Curve 193
Fig. 4.14: A typical Flow-Mass Curve 197
Fig. 5.1: Classification and distribution of subsurface water. 226
Fig. 5.2: Type of Aquifers. 229
Fig. 5.3: Pore space in various geologic formations. 230
Fig. 5.4: Unconfined aquifer case for derivation of Thiem's formula. 241
Fig. 5.5: Drawdown curve in a confined aquifer. 243
Fig. 5.6: Unconfined aquifer cases for derivation of Dupuit's formula. 244
Fig. 5.7: Confined aquifer case for derivation of Dupuit's formula. 245
Fig. 5.8: Flow lines in a well fully and partially penetrating the aquifer. 247
Fig. 5.9: Spherical flow in a well. 248
Fig. 5.10: Well loss 252
Fig. 5.11: Giving values of C. 252
Fig. 5.12: Yield-drawdown curve. 253

Fig. 5.13: Map of the aquifer test site. 257
Fig. 5.14: Change of water level in Well B. 258
Fig. 5.15 Theis type curve. 260
Fig. 5.16: Matching of type curve. 261
Fig. 5.17: Depression cone during pumping from a confined aquifer. 263
Fig. 5.18: Steady and unsteady state of depression cone during pumping from a confined aquifer. 264
Fig. 5.19: Steady state condition during pumping from a confined aquifer. 264
Fig. 5.20: Time drawdown graph. 265
Fig. 5.21: Cluster of pumping and observation wells in a confined aquifer. 266
Fig. 5.22: Distance-drawdown graph. 267
Fig. 5.23: Schematic diagram of pumping test of a confined aquifer. 269
Fig. 5.24: Time Drawdown curve. 270
Fig. 5.25: Pumping rate and drawdown relationship. 271
Fig. 6.1: (a) Tehri Dam; (b) Powerhouse of Tehri Hydropower Project 285
Fig. 6.2: Levees 291
Fig. 6.3: Dam & reservoir 291
Fig. 6.4: Flood wall 292
Fig. 6.5: Typical sketch of floodplain, levees and channel 293
Fig. 6.6: Flood management components and approaches (National Commission on Floods, 1980)295
Fig. 6.7: Triangular texture diagram 305
Fig. 6.8: Surface irrigation 315
Fig. 6.9: Drip/Trickle irrigation 318
Fig. 6.10: Sprinkler irrigation 319
Fig. 7.1: A Schematic layout of a typical canal system 337
Fig. 7.2: Various types of canal alignments. 341
Fig. 8.1: Typical Sections of Homogenous Earth Dam 398
Fig. 8.2: Typical Section of a Rock-fill Dam 398
Fig. 8.3: Flownet shape change on account of regions of different permeability 403
Fig. 8.4:Comparison of Flow net in the (I) Transformed and (II) Original Section. 405
Fig. 8.5: General Conditions for Line of Seepage 406

Fig. 8.6: Embankment Dam with Kozeny's Conditions 407
Fig. 8.7: Usual Loading Combination for a Gravity Dam 409
Fig. 8.8: Modification in Uplift Force due to Drain 411
Fig. 8.9:Variation of c_m with Upstream Face Inclination 415
Fig. 8.10: Resultant Force on Gravity Dam 420
Fig. 8.11: Normal Stresses on the Base of a Gravity Dam 421
Fig. 8.12:Principal Stresses in a Gravity Dam 422
Fig. 8.13: Elementary profile of a Gravity Dam 425
Fig. 8.14:Arch Dam 428
Fig. 8.15: Buttress Dams 429
Fig. 8.16: Raising Height of Buttress Dam 430
Fig. 8.17: Flood Hydrograph of Inflow 439
Fig. 8.18: Mass Inflow Curve 439
Fig. 8.19: Demand Curves 440
Fig. 8.20: Determination of Reservoir Capacity 441
Fig. 8.21: Determination of Yield from Reservoir of Specified Capacity 442
Fig. 8.22: Brune's curve for trap efficiency 444
Fig. 8.23: Churchill curve for trap efficiency 445
Fig. 8.24: Generalised Sediment Deposition Zones in a Reservoir. 450
Fig. 8.25: Longitudinal Patterns of Sediment Deposition in Reservoirs. 451
Fig. 8.26: Sketch for Problems 1 and 2. 458
Fig. 8.27: Sketch for Problem 3. 458
Fig. 8.28: Sketch for Problem 5. 459
Fig. 8.29: Sketch for Problem 8. 460

LIST OF TABLES

Table 1.1: World Water Reserves. 16

Table 1.2: Global water balance 17

Table 1.3: Water balance of the oceans 18

Table 1.4: Annual Water Balance of Continents 18

Table 2.1: Intensity of different types of rain 38

Table 2.2: Evaluation of Double-Mass curve of Example 2.4. 59

Table 2.3: Calculation of weighted rainfall by Thiessen polygon method 66

Table 3.1: Value of pan coefficients C_p in use for different pans 99

Table 3.2: Saturation vapour pressure of water 101

Table 3.3: Surface Areas and Capacities of Some Indian Reservoirs 109

Table 3.4: Mean monthly solar radiation at the top of the atmosphere (Ha), measured in millimetres of evaporable water 117

Table 3.5: Mean monthly values of possible sunshine hours (N) 117

Table 3.6: Values of reflection coefficient (r) 118

Table 3.7: Classification of soil based on the infiltration capacity 140

Table 4.1: Description of hydrologic groups 179

Table 4.2: Antecedent Moisture (AMC) for Obtaining the CN 180

Table 4.3: Curve numbers for three antecedent moisture conditions 181

Table 4.4: Runoff curve numbers for Hydrological cover conditions 184

Table 4.5: India's water resources at a glance 200

Table 4.6: Surface water resources potential of river basins (km^3) of India 201

Table 5.1 Selected values of W(u) for the given value of 1/u 258

Table 5.2: Drawdown of the water level in an observation well (r = 60.96 m) 274

Table 6.1: Basic water requirements for human needs 283

Table 7.1: Values of n for excavated open channels 361

Table 7.2: Values of Rugosity coefficient recommended by Buckley for alluvial channels 362

Table 7.3: Values of C.V.R. for different types of soils 365

Table 8.1: Thickness and size of dumped rock riprap 408

Table 8.2: Observed trap efficiency of Indian reservoirs 444
Table 8.3: Coefficient of proportionality (K) 447
Table 8.4: Calculation of Trap Efficiency and Sediment Volume 452
Table 8.5: Elevation-Area-Capacity and Sediment Deposition Computation by Empirical Area Reduction Method 454

TABLE OF CONTENTS

Foreword iii
Acknowledgement v
Preface vi
Outcome Based Education vii
Course Outcomes ix
Guidelines for Teachers xi
Guidelines for Students xii
Abbreviations and Symbols xiii
List of Figures xxii
List of Tables xxvi

Unit-1: Introduction **1-35**
Unit specifics 1
Rationale 1
Pre-requisite 1
Unit outcomes 1
1.1 Introduction 2
1.2 Hydrology 3
1.3 Catchment Area 3
1.4 Hydrological Cycle 5
1.5 The Water Balance Equation 9
1.6 World Water Balance 15
1.7 Global Freshwater Resources 18
1.8 History of Hydrology 20
1.9 Applications in Engineering 21
1.10 Sources of Data 23
Summary 25
Exercise 26
Know More 34
Bibliography 34

Unit-2: Precipitation **36-91**

Unit specifics 36

Rationale 36

Pre-requisite 36

Unit outcomes 36

2.1 Introduction 37

2.2 Forms of Precipitation 38

2.3 Characteristics of Precipitation in India 41

2.4 Seasons in India 42

2.5 Monsoons 43

2.6 Annual Precipitation in India 45

2.7 Measurement of Precipitation 47

2.7.1 Measurement of rainfall 47

2.7.2 Measurement of snowfall 51

2.8 Measurement of Rainfall 52

2.8.1 Adequacy of Rain Gauge Stations 52

2.8.2 Preparation of data 54

2.8.3 Test for Consistency of Record 57

2.9 Presentation of Rainfall Data 61

2.9.1 Mass Curve of Rainfall 61

2.9.2 Hyetograph 62

2.9.3 Point Rainfall 63

2.10 Mean Precipitation Over an Area 63

2.10.1 Arithmetical-Mean technique 63

2.10.2 Thiessen Polygon Method 63

2.10.3 Isohyet Method 66

2.11 Storm Analysis 68

2.11.1 Depth Area Relation 68

2.11.2 Depth Area Duration Analysis 70

2.11.3 Depth-Duration Curve 73

2.11.4 Rainfall - Depth - Duration Frequency Curve 74

2.11.5 Depth-Area-Duration Curve 75
2.12 Probable Maximum Precipitation (PMP) 75
2.13 Rainfall Data in India 76
Summary 77
Exercise 78
Know More 90
Bibliography 91
Unit-3: Abstraction from Precipitation **92-153**
Unit specifics 92
Rationale 92
Pre-requisite 92
Unit outcomes 92
3.1 Evaporation Process 93
3.2 Evaporimeters 95
3.2.1 Types of Evaporimeters 95
3.2.2 Limitations and drawbacks 98
3.2.3 Pan coefficient Cp 99
3.3 Empirical Evaporation Equations 100
3.3.1 Dalton type equation 100
3.3.2 Meyer's formula 100
3.3.3 Rohwer's formula 102
3.4 Analytical Methods for Evaporation Estimation 103
3.4.1 Water-Budget Method 103
3.4.2 Energy-Budget Method 104
3.4.3 Mass-Transfer Method 105
3.5 Reservoir Evaporation and Methods for its Reduction 106
3.5.1 Methods to Reduce Evaporation Losses 107
3.6 Transpiration 109
3.7 Evapotranspiration 110
3.7.1 Measurement of Evapotranspiration 113

3.7.2 Evapotranspiration Equations 115
3.8 Potential Evapotranspiration Over India 121
3.9 Actual Evapotranspiration (AET) 123
3.10 Initial Loss 125
3.10.1 Interception 126
3.10.2 Depression storage 128
3.11 Infiltration 128
3.12 Infiltration Capacity 131
3.13 Measurement of Infiltration 133
3.13.1 Simple (Tube type) infiltrometer 133
3.13.2 Double-ring infiltrometer 133
3.13.3 Rainfall Simulators 134
3.13.4 Hydrograph Analysis 136
3.14 Modelling Infiltration Capacity 136
3.14.1 Horton's Equation (1933) 137
3.14.2 Philip's Model 138
3.14.3 Kostiakov Model 139
3.14.4 Green–Ampt Model 139
3.15 Classification of Infiltration Capacities 139
3.16 Infiltration Indices 140
3.16.1 Φ-index 140
3.16.2 W- Index 142
Summary 142
Exercise 143
Know More 153
Bibliography 153
Unit-4: Runoff **154-223**
Unit specifics 154
Rationale 154
Pre-requisite 154

Unit outcomes 154

4.1 Introduction 155

4.1.1 Factors Influencing Runoff 156

4.1.2 Need for the study about Runoff 157

4.1.3 Runoff Classification 158

4.1.4 Natural Flow 159

4.1.5 Runoff Volume (Yield) 161

4.2 Hydrograph 162

4.2.1 Hydrograph generation 163

4.2.2 Components of a natural hydrograph 164

4.2.3 The contribution to base flow to stream discharge 166

4.2.4 Factors Affecting Hydrographs 167

4.3 Base Flow Separation 170

4.3.1 Factors affecting Base flow generation 170

4.3.2 Methods for base flow generation 171

4.4 Effective Rainfall 174

4.5 SCS-CN Method of Estimating Runoff Volume 176

4.5.1 Soils 179

4.5.2 Antecedent Moisture Condition (AMC) 180

4.5.3 Land Use 182

4.5.4 SCS-CN Equation for Indian Conditions 188

4.5.5 Procedure for Estimating Runoff Volume from a Catchment 189

4.5.6 Advantages of SCS-CN method 190

4.5.7 Limitations 190

4.6 Flow-Duration Curves 192

4.6.1 Theoretical Foundations 193

4.6.2 Data Collection and Analysis 194

4.6.3 Curve Construction 194

4.6.4 Interpretation of Curve Segments 195

4.6.5 Geological Influences on Streamflow 195

4.6.6 Applications in Water Resources Engineering 196

4.6.7 Advanced Topics in Flow-Duration Curves 196
4.7 Flow-Mass Curve 197
4.7.1 Calculation of Storage Volume 198
4.8 Surface Water Resources of India 199
4.9 Environmental Flows 202
4.9.1 Trade-offs in Development and Conservation 203
4.9.2 Estimation of Environmental Flows 203
4.9.3 Environmental Flow Assessment Methodologies 204
4.9.4 Implementation of EF 207
4.9.5 Future Challenges 209
4.10 Unit Hydrograph Approach 209
4.10.1 Method of Constructing a Unit Hydrograph 210
4.10.2 Uses of Unit Hydrographs 210
4.10.3 Synthetic Unit Hydrograph Parameters 211
Summary 214
Exercise 214
Know More 223
Bibliography 223
Unit-5: Groundwater and Well Hydrology 224-280
Unit specifics 224
Rationale 224
Pre-requisite 224
Unit outcomes 224
5.1 Introduction 225
5.2 Definition of Groundwater 225
5.3 Forms of Subsurface Water 225
5.4 Types of Subsurface Formation 227
5.5 Aquifer Types and their Hydraulic Properties 228
5.6 Darcy's Law for Groundwater Velocity 235
5.7 Empirical Formulas for Estimating Groundwater Velocity 237

5.7.1 Field Measurement of the Velocity of Groundwater 238
5.8 Groundwater Hydraulics 240
5.8.1 Analytical methods for estimation of Yield from wells and tube wells 240
5.8.2 Dupuit's formulae for estimating well yield under steady state condition 244
5.9 Partial Penetration of an Aquifer by a Well: From Equation 246
5.9.1 Interference among wells 248
5.10 Well Loss and Specific Capacity of Wells 251
5.11 Aquifer Test 256
5.11.1 Analysis of Aquifer Test Data 259
5.12 Time Drawdown Analysis 262
5.13 Distance Drawdown Analysis 266
5.14 Single Well Test 268
Summary 271
Exercise 271
Know More 279
Bibliography 280
Unit-6: Water Withdrawals and Uses **281-333**
Unit specifics 281
Rationale 281
Pre-requisite 281
Unit outcomes 281
6.1 Water Demand 282
6.1.1 Domestic Water Supply 282
6.1.2 Water for Agriculture 284
6.1.3 Water for Hydropower 284
6.1.4 Water for Electric Production 286
6.2 Flood and its Management 287
6.2.1 Types of Floods 287
6.2.2 Causes of Floods 288
6.2.3 Impacts of Floods: 288

	6.2.4	Flood Management Measures	289
	6.2.5	Flood Control Strategies	290
6.3	Analysis on Surface Water Supply		294
	6.3.1	Quantity Analysis	294
	6.3.2	Quality Analysis	294
	6.3.3	Sustainability and Management	295
	6.3.4	Infrastructure Analysis	296
	6.3.5	Community and Stakeholder Engagement	296
	6.3.6	Regulatory Compliance	296
6.4	Water Requirement of Crops		297
	6.4.1	Factors Affecting Water Requirements of Crops	297
6.5	Cropping Pattern		298
	6.5.1	Crop and Crop Season in India	300
	6.5.2	Duty and Delta	300
6.6	Quality of Irrigation Water		302
	6.6.1	Physical Characteristics	302
	6.6.2	Chemical Characteristics	302
	6.6.3	Biological Characteristics	303
	6.6.4	Other Considerations	303
	6.6.5	Water Quality Management Strategies	303
6.7	Soil-Water Relationship		304
	6.7.1	Soil Properties	304
	6.7.2	Root Zone Soil Water	306
	6.7.3	Infiltration	308
6.8	Consumptive Use		310
6.9	Irrigation Requirement		311
	6.9.1	Factors Affecting Irrigation Requirement	311
	6.9.2	Frequency of Irrigation	312
	6.9.3	Methods of Irrigation	314
Summary			322
Exercise			323

Know More 332
Bibliography 332

Unit-7: Distribution Systems **334-394**
Unit specifics 334
Rationale 334
Pre-requisite 334
Unit outcomes 334
7.1 Introduction 335
7.2 Canal Distribution Systems 335
7.2.1 Classification of Canals 336
7.2.2 Categorization of Canal Systems 337
7.3 Command Area 339
7.3.1 Gross Command Area (GCA) 339
7.3.2 Cultivable Command Area (CCA) 339
7.4 Alignments of Canals 339
7.5 Canal Losses 342
7.5.1 Causes of Canal Losses 342
7.5,2 Factors Affecting the Canal Losses 343
7.6 Discharge in Canals 343
7.6.1 Factors affecting Canal discharge 344
7.6.2 Head discharge and cultivable commanded area (CCA) 344
7.7 Design Discharge 345
7.7.1 Factors Influencing Design Discharge 345
7.7.2 Applications of Design Discharge 345
7.8 Design Discharge Estimation 346
7.8.1 Methods for Design Discharge Estimation 348
7.8.2 Natural Channels 356
7.8.3 Rigid boundary channels 357
7.8.4 Alluvial channels 358
7.9 Open Channel Flow Hydraulics 360

7.10 Kennedy's Silt Theory for Steady Regime Channels 364
7.10.1 Channel design by Kennedy's theory 365
7.11 Lacey's Theory 368
7.11.1 Lacey's silt theory 368
7.11.2 Channel design using Lacey's theory 369
7.11.3 Factors for channel design consideration 371
7.12 Canal Outlets 373
7.12.1 Factors Affecting Canal Outlets 374
7.12.2 Modular Canal Outlets 375
7.12.3 Non-modular canal outlets 376
7.12.4 Semi-modular canal outlets 377
7.13 Water Logging 378
7.14 Lining of Canals 379
7.14.1 Necessity of Drainage in Irrigated Lands 381
7.14.2 Methods of Drainage in Irrigated Lands: 382
Summary 383
Exercise 384
Know More 393
Bibliography 394
Unit-8: Dams and Spillways **395-467**
Unit specifics 395
Rationale 395
Pre-requisite 395
Unit outcomes 395
8.1 Introduction 396
8.2 Embankment Dams 396
8.2.1 Classification 397
8.2.2 Design Considerations 398
8.2.3 Factors Influencing the Design of an Embankment Dam 399
8.2.4 General Design Criteria for Embankment Dams 401

	8.2.5	Estimation and Control of Seepage	402
8.3	Gravity Dams		408
	8.3.1	Forces on a Gravity Dam	409
	8.3.2	Causes of Failure of a Gravity Dam	416
	8.3.3	Stress Analysis of Gravity Dams	418
	8.3.4	Elementary Profile of a Gravity Dam	424
	8.3.5	Design of a Gravity Dam	427
8.4	Arch Dams		427
	8.4.1	Advantages of Arch Dams	427
	8.4.2	Disadvantages	428
8.5	Buttress Dams		428
	8.5.1	Advantages of Buttress Dams	429
	8.5.2	Disadvantages of Buttress Dams	431
8.6	Spillways		431
	8.6.1	Components of Spillway	432
8.7	Types of Gates for Spillway Crests		433
	8.7.1	Flashboards and Stoplogs	434
	8.7.2	Vertical Lift Gates	435
	8.7.3	Radial (or Tainter) Gates	435
	8.7.4	Drum Gates	435
8.8	Reservoirs		436
	8.8.1	Types of Reservoirs	436
8.9	Reservoir Capacity and Yield of Reservoirs		437
	8.9.1	Yield	437
	8.9.2	Safe yield or firm yield	438
	8.9.3	Secondary yield	438
	8.9.4	Average yield	438
	8.9.5	Mass inflow curve	438
	8.9.6	Demand curve	439
	8.9.7	Calculation of Reservoir Capacity for a Specified Yield from the Mass Inflow Curve Procedure (Figure 8.20):	440

8.9.8 Determination of Safe Yield from a Reservoir of a Given Capacity 441
8.10 Sedimentation 442
8.10.1 Silting Rate 443
8.10.2 Description of Terms in Churchill's Curve 444
8.10.3 Prediction of Rate of Reservoir Sedimentation 446
8.10.4 Prediction of Sediment Distribution 449
8.10.5 Elevation Area Capacity 453
8.10.6 Computation of Revised Area and Capacity after 100 Years 453
8.11 Selection of Suitable Site 454
Summary 456
Exercise 456
Know More 466
Bibliography 467
References for Further Learning 468-469
CO and PO Attainable 470
Index 471-487

1 Introduction

UNIT SPECIFICS

This unit introduces the subject as a scientific study of water and discusses the components of the hydrological cycle, including, precipitation, abstraction losses, and runoff as surface & groundwater. The unit also deals with the water balance equation, a fundamental condensation concept in hydrology that represents the accounting of water inputs, outputs, and changes in storage within a system over a specified period. The global water availability on the Earth, oceans, and continents is presented with the water balance components. The unit also highlights how water interacts with the atmosphere, land, and living organisms, which is crucial in understanding and managing water resources on Earth. Finally, the historical background, applications in water resources engineering and the sources of hydrological data is highlighted.

RATIONALE

To understand the fundamentals of the hydrology cycle, water-budget and water balance.

PRE-REQUISITE

Nil

UNIT OUTCOMES

The list of outcomes of this unit is as follows:

U1-O1: Understand the fundamentals of the hydrological cycle

U1-O2: Understand the water budget and water balance

U1-O3: Applications in engineering and socially relevant problems

Unit Outcomes	Expected Mapping with Course Outcomes (1- Weak Correlation; 2- Medium correlation; 3- Strong Correlation)							
	CO-1	CO-2	CO-3	CO-4	CO-5	CO-6	CO-7	CO-8
U1-O1	3	3	3	3	3	3	3	3
U1-O2	3	3	3	3	3	3	1	3
U1-O3	1	1	3	3	3	3	3	3

1.1 INTRODUCTION

Earth is frequently called the blue planet because water occupies more than three-fourths of the planet's surface. Estimates place the entire volume of water at 1.4 billion cubic kilometers on Earth, and if this water were to be evenly distributed as a layer throughout the planet, it would be close to 3 kilometers thick layer. About 97% of the total available water on the earth is found in oceans and sea, and only 2.7% of the total water is freshwater available on the Earth. About 22.6% of this is buried as groundwater, while the remainder, about 75.2% of freshwater, is frozen in polar regions. Static water, a sizable portion of groundwater, is buried too deeply for convenient extraction. Rivers, lakes, soil, and other bodies of water contain a tiny amount of the residual water. Therefore, humans can only use a tiny portion of the water on Earth.

On a global basis, over 3,240 km^3 of Freshwater is used each year. Around the world, sector-wise, considerable variation is found in the use of water. Agriculture is the dominant water user in Asia, whereas in most parts of Europe and North America, industries and municipalities are the biggest users in the water sector. Generally, 69% of water use is accounted for by agriculture, with the remaining 23% going to industry and 8% going to residential use. Over the past century, there has been a notable increase in the usage of water for numerous reasons, and this tendency is still present now. As a result, attention is now concentrated on maintaining environmental quality and using water resources sustainably.

Information on the water balance of river and lake basins across short time intervals (week, day, and season) is used in operational management. For instance, India receives 117 cm of rain on average every year, which is equivalent to about 3,840 × 10^9 m^3 of water. A little over half of the yearly rainfall is lost to evaporation and transpiration, while the remaining 49% is transformed into surface runoff and Groundwater.

1.2 HYDROLOGY

The study of water science is known as hydrology. It is the area of study that deals with how water is distributed, occurs, and moves through the atmosphere of the earth. Water in lakes, reservoirs and streams, precipitation and snowfall, snow and ice on the earth, and water found in the pores of rocks and soil are all covered in this subfield of earth science. In general, the study of hydrology is an interdisciplinary field that incorporates elements of meteorology, geology, statistics, chemistry, physics, and fluid mechanics, among other related sciences. Hydrology is essentially an applied science. Sometimes, the topic is classified to emphasize how relevant it is. The topic is occasionally classified as (i) Scientific hydrology and (ii) Applied or engineering hydrology. Scientific hydrology is a field of research that focuses mostly on academic issues. Meanwhile, applied or engineering hydrology is known as the study of engineering applications. Broadly speaking, engineering hydrology is concerned with (i) water resource estimation, (ii) the analysis of elements such as runoff, precipitation, evapotranspiration, and their interactions, and (iii) disaster issues like floods and droughts and ways to mitigate them.

1.3 CATCHMENT AREA

A catchment area, which is also referred to as a watershed, drainage basin, or river basin, is a stretch of land that is divided by ridges, hills, and mountains on its terrain and from which all surface water flows to a common outlet point, which may be a lake, a river, or an ocean. It is essentially a natural basin or bowl-like area that collects and channelize rainfall and runoff towards a central point or outlet.

Catchment areas play a critical role in the hydrological cycle and are fundamental units for studying and managing water resources at the local, regional, and global scales. They serve as interconnected systems that link precipitation, surface water, groundwater, and the environment, shaping the distribution and availability of water resources within a given geographic area. Here are some of the key characteristics and concepts related to catchment areas:

a. **Boundary**: The boundary of a catchment area is delineated by the highest points of elevation, known as divides or drainage divides, which separate one catchment area from another. Water falling within the boundary of a catchment area will ultimately flow to the same outlet point.

b. **Hydrological Processes**: Within a catchment area, various hydrological processes occur, including precipitation, infiltration, runoff, evaporation, and groundwater flow. These processes interact to determine the movement and distribution of water within the catchment.

c. **Size**: Catchment areas can vary in size from small headwater streams that drain a few hectares to large river basins that span thousands of square kilometers. The size of a catchment area influences the magnitude and timing of streamflow and the overall hydrological response to precipitation events.

d. **Topography**: The topography of a catchment area, including slope, elevation, and land cover, influences the distribution of precipitation, the rate of runoff, and the types of landforms and drainage patterns within the catchment.

e. **Land Use**: Human activities within a catchment area, such as agriculture, urbanization, deforestation, and land development, can significantly impact hydrological processes and water quality. Changes in land use can alter the natural flow paths of water, increase runoff, and lead to erosion, sedimentation, and pollution of water bodies.

f. **Management**: Understanding the characteristics and behavior of catchment areas is essential for effective water resources management and planning. Watershed management strategies aim to protect and enhance water quality, mitigate flood risks, sustainably manage water resources, and preserve the ecological health of catchment areas.

A river's catchment area and its tributaries' sub-catchment areas are depicted in Figure 1.1. The locations of field water measurements used to calculate runoff are shown by the red points.

1.4 HYDROLOGICAL CYCLE

The continuity of mass can be quantitatively expressed using an equation for the hydrological cycle. This expression is also referred to as the water budget equation or the water balance equation, depending on the aim of the computation. An assessment of water resources and their changes can be done quantitatively to the water balance approach. The study of the water balance of lakes, river basins, unsaturated zones, and ground-water basins (e.g., inter-basin transfer, stream flow management, irrigation scheduling, etc.) provides a basis for project planning and the prudent use of water resources in space and time.

The hydrological cycle plays a crucial role in regulating the Earth's climate and sustaining life by redistributing water across various ecosystems and maintaining a balance of freshwater resources. The hydrological cycle is a continuing occurrence that characterizes the flow of water on Earth's surface as well as below it. It involves the circulation of water in various forms - liquid, solid (ice), and gas (water vapour) - through different reservoirs, including oceans, rivers, lakes, groundwater, atmosphere, and glaciers.

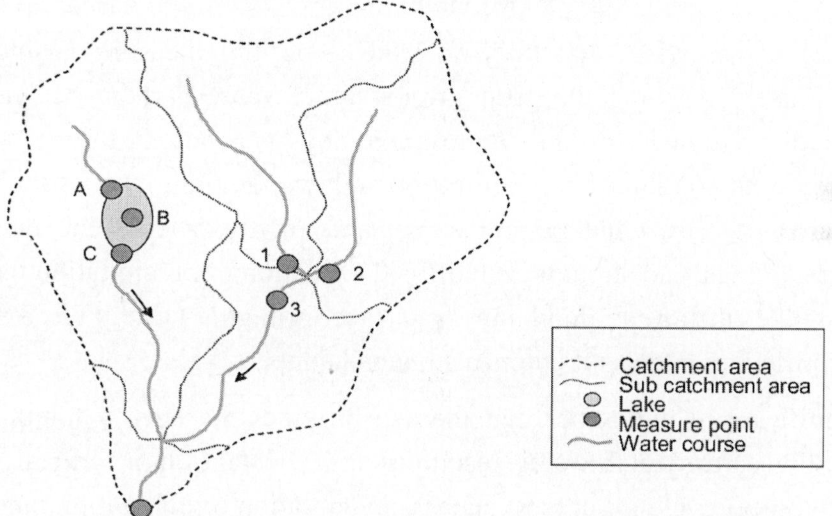

Figure 1.1: Catchment area of a river

Figure 1.2 depicts a typical descriptive illustration of a hydrological cycle. Horton proposed a qualitative depiction of the hydraulic cycle, which is referred to as Horton's representation of the hydrological cycle and is shown in Figure 1.3.

The hydrological cycle is a highly intricate system. The oceans provide a handy place to start when describing the cycle. The ocean's water evaporates because of the thermal energy generated by solar radiation. Clouds are created as the rising water vapour ascends. While most clouds condense and return to the oceans as rain, some clouds are driven by winds to the land.

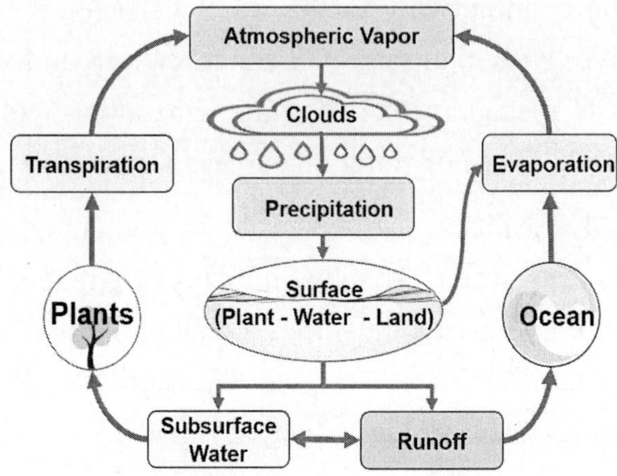

Figure 1.2: Hydrological Cycle

A portion of the water that falls on land seeps into the soil, raising its moisture content and ultimately reaching the groundwater table. Some part of the water below the earth's surface is returned to the atmosphere by vegetation through transpiration. Precipitation meets the conditions for infiltration and evaporation, reaches the surface of the land, moves across it, down the natural slope, and finally reaches the ocean through a network of rivers, gullies, and streams. Runoff is the fraction of precipitation that travels both above and below the earth's stream channel's surface through a variety of pathways. Runoff changes to streamflow as soon as it enters a stream channel.

It is significant to remember that the sun provides the energy for the hydrological cycle and that the earth's total water resources are constant. It is crucial to understand numerous hydrological cycle processes, such as evaporation, precipitation, and groundwater flow, in order to understand hydrology as a scientific field with rigorous investigation. It also becomes evident that almost every component of the hydrologic cycle, for example, artificial precipitation, suppressing evaporation, altering the land's vegetation and use, extracting groundwater, etc., can be influenced by people. Interference at one point in the cycle can have major consequences at a later point.

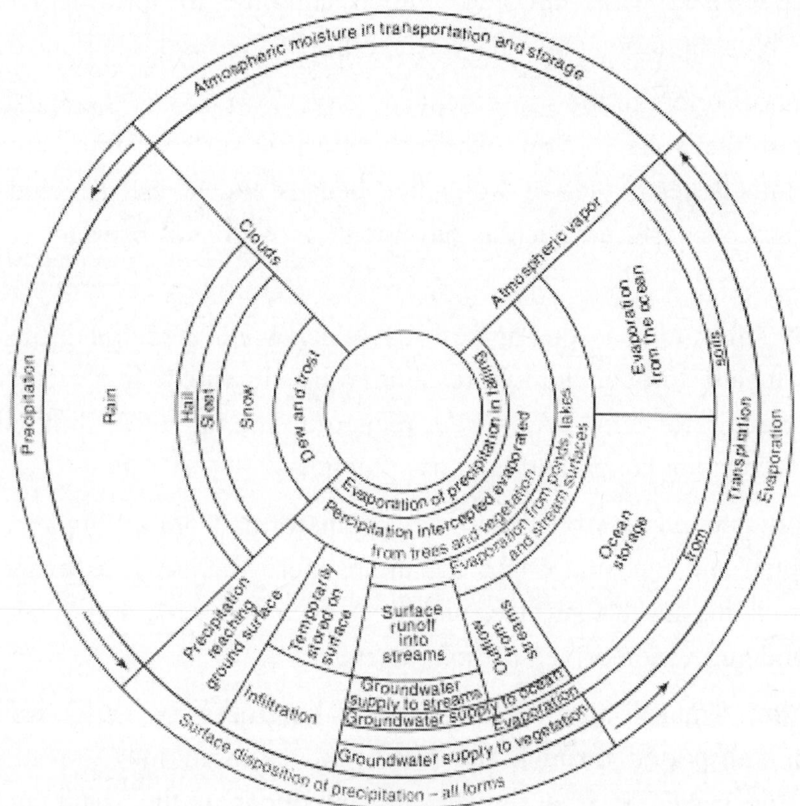

Figure 1.3: Horton's depiction of the water cycle

Numerous disciplines, including geography, forestry, agriculture, economics, sociology, and politics, are significantly impacted by the hydrological cycle. A few engineering applications where knowledge of the hydrological cycle and hydrological issues are used are water supply, irrigation, drainage, water power, flood control, navigation, coastal works, salinity control, and recreational application. The hydrological cycle consists of several key processes described below:

a. **Condensation**: Water vapour in the atmosphere cools and condenses into tiny droplets or ice crystals, forming clouds. Condensation is also responsible for the formation of dew and fog.

b. **Precipitation**: Precipitation is the primary input of water into the Earth's hydrological system. It includes rainfall, snowfall, sleet, and hail. Precipitation occurs when Clouds comprise condensed water droplets, which combine to form larger droplets or ice crystals. Such precipitation causes crystals to fall to the Earth's surface due to gravity.

c. **Runoff:** Runoff refers to the movement of water from excess precipitation that does not infiltrate into the soil; it flows over the Earth's surface, into streams, rivers, and eventually into larger bodies of water like oceans and lakes. It includes surface runoff from impermeable surfaces, such as pavement, as well as subsurface runoff through soil layers.

d. **Infiltration**: Infiltration is the process by which water penetrates the soil surface and moves downward through soil pores and rock fractures. It replenishes groundwater aquifers and contributes to groundwater recharge. The rate of infiltration depends on soil properties, vegetation cover, and land use practices.

e. **Evaporation**: The process by which water transforms from a liquid into a gaseous state (water vapour) and then back into the atmosphere is known as evaporation. It occurs primarily from the Earth's surface, including oceans, rivers, and lakes, as well as from moist soil and vegetation driven by solar energy.

f. **Transpiration**: Transpiration is referred to as the process by which water is taken up by plant roots, transported through the plant, and released into the atmosphere through stomata in the leaves. It is a significant component of the water cycle, especially in terrestrial ecosystems, where it influences humidity levels and contributes to cloud formation. Transpiration contributes to the water vapour content in the atmosphere. Transpiration, the release of water vapour from plant leaves, also contributes to evaporation.

g. **Storage**: Water is stored in various reservoirs within the Earth's hydrological system, including oceans, glaciers, ice caps, lakes, rivers, soil moisture, and groundwater aquifers. These reservoirs play critical roles in regulating water availability, sustaining ecosystems, and supporting human activities.

h. **Groundwater Flow**: The transport of water via aquifers, which are saturated in subsurface formations, is referred to as groundwater flow. Groundwater provides a vital source of freshwater for drinking, irrigation, and industrial activities. Through springs or base flow into rivers and streams, groundwater seepage supplies water to surface water bodies.

i. **Sublimation**: This is the direct, solid (ice or snow) to vapour transformation of water without converting into liquid form. It occurs primarily in Polar Regions and at high altitudes.

j. **Melting:** Solid ice and snow can melt into liquid water when exposed to warmer temperatures, contributing to surface runoff, or infiltrating into the ground.

1.5 THE WATER BALANCE EQUATION

The water balance equation is a fundamental concept in hydrology that represents the accounting of water inputs, outputs, and changes in storage within a defined system over a specified period. The basic principle of the water balance equation is the conservation of mass. The equation is also known as the continuity equation. The water balance equation can be applied at various spatial and temporal scales, ranging from small watersheds to large river basins and from daily to annual time periods. It serves as a fundamental tool for understanding the hydrological processes within a given area, assessing water availability, and managing water resources sustainably. According to this equation, the change in water storage within a volume is equal to the difference between the entire input and output for that specific volume of space and time. The water continuity equation (Equation 1.1) for a given catchment at a time interval t can be expressed as follows:

Mass Inflow – Mass Outflow = Change in Mass Storage **.............(1.1)**

The general form of the expression for the water budget equation of a catchment within a time interval t is given as Equation (1.2a); however, it should be emphasized that all the terms in a hydrological water budget may not be determined with the same degree of accuracy. This equation states that the change in storage ΔS in each catchment area represents the difference between the water stored within the system at the beginning and end of the specified period after accounting for changes in groundwater levels, soil moisture, snowpack, and other forms of water storage. Based on Equation (1.2a), change in storage ΔS is equal to the subtraction of evapotranspiration E, runoff R, transpiration T and Groundwater flow G from total precipitation P. The components of a water balancing equation can be expressed as the mean depth of water for the control volume (mm) or as the volume of water (m^3). Based on the above, Equation 1.1 can be rearranged as follows.

Change in Mass Storage = Mass Inflow – Mass Outflow

$\Delta S = P - (E + R + G + T)$ ....(1.2 a)

where:

- P represents precipitation (input)
- E represents evaporation (output)
- R represents runoff (output)
- G represents groundwater flow out of the catchment (output)
- T represents transpiration (output)
- ΔS represents the change in storage (change in water stored within the system)

In a time interval Δt, the water budget equation (1.1) for the lake/reservoir can also be written as Equation 1.2b:

Input volume - Output volume = Change in storage of the lake

$\Delta S = (\bar{I} \times \Delta t + P \times A) - (\underline{Q} \times \Delta t + E \times A)$ ....(1.2 b)

where,

- $\bar{I}$ = Mean inflow rate of water into the lake/reservoir,
- $\underline{Q}$= Mean outflow rate from the lake/reservoir,
- P = precipitation,
- E = evaporation,
- A = average surface area of the lake/reservoir and
- ⊗S = change in storage volume of the lake/reservoir.

In the water budget equation described above, storage S consists of three components as:

S_s = Surface water storage

S_{sm} = Water storage as soil moisture

S_g = Water storage as Groundwater

Thus, in equation (1.2a), ΔS can be represented as:

$$\mathbf{\Delta S = \Delta S_s + \Delta S_{sm} + \Delta S_g} \qquad \textbf{.........(1.2c)}$$

The hydrologic budget of a drainage basin contains the mathematical expression for its hydrologic cycle, commonly referred to as its water balance. It is expressed as the rate of change of storage within the drainage basin, S, over a specific time interval, t, equal to the difference between inflow, I, and outflow, O. The basin's hydrologic budget can be stated as follows (Equation 1.3a, 1.3b) when seen as a reservoir or as a black-box system:

$$\frac{\Delta S}{\Delta t} = \bar{I} - \bar{O} \qquad \textbf{.............(1.3a)}$$

$$\frac{S_2 - S_1}{\Delta t} = \frac{I_2 + I_1}{2} - \frac{O_2 + O_1}{2} \qquad \textbf{.............(1.3b)}$$

where,

$\bar{I}$ = average inflow for the time interval Δt

$\bar{O}$ =average outflow for the time interval Δt.

ΔS = rate of change of storage

Δt = given period of time

Subscripts 1 and 2 = values of the variables at the start and end of the time interval $\Delta t = t_2 - t_1$. If I and O vary continuously with time t, then Equation (1.3a, 1.3b) can be written as Equation (1.4).

$$\frac{dS(t)}{dt} = I(t) - O(t) \qquad \textbf{.............(1.4)}$$

It is implied in equations (1.3a), (1.3b), and (1.4) that I, O, and S are spatially uniform or do not vary in space. Eq. (1.4) is also denoted as the spatially lumped continuity equation or water budget equation. In terms of a rainfall-runoff relationship, the value of runoff in equation (1.2a) can be represented as Equation 1.5:

$$R = P - L \qquad \textbf{.............(1.5)}$$

where L = Losses = Water received from precipitation is not available for runoff due to several losses, i.e., evaporation, transpiration, infiltration, and surface storage.

Example 1.1: The yearly flow from a watershed with a 12 km^2 catchment area is 0.2x10^{7}m^3. What will the depth of water correspond to the given stream flow?

Solution:

$$\text{Depth of water} = \frac{Volume\ of\ water}{Catchment\ Area\ of\ watershed}$$

$$\text{Depth of water} = \frac{0.2x10^7}{12x10^6}$$

$$= 2/12 \text{ meter}$$

$$= 16.67 \text{ cm}$$

Example 1.2: At the start of a particular month, the water surface elevation of a reservoir was 105.200 meters above the datum. Surface runoff sources contributed an average of 8.0 m^3/s to the lake's inflow during that month. During that same time frame, the lake's discharge averaged 7.6 m^3/s. Additionally, the lake saw 175 mm of rainfall in that month, and 7.15 cm of evaporation was measured from the lake's surface. At the end of the month, evaluate the lake's water surface elevation and write down the water budget equation. 5,500 hectares might be considered the typical lake surface area. Assume that the groundwater reservoir receives no input at all.

Solution

In a time interval Δt, the water budget for the lake can be written as:

Input volume - Output volume = Change in storage of the lake

$$(\bar{I} \times \Delta t + P \times A) - (\overline{O} \times \Delta t + E \times A) = \Delta S$$

where $\bar{I}$ = Mean inflow rate of water into the lake,

$\overline{O}$ = Mean outflow rate from the lake,

P = Precipitation,

E = Evaporation,

A = Average surface area of the lake and

ΔS = Change in storage volume of the lake.

Here Δt = 1 month = 30 x 24 x 60 x 60 = 2.592 X 10^6 s = 2.592 Ms

In one month:

Inflow volume = $\bar{I} \times \Delta t$ = 8.0 x 2.592 = 20.736 M m^3

Outflow volume = $\bar{O} \times \Delta t$ = 7.6 x 2.592 = 19.699 M m^3

Input due to precipitation = $PA = \frac{175 \times 5500 \times 100 \times 100}{1000X\ 10^6} = \frac{175 \times 55}{1000}$ M m^3 = 9.625 M m^3

Outflow due to evaporation $= EA = \frac{7.15 \times 5500 \times 100 \times 100}{100X\ 10^6} = \frac{7.15 \times 55}{100}$ =3.93 M m^3

Hence $\Delta S = (\bar{I} \times \Delta t + PA) - (\bar{O} \times \Delta t + EA)$

= 20.736 + 9.625-19.699-3.93

= 6.732 M m^3

Change in elevation $\Delta z = \frac{S}{A} = \frac{6.732x10^6}{5500 \times 100 \times 100}$ = 0.1224 m

New water surface elevation at the end of the month = 105.200 + 0.1224

= 105.3224 m above the datum.

Example 1.3: A storm dropped 15.5 cm of rain in 2 hours over a tiny watershed of 250 ha. Prior to the storm, the stream at the catchment's outlet was dry and had an average discharge of 1.8 m^3/s over a 12-hour period. When the runoff event ended, the stream was once again dry. (a) How much water was unavailable for runoff as a result of the combined effects of transpiration, evaporation, and infiltration? (b) What is the precipitation to runoff ratio?

Solution

The water-budget equation for the watershed in a time duration Δt is

R = P - L

where,

L = losses = Rainfall-generated water is unavailable for runoff because of evaporation, transpiration, infiltration, and surface storage, among other losses.

Δt = duration of the runoff = 12 hours.

Note: Rainfall received in the first 2 hours only, and for the rest 10 hours, the precipitation was zero.

(a) *P*= Input due to precipitation in 12 hours

$= 250 \times 100 \times 100 \times (15.5/100) = 387{,}500 \text{ m}^3$

R= Runoff volume = outflow volume at the catchment outlet in 12 hours

$= 1.8 \times 12 \times 60 \times 60 = 77{,}760 \text{ m}^3$

Hence,

Losses $L = 387{,}500 - 77{,}760 = 309{,}740 \text{ m}^3$

(b) Runoff/rainfall = 77,760/387,500 = 0.20

(This ratio is known as *runoff coefficient*)

Example 1.4: A watershed's catchment area is 126 km^2, and it obtains 112 cm of precipitation annually. At the watershed's outlet, the average outflow rate for the stream that drains the catchment was found to be 1.76 m^3/s for the first four months, 2.2 m^3/s for the next seven months, and 3.7 m^3/s for the last month. Find out: (i) what is the runoff coefficient for the watershed? (ii) If the watershed's afforestation lowers the runoff coefficient to 0.30, what is the increase in the abstraction from precipitation owing to infiltration, evaporation, and transpiration for the same annual rainfall of 112 cm?

Solution:

(i) Before Afforestation:

Consider a period = Δt = 1 year

Input volume to the catchment through precipitation = P × A

$$=126 \times 10^6 \times \left(\frac{112}{100}\right)$$

$$=141.12 \text{ Mm}^3$$

Runoff = Output volume = $(1.76 \times 4) + (2.2 \times 7) + (3.7 \times 1) = 26.14 \text{ Mm}^3$.month

$$= 26.14 \times \left(\frac{365}{12}\right) \times 24 \times 60 \times 60$$

$= 68.69 \text{ Mm}^3$

Runoff coefficient = Runoff / Rainfall

$= 68.69/141.12 \quad = \quad 0.4867$

Abstraction volume = 141.12 - 68.69 = 72.43 Mm^3

(ii) After Afforestation

Runoff = 0.30 x 141.12 = 42.336 Mm^3

New abstraction volume = 141.12 - 42.336 = 98.784 Mm^3

Increase in abstraction = 98.784 - 72.43 = 26.354 Mm^3

1.6 WORLD WATER BALANCE

The constant movement of water on, above, and below the surface of the Earth is described by the global water balance, commonly referred to as the water cycle or the global hydrological cycle. It includes processes such as groundwater movement, runoff, infiltration, condensation, and precipitation. The water balance is crucial for maintaining the Earth's ecosystems, climate regulation, and providing freshwater resources for human activities.

The world water balance is dynamic and interconnected, with water constantly cycling between different reservoirs and undergoing various physical processes. Changes in climate, land use, and human activities can impact the water balance, leading to shifts in precipitation patterns, changes in runoff dynamics, alterations in groundwater recharge rates, and fluctuations in water availability. Understanding the world water balance is essential for sustainable water management and addressing water-related challenges, such as water scarcity, droughts, floods, and ecosystem degradation.

Approximately 1,357.5 million cubic kilometers (M km^3) of water exist in the world. Over 97% of the world's water is found in the oceans and seas, where it is too salty for its most useful uses. Two third of the balance water is present in the form of glaciers, ice caps, permafrost, underground aquifers, and marshes. An estimated 108,000 km^3 of precipitation falls over Earth's surface annually. By evaporation, around 60% (61,000 km^3) of the water returns directly to the sky, leaving 47,000 km^3 to flow towards the sea. Thus, the globe only has 35 M km^3 of pure water accessible. Of this, 10.5 M km^3 of fresh Groundwater and 21.6

M km^3 of Antarctic ice are present. Table 1.1 shows where Freshwater is distributed on Earth.

The earth receives about 0.42 M km^3 of water in the form of precipitation. Out of these, 0.32 M km^3 is received in the oceans and seas, while the remaining 0.10 M km^3 is received on the earth. Every year, the rivers, and springs transport 0.038 million cubic kilometers of water to the sea. Nevertheless, only 4% of the entire river flow is utilized for irrigation, and the remaining portion flows uselessly into the sea. Table 1.2 represents the Global water balance from different sources.

Table 1.1: World Water Reserves.

SN	Item	Area (M km^2)	Volume (M km^3)	Percent total water	Percent Freshwater
1.	Oceans	361.3	1338.0	96.5	-
2.	Groundwater				
2a.	Fresh Groundwater	134.8	10.530	0.76	30.1
2b.	Saline Groundwater	134.8	12.870	0.93	-
3.	Soil moisture	82.0	0.0165	0.0012	0.05
4.	Ice caps and glaciers	16.0	24.0235	1.7	68.6
5.	Other ice and snow	0.3	0.3406	0.025	1.0
6.	Lakes				
6a.	Fresh Lakes	1.2	0.0910	0.007	0.26
6b.	Saline Lakes	0.8	0.0854	0.006	-
7.	Wetlands	2.7	0.01147	0.0008	0.03
8.	Rivers	148.8	0.00212	0.0002	0.006
9.	Biological water	510.0	0.00112	0.0001	0.003
10.	Atmospheric water	510.0	0.01290	0.001	0.04
Total	All kinds of water	510.0	1386.0	100.0	
	Freshwater	148.8	35.0	2.5	100

Source: UNESCO (1978).

The oceans' water balance is shown in Table 1.3. The table illustrates the considerable water transfer between the oceans as well as the variations in precipitation and evaporation levels between them.

Table 1.2: Global water balance

<table>
<tr><th>SN</th><th colspan="2">Item</th><th>Unit</th><th>Ocean</th><th>Land</th></tr>
<tr><td>1</td><td colspan="2">Area</td><td>M km^2</td><td>361.30</td><td>148.8</td></tr>
<tr><td rowspan="2">2</td><td colspan="2" rowspan="2">Precipitation</td><td>km^3/year</td><td>4,58,000</td><td>1,19,000</td></tr>
<tr><td>mm/year</td><td>1270</td><td>800</td></tr>
<tr><td rowspan="2">3</td><td colspan="2" rowspan="2">Evaporation</td><td>km^3/year</td><td>5,05,000</td><td>72,000</td></tr>
<tr><td>mm/year</td><td>1400</td><td>484</td></tr>
<tr><td rowspan="4">4</td><td rowspan="2">Runoff to Ocean</td><td>Rivers</td><td>km^3/year</td><td></td><td>44,700</td></tr>
<tr><td>Groundwater</td><td>km^3/year</td><td></td><td>2,200</td></tr>
<tr><td colspan="2" rowspan="2">Total runoff</td><td>km^3/year</td><td></td><td>47,000</td></tr>
<tr><td>mm/year</td><td></td><td>316</td></tr>
</table>

Source: UNESCO (1978).

According to Shiklomanov (2000), The worldwide estimated mean amount of renewable water resources is 42,750 km^3/year. Table 1.4 displays the continental land mass's water balance. The table shows that while Africa is the driest continent in the world, with only 20% of precipitation turning into runoff, North America and Europe have the largest amounts of runoff. Asia has the most plentiful water resources worldwide.

The world's rivers release over 40,000 km^3 of water into the ocean annually. Four major rivers, the Amazon, Congo, Orinoco, and Parana, flow into the Atlantic Ocean, accounting for about half of its discharge (20,000 km^3). The world's rivers have an average yearly flow of 1.2 million m^3/s. The average discharge of the Amazon, the greatest river in the world, is 200,000 m^3/s. The Brahmaputra and the Ganges are India's two largest rivers, with average discharges of 16,200 m^3/s and 15,600 m^3/s, respectively.

Table 1.3: Water balance of the oceans

Ocean	Area (M km²)	Precipitation	Inflow from adjacent continents	Evaporation	Water exchange with other oceans
Atlantic	107	780	200	1040	-60
Arctic	12	240	230	120	350
Indian	75	1010	70	1380	-300
Pacific	167	1210	60	1140	130

Table 1.4: Annual Water Balance of Continents

Continent	Area (M km²)	Precipitation	Total runoff	Runoff as % of precipitation	Evapo-transpiration
Africa	30.3	686	139	20	547
Asia	45.0	726	293	40	433
Australia	8.7	736	226	30	510
Europe	9.8	734	319	43	415
North America	20.7	670	287	43	383
South America	17.8	1648	583	35	1065

1.7 GLOBAL FRESHWATER RESOURCES

The distribution of global freshwater resources is highly uneven, with some regions having abundant water supplies while others face water scarcity. Here is an overview of the distribution of freshwater resources globally:

a. **Surface Water**: Surface water refers to water bodies such as rivers, lakes, and reservoirs. The distribution of surface water is influenced by factors such as precipitation patterns, topography, and geology. Regions with high rainfall, such as tropical rainforests and temperate zones, tend to have abundant surface water resources. Conversely, arid, and semi-arid regions receive less precipitation and have limited surface water availability.

b. **Groundwater**: Groundwater refers to water stored underground in aquifers. The distribution of groundwater resources is influenced by geological factors such as rock permeability and aquifer recharge rates. Groundwater is often more abundant in regions with porous, permeable rock formations and high rates of recharge, such as alluvial plains and coastal areas. However, overexploitation of groundwater resources can lead to depletion and groundwater depletion, particularly observed in densely populated areas and regions with intensive agriculture.

c. **Glaciers and Ice Caps**: Glaciers and ice caps store vast quantities of freshwater, primarily in polar regions and mountainous areas. These ice masses slowly release water over time, contributing to surface water runoff and maintaining river flow during dry seasons. However, climate change is causing glaciers and ice caps to melt at an accelerated rate, leading to concerns about water availability in downstream regions that rely on glacier-fed rivers for freshwater supply.

d. **Permafrost**: Permafrost refers to permanently frozen ground found in polar and high-latitude regions. Permafrost stores significant amounts of freshwater in the form of ice and groundwater. However, thawing permafrost due to climate change can release stored water, altering hydrological systems and impacting ecosystems and infrastructure.

e. **Distribution Challenges**: Despite the overall abundance of freshwater on Earth, access to clean and reliable freshwater resources remains a challenge for many regions due to uneven distribution, population growth, pollution, and water management issues. Additionally, water scarcity and competition for water resources are exacerbated by factors such as agricultural irrigation, industrial development, urbanization, and climate variability.

In summary, freshwater resources are distributed unevenly across the globe, with some regions having abundant surface water, groundwater, glaciers, and ice caps, while others face water scarcity and challenges in accessing clean and reliable freshwater supplies. Sustainable management and conservation of freshwater resources are essential to ensure water security and meet the needs of growing populations while preserving ecosystems and maintaining environmental sustainability.

1.8 HISTORY OF HYDROLOGY

Since water is a basic necessity for life, humans have always tried to make the most of the water resources that are accessible to them. The field of hydrology has a long history, having developed from primitive water management techniques and early human observations to the highly developed scientific field it is today. There are historical examples of civilizations that thrived while there were stable water sources and then fell apart when those supplies ran out. Vedic literature has numerous allusions regarding groundwater supply and its practical uses. Archaeological excavations at Mohenjodaro have indicated that the Indus Valley culture was aware of groundwater development through wells as early as 3000 BC. Quotations from old Hindu scriptures imply that knowledge of the hydrologic cycle pertains back from the Vedic period. The Arthashastra, written by Chanakya circa 300 BC, contains the earliest explanation of the rain gauge and its application. The Brihatsamhita by Varaharmihira (c. AD 505–587) includes explanations of the rain gauge, wind vane, and rainfall forecast techniques. Egyptians were aware of the value of measuring a river's stage; documents detailing the Nile's stages from 1800 BC have been found. Europe did not learn about the hydrologic cycle until much later, in the year 1500.

Here is a brief overview of key developments in the history of hydrology:

a. **Ancient Civilizations**: The earliest known civilizations, such as those in Mesopotamia, Egypt, India, and China, developed rudimentary forms of hydrological knowledge and water management techniques. They constructed irrigation systems, reservoirs, and aqueducts to control the water flow for agriculture, domestic use, and flood mitigation.

b. **Classical Period**: Greek and Roman scholars made significant contributions to hydrological understanding. Greek philosophers such as Thales and Aristotle pondered the nature of water, while engineers like Archimedes developed principles of fluid mechanics. The Romans constructed elaborate aqueducts and drainage systems to supply water to cities and manage wastewater.

c. **Middle Ages**: During the Middle Ages in Europe, Islamic scholars preserved and expanded upon classical knowledge. Islamic engineers developed sophisticated irrigation systems, water clocks, and water-powered machines, contributing to advancements in hydrological engineering.

d. **Renaissance and Early Modern Period**: The Renaissance saw renewed interest in scientific inquiry, leading to advancements in hydrological measurement and understanding. Leonardo da Vinci conducted studies on water flow, river morphology, and sediment transport. In the 17th and 18th centuries, scientists like Edme Mariotte and Daniel Bernoulli made significant contributions to hydrodynamics and fluid mechanics.

e. **18th and 19th Centuries**: The Enlightenment era brought further advancements in hydrological science. Hydrological measurement techniques, such as stream gauging and rainfall observation, began to be standardized. Engineers like Pierre Perrault and Henry Darcy conducted pioneering studies on hydrological processes, including evaporation, infiltration, and groundwater flow.

f. **20th Century**: The 20th century saw rapid advancements in hydrological science and engineering, driven by technological innovations and an increased understanding of complex hydrological processes. The development of hydrological models, remote sensing technologies, and computer simulations revolutionized the field. Hydrologists made significant contributions to understanding climate change impacts, water resource management, flood forecasting, and environmental conservation.

g. **Contemporary Period**: In the 21st century, hydrology continues to evolve in response to emerging challenges such as climate change, population growth, and urbanization. Integrated approaches to water management, incorporating principles of sustainability and ecosystem management, are gaining prominence. Advances in data collection, modelling, and computational techniques are enhancing our understanding of the hydrological processes and improving water management practices worldwide.

Throughout history, hydrology has been essential for human survival and development, shaping civilizations and influencing the management of water resources and the environment. Today, hydrology remains a vital scientific discipline with interdisciplinary applications in engineering, agriculture, environmental science, and policy-making.

1.9 APPLICATIONS IN ENGINEERING

Hydrology plays a crucial role in various engineering disciplines. Some of the key applications of hydrology in engineering include:

a. **Water Resources Management**: Hydrology helps in assessing the availability, distribution, and sustainable use of water resources. Engineers use hydrological data to design systems for water supply, irrigation, hydropower generation, and flood control.

b. **Hydraulic Structure Design**: Engineers use hydrological analysis to design hydraulic structures such as dams, levees, spillways, and canals. Understanding the flow characteristics of water bodies helps in designing structures that can withstand various hydraulic conditions and efficiently manage water flow.

c. **Flood Risk Assessment and Management**: Hydrological studies are essential for assessing flood risks in a given area. Engineers use hydrological models to simulate rainfall-runoff processes, estimate flood volumes, and determine flood inundation extents. This information is crucial for designing flood control measures, such as levees, flood walls, and Stormwater management systems, to mitigate flood risks and protect communities.

d. **Stormwater Management**: Hydrology is essential in designing Stormwater management systems for urban areas. Engineers use hydrological principles to estimate peak flows and runoff volumes from rainfall events, design drainage systems, and implement best management practices for Stormwater control, such as detention ponds, green infrastructure, and permeable pavements.

e. **Erosion and Sediment Control**: Hydrology helps assess the erosive potential of water and design erosion and sediment control measures for construction sites, agricultural lands, and riverbanks. Engineers use hydrological data to predict sediment transport, erosion rates, and sediment deposition and develop erosion control structures and land management practices to minimize soil erosion and protect water quality.

f. **Groundwater Management**: Hydrology plays a critical role in managing groundwater resources sustainably. Engineers use hydrological models to simulate groundwater flow and recharge processes, assess aquifer characteristics, and design well fields and groundwater extraction systems for water supply, irrigation, and industrial use.

g. **Environmental Engineering**: Hydrology is integral to environmental engineering projects aimed at protecting and restoring aquatic ecosystems. Engineers use hydrological data to assess the impacts of human activities on water quality, habitat availability, and ecosystem health and develop strategies for water quality improvement, habitat restoration, and watershed management.

h. **Climate Change Adaptation**: Hydrology is increasingly important in adapting engineering systems to the impacts of climate change, such as changes in precipitation patterns, temperature, and sea level rise. Engineers use hydrological models to assess future water availability, flood risks, and drought vulnerabilities and design resilient infrastructure and adaptive management strategies to cope with changing hydrological conditions.

These applications demonstrate the significant role of hydrology in engineering practices aimed at managing water resources, mitigating risks, and promoting sustainable development.

1.10 SOURCES OF DATA

A hydrologist would need information about the several crucial stages of the hydrological cycle affecting the problem catchment, depending on the scope of the problem. Temperature, humidity, wind speed, precipitation, stream flow, evaporation and evapotranspiration, infiltration characteristics, soil characteristics, land use and land cover, groundwater characteristics, physical and geological characteristics, water quality, etc., are among the parameters that are usually required for studies. The different sources to obtain the required data mentioned above are given below.

a. **Government Agencies**: Many countries have government agencies responsible for monitoring water resources and providing hydrologic data. For example, In India:

- Hydro-meteorological data, including temperature, humidity, wind speed, precipitation, etc., is gathered by the India Meteorological Department (IMD) and a few state government organizations.

- River flow data for all major rivers of the country have been monitored by the Central Water Commission (CWC).

- The State Water Resources/Irrigation Department collects stream-flow data of various rivers and streams in India.
- Groundwater data is gathered by the Central Groundwater Board (CGWB) and state government groundwater agencies.
- Data on soil, evapotranspiration and infiltration properties of soil is provided by state government agencies such as the Department of Agriculture, the Department of Watershed Development, and the Department of Irrigation.
- The topographical maps that are provided by the Survey of India must be analyzed in order to determine the physical characteristics of the study area.
- The State Geology Directorate and the Geological Survey of India would have information on the geological features of the basin under study.
- A location's soil characteristics can be obtained on pertinent maps from the National Bureau of Soil Survey and Land Use Planning (NBSS&LUP), published in 1996.
- The State Watershed Development Department and the State Agriculture Department have more comprehensive data available.
- State Remote Sensing Agencies often provide data on land use and land cover.
- The National Remote Sensing Agency (NRSA) of the Government of India will provide multispectral and multi-temporal satellite imagery, from which specific details must be gathered.
- Data on water quality is gathered by the Central and State Pollution Control Boards (CPCB/SPCB).

b. **International Organizations**: Organizations like the World Meteorological Organization (WMO) and the United Nations Educational, Scientific and Cultural Organization (UNESCO) often gather and share hydrologic data globally.

c. **Academic Institutions**: Universities and research institutions conduct hydrological studies and often make their data available to the public.

d. **Non-Governmental Organizations (NGOs)**: NGOs focused on environmental issues may collect hydrologic data for specific regions or water bodies.

e. **Remote Sensing**: Significant data for hydrological studies, including information on precipitation, soil moisture, and surface water can be provided through satellite imagery and remote sensing technologies.

f. **Weather Stations**: Weather stations frequently gather information on precipitation, temperature, humidity, and wind speed etc, which can be used in hydrological analyses.

g. **Private Companies**: Some private companies specialize in hydrological monitoring and may offer data services for a fee.

h. **Published Literature**: Hydrological research studies often publish their findings in academic journals, which may include datasets as supplementary material.

i. **Online Databases**: Various online databases aggregate hydrological data from multiple sources and make it accessible to researchers, policymakers, and the public.

j. **Field Measurements**: Hydrological data can also be collected through field measurements using instruments such as stream gauges, rain gauges, and groundwater wells.

SUMMARY

Hydrology is the scientific study of water in the environment, focusing on its distribution, movement, and properties. It examines how water interacts with the atmosphere, land, and living organisms, playing a crucial role in understanding water resources, flood forecasting, and managing water-related challenges. Water resources management involves the planning, developing, and managing of water resources to meet various human needs and environmental requirements. This unit describes the main elements of the hydrological cycle, including condensation, precipitation, abstraction losses, surface water & groundwater. It has been described in detail which are affected by the watershed/catchment area having different vital characteristics. The unit also deals with the water balance equation, a fundamental concept in hydrology that represents the accounting of water inputs, outputs, and changes in storage within a defined system over a specified period. The global water availability on the earth, oceans, and continents is presented in this unit. In the end, the water sources and the applicable fields of hydrology are also described.

EXERCISE

Revision Questions

1. What do you understand by the term "Hydrology"? Explain key applications of hydrology.
2. Describe the term "catchment area" with the sketch. Explain key characteristics and concepts related to catchment areas.
3. Draw a sketch indicating Horton's representation of the hydrological cycle.
4. Explain the hydrologic cycle. Describe briefly how human influence affects different stages of this cycle.
5. Describe key processes of the hydrologic cycle.
6. Explain the hydrological water budget using examples.
7. What are the noteworthy aspects of global water-balance studies?
8. Give a brief explanation of the distribution of freshwater resources around the world.
9. Enumerate the principal applications in which hydrological studies are important.
10. Give a brief overview of sources of hydrological data in India.

NUMERICAL PROBLEMS

1. For a 210 km^2 area, three and a half cm of rain each day is equivalent to an average rate of input of how many cubic meters of water per second?
2. The total surface area of the watershed at a stream location is 80,000 hectares. At this point, 980 mm of precipitation occurs on average each year. (i) Calculate the mean annual flow rate of the stream in m^3/s, if 30% of the rainfall reaches the basin outlet as stream flow. (ii) For the same mean annual rainfall of 980 mm, if watershed restoration, including afforestation, reduces the runoff rate to 25%, what will be the increase in the volume of abstraction from all causes?
3. At the start of a specific month, the water surface elevation of a lake was 104.500 meters above the datum. The surface runoff sources contributed an average of 6.5 m^3/s to the lake's inflow during that month. During that same time frame, the lake's discharge

averaged 7.0 m^3/s. Additionally, the lake experienced 156 mm of rainfall in that month, and 7.20 cm of surface evaporation was calculated. At the end of the month, compute the lake's water surface elevation and write down the water-budget equation. One can estimate the average lake surface area to be 5150 hectares. Assume that the groundwater reservoir receives no input at all.

4. A storm produced 11.6 cm of rainfall in 75 minutes on a small 170-hectare watershed. Prior to the storm, the river that drains the catchment at the exit was dry, and during the nine-hour runoff, the average discharge was 1.6 m3/s. When the runoff event ended, the river was once again dry. (A) How much water was prevented from flowing into the stream as a result of transpiration, evaporation, and infiltration working together? What is the precipitation to runoff ratio?
5. A watershed's catchment area is 160 km^2. Rainfall in the watershed is 150 cm per year. The average runoff rates at the watershed's outlet were determined to be (i) 1.6 m^3/s for the first three months, (ii) 2.5 m^3/s for the next six months, and (iii) 4.5 m^3/s for the final three months in the stream that drains the catchment. What will the catchment's runoff coefficient be? (ii) For a given annual rainfall of 150 cm, if the catchment's afforestation decreases the runoff coefficient to 0.35, what is the increase in precipitation abstraction from infiltration, evaporation, and transpiration?
6. Determine the steady rate of withdrawal from a 1295-hectare reservoir during the duration of a 30-day month, where the reservoir's average daily intake was 0.6 Mm^3/day, but its level decreased by 0.74 m. Over the course of the month, the reservoir observed an average seepage loss of 2.7 cm, a total of 19.5 cm of precipitation, and 9.8 cm of evaporation.
7. A flood wave was moving through a river stretch. The estimated amount of water stored in the reach at any given time was 16.6 hectare-meter. If, over a three-hour period, the average inflow and outflow are 11.2 m^3/s and 13.9 m^3/s, respectively, what would be the storage in the reach?
8. A catchment has four sub-areas. The following table shows the yearly precipitation and evaporation from each of the sub-areas. Calculate the yearly average values of (i) precipitation and (ii) evaporation for the entire watershed, assuming that there is no annual change in the groundwater storage. What are the annual runoff coefficients for the sub-areas and for the total catchment taken as a whole?

Sub-area	Area (Mm^2)	Annual precipitation (mm)	Annual evaporation (mm)
A	11.2	997	510
B	3.5	850	440
C	8.1	890	425
D	16.9	1320	670

9. Estimate the residence time of the following by using Tables 1.1 and 1.2.
   a. Global atmospheric moisture
   b. Global groundwater by assuming that only the fresh groundwater runs off the oceans
   c. Ocean water

Multiple Choice Questions

1. What is hydrology primarily concerned with?

A) The study of climate
B) The distribution and movement of water on Earth
C) The analysis of soil properties
D) The monitoring of weather patterns

2. Which process describes the movement of water from the surface into the soil?

A) Evaporation B) Transpiration
C) Infiltration D) Runoff

3. What is the term for water that flows over the ground surface after precipitation?

A) Infiltration B) Groundwater
C) Runoff D) Evaporation

4. Which of the following is a key component of the hydrological cycle?

A) Evaporation B) Filtration
C) Sedimentation D) Photosynthesis

5. Which term describes the area of land that drains water to a specific point, such as a river or lake?

A) Aquifer B) Watershed
C) Estuary D) Basin

6. What is the role of transpiration in the hydrological cycle?

A) It contributes to surface runoff

B) It helps in the movement of groundwater

C) It releases water vapor from plants into the atmosphere

D) It facilitates water absorption by soil

7. Groundwater can be stored in which of the following?

A) Aquifers
B) Reservoirs

C) Rivers
D) Lakes

8. What is the primary process that transforms liquid water into water vapor?

A) Condensation
B) Precipitation

C) Evaporation
D) Infiltration

9. Which of the following processes is part of the hydrological cycle?

A) Transpiration
B) Photosynthesis

C) Respiration
D) Nitrogen fixation

10. During which phase of the hydrological cycle does water return to the Earth's surface?

A) Evaporation
B) Condensation

C) Precipitation
D) Runoff

11. What role do rivers and streams play in the hydrological cycle?

A) They contribute to evaporation only

B) They transport water from the land to the ocean

C) They primarily store water

D) They have no role in the cycle

12. In which part of the hydrological cycle does infiltration occur?

A) When water vapor condenses into clouds
B) When rainwater soaks into the ground

C) When snow melts
D) When water evaporates from lakes

13. What is the term for water vapor that plants release into the atmosphere?

A) Precipitation
B) Evaporation

C) Transpiration
D) Infiltration

14. How does climate change impact the hydrological cycle?

A) It has no impact
B) It can increase evaporation rates
C) It can decrease precipitation
D) Both B and C

15. What is the term for the continuous movement of water in the environment?

A) Water cycle
B) Hydrological cycle
C) Aquatic cycle
D) Atmospheric cycle

16. What is a catchment area?

A) An area where water is stored
B) A region from which water drains into a specific water body
C) A designated land area for irrigation
D) A type of water treatment facility

17. What is another term commonly used for catchment area?

A) Watershed
B) Reservoir
C) Estuary
D) Basin

18. Which of the following factors can affect the size of a catchment area?

A) Topography
B) Land use
C) Vegetation cover
D) All of the above

19. How does vegetation in a catchment area impact water flow?

A) It increases evaporation only
B) It decreases infiltration rates
C) It reduces runoff and enhances water absorption
D) It has no effect on water flow

20. What is the primary purpose of studying catchment areas in hydrology?

A) To assess soil quality
B) To monitor air quality
C) To manage water resources and predict flooding
D) To analyse climate patterns

21. Which of the following best describes a "sub-catchment"?

A) A smaller area within a larger catchment
B) An area that only collects groundwater
C) A man-made structure for water storage
D) A type of irrigation system

22. What role do catchment areas play in the water cycle?

A) They only store surface water

B) They facilitate water movement and distribution

C) They prevent evaporation

D) They eliminate the need for precipitation

23. What is a common challenge associated with urban catchment areas?

A) Increased vegetation growth
B) Reduced water runoff

C) Increased runoff and pollution
D) Enhanced groundwater recharge

24. What does the water balance equation typically represent?

A) The amount of precipitation only
B) The input and output of water in a system

C) The temperature variations in a watershed
D) The distribution of water in oceans

25. Which of the following components is NOT usually included in the water balance equation?

A) Precipitation (P)
B) Evapotranspiration (ET)

C) Groundwater recharge (G)
D) Solar radiation (SR)

26. The simplified water balance equation can be expressed as:

A) $P = ET + R + D$
B) $P = ET + Q + \Delta S$

C) $P + ET = Q + \Delta S$
D) $P + Q = ET + \Delta S$

27. In the water balance equation, what does ΔS represent?

A) The amount of precipitation
B) The change in storage

C) The surface runoff
D) The total evaporation

28. Which factor can lead to a negative water balance in a catchment area?

A) High precipitation rates
B) Increased evapotranspiration

C) Reduced groundwater recharge
D) Both B and C

29. When assessing water balance, what does "Q" typically refer to?

A) Water quality
B) Groundwater levels

C) Streamflow or discharge
D) Water temperature

30. Which of the following scenarios could result in a positive water balance?

A) Extended drought periods

B) Increased precipitation and reduced evaporation

C) High rates of evapotranspiration

D) Increased urbanization

31. What is the importance of understanding water balance in hydrology?

A) It helps to predict weather patterns

B) It aids in water resource management and planning

C) It determines soil fertility

D) It measures atmospheric pressure

32. What percentage of the Earth's surface is covered by water?

A) 50%
B) 70%

C) 90%
D) 30%

33. What proportion of the Earth's water is freshwater?

A) About 2.5%
B) About 10%

C) About 25%
D) About 50%

34. Where is the majority of the world's freshwater stored?

A) Rivers
B) Lakes

C) Glaciers and ice caps
D) Aquifers

35. What is the main source of freshwater for human consumption?

A) Surface water
B) Desalination

C) Groundwater
D) Atmospheric moisture

36. In the global water balance, what does "P" typically represent?

A) Precipitation
B) Pollution

C) Permeability
D) Pressure

37. What is a significant contributor to the loss of freshwater resources globally?

A) Increased vegetation
B) Urbanization and industrialization

C) Natural climate variability
D) Improved water management

38. Which of the following is NOT a major use of freshwater globally?

A) Agriculture
B) Industrial processes

C) Hydropower generation
D) Solar power generation

39. What does the term "virtual water" refer to?

A) Water stored in glaciers

B) Water used in the production of goods and services

C) Water that is recycled

D) Water that cannot be quantified

40. What is one of the primary applications of hydrology in agriculture?

A) Pest control
B) Soil fertilization
C) Irrigation management
D) Crop selection

41. How is hydrology applied in urban planning?

A) To determine building materials

B) To assess water supply and drainage systems

C) To forecast population growth

D) To design recreational parks

42. Which application of hydrology helps in flood risk assessment?

A) Streamflow monitoring
B) Soil quality testing
C) Weather forecasting
D) Wildlife conservation

43. In the context of water quality management, hydrology is used to:

A) Analyse soil erosion
B) Measure atmospheric pressure
C) Monitor pollutants in water bodies
D) Evaluate plant growth

44. What role does hydrology play in climate change studies?

A) It measures solar radiation levels

B) It assesses changes in precipitation patterns and water availability

C) It predicts wind patterns

D) It analyses soil temperature

45. How is hydrology utilized in water resource management?

A) To design new agricultural techniques
B) To allocate water rights and usage
C) To build new cities
D) To develop renewable energy sources

46. What is a key application of hydrology in the field of ecology?

A) Designing irrigation systems

B) Assessing the health of aquatic ecosystems

C) Creating urban landscapes

D) Predicting natural disasters

47. In what way is hydrology important for engineering projects?

A) It provides aesthetic designs

B) It ensures structures can withstand water-related impacts

C) It helps choose construction materials

D) It enhances project marketing

Answer: 1-B; 2-C; 3-C; 4-A; 5-B; 6-C; 7-A; 8-C; 9-A; 10-C; 11-B; 12-B; 13-C; 14-D; 15-B; 16-B; 17-A; 18-D; 19-C;20-C; 21-A; 22-B; 23-C; 24-B; 25-D; 26-B; 27-B; 28-D; 29-C; 30-B; 31-B; 32-B; 33-A; 34-C; 35-C; 36-A; 37-B; 38-D; 39-B; 40-C; 41-B; 42-A; 43-C; 44-B; 45-B; 46-B; 47-B

KNOW MORE

For more information related to this topic scan the QR code.

OR

Type this link in your browser

https://drive.google.com/file/d/1b7TXl3OIfOsLVWnBx9VUQmXulqReFCqv/view?usp=sharing

BIBLIOGRAPHY

1. Asawa, G.L. (2005), "Irrigation and Water Resources Engineering, New Age International (P) Ltd., ISBN-13: 978-8122416732
2. Chow. V.T, (Ed), (1964), "Handbook of Applied Hydrology", McGraw-Hill, New York, NY, 1964.
3. IMD (1971), "Scientific Report No. 136, India Meteorological Department,1971, Government of India.

4. Jain S. K., Agarwal, P. K. And Singh, V. P. (2007), Hydrology and Water Resources of India, Water Science and Technology Library, Vol 57, Springer, The Netherland, ISBN-10 1-4020-5179-4(HB).
5. Schendel, V, (1975), “The world's water resources and water balance”, Natural Resources and Development, Inst. for Sci. Coop, Hannover, Germany, pp. 8-14, Vol. 1, 1975.
6. Shiklomanov (2000), “Appraisal and assessment of world water resources”, Water Int (25(1):11-32.
7. Singh, V. P. (1992). Elementary hydrology. Prentice Hall, New Jersy, 973.
8. Subramanya, K. (2015), “Engineering Hydrology”, McGraw-Hill Education (India) Private Limited, New Delhi, 2015, ISBN(13): 978-1-25902997-4.
9. UNESCO, (1978), “World Water Balance and Water Resources of the Earth”, Studies and Reports in Hydrology, 25, UNESCO, Paris, France, 1978.
10. Van der Leeden, (1975), “Water Resources of the World”, Water Information Center, Port Washington, N.Y., USA, 1975.

2 Precipitation

UNIT SPECIFICS

In engineering hydrology, precipitation is the primary source of water supply, and it can take different forms, including rain, snow, hail, and dew. The precipitation types, availability, measurement, and use are considered the basic requirements of a hydrological investigation. This unit describes the different forms of precipitation and their intensity and impact on water resources, agriculture, and ecosystems. In this unit, the characteristics of precipitation and different seasons in India have been discussed. Different measurement techniques of rainfall/snowfall have been described for the annual precipitation in India. In addition, methods to estimate missing data and data consistency analysis have also been discussed in the unit. Different methods for obtaining mean rainfall over a watershed are presented. Furthermore, the Depth Area Duration (DAD) curve, Probable Maximum Precipitation (PMP) and Rainfall data availability in India have been discussed.

RATIONALE

To understand the different forms of precipitation, mean precipitation over an area, Depth-area-duration relationship, maximum intensity/ depth-duration-frequency relationship, and Probable Maximum Precipitation for the design of various hydrological problems.

PRE-REQUISITE

Nil

UNIT OUTCOMES

The list of outcomes of this unit is as follows:

U2-O1: To understand different forms and characteristics of precipitation and its measurement

U2-O2: Presentation of rainfall data and measurement of mean precipitation over an area

U2-O3: To understand the intensity-duration-frequency and depth-area-duration curves

U2-O4: To understand the Probable Maximum Precipitation (PMP) over an area

Unit Outcomes	Expected Mapping with Course Outcomes (1- Weak Correlation; 2- Medium correlation; 3- Strong Correlation)							
	CO-1	CO-2	CO-3	CO-4	CO-5	CO-6	CO-7	CO-8
U2-O1	3	3	3	3	3	3	3	3
U2-O2	3	3	3	3	3	3	1	3
U2-O3	1	1	3	3	3	3	3	3
U2-O4	1	3	1	2	2	2	3	3

2.1 INTRODUCTION

Precipitation refers to any type of water, whether liquid or solid, that descends from the atmosphere and falls on the earth. Rain, snow, sleet, and hail are the different forms of precipitation. Precipitation is an important component of the Earth's water cycle, playing a key role in replenishing freshwater sources such as rivers, lakes, and groundwater. It is driven by various atmospheric processes, including condensation, which occurs when water vapour in the atmosphere cools and forms droplets or ice crystals that eventually grow heavy enough to descend to the ground under the influence of gravity. Different climates and areas have quite distinct patterns of precipitation, which are determined by elements like air pressure, temperature, humidity, and physical features.

The main source of water supply is precipitation, which can take different forms, including rain, snow, hail, dew, and so forth. Based on the origin, the precipitation can be classified into three types as orographic, convective, and cyclonic. During the monsoon season, orographic precipitation accounts for the majority of precipitation in India. Rain makes up the majority of precipitation, with relatively little coming from other forms like snow, hail, dew, etc. It is desirable to know the amount of rainfall in a specific area in order to determine the available water resources. Rain gauges are devices used to measure the amount of rain that falls. The Standard Non-Recording Rain Gauge has been adopted by the Indian Meteorological Department, which typically records rainfall at all stations at 8:30 AM.

The other types of rain gauges are the self-recording and tipping bucket models. Rainfall that falls during a 24-hour period is referred to as daily rainfall, whereas rainfall that falls within a year is referred to as annual rainfall. Rainfall data should be gathered annually for a duration of 35 to 40 years in order to calculate the average yearly rainfall. The formula for the index of wetness (Equation 2.1) can be given as:

$$Index\ of\ wetness = \frac{Actual\ ranfall\ in\ a\ given\ year\ at\ the\ given\ place}{Average\ aanual\ rain\ fall\ of\ that\ place} \quad \text{.............(2.1)}$$

A year's worth of wetness can be estimated using the index of wetness. For instance, a nearly 60% index value indicates a 40% shortfall. A substantial deficiency is defined as 30-45%, a major deficiency as 45–60%, and a calamitous deficiency as greater than 60%. A year with rainfall that is equal to or greater than the average is referred to as a normal year, whereas one with less rainfall is referred to as a sub-normal year.

2.2 FORMS OF PRECIPITATION

Depending on atmospheric conditions, rain, snow, sleet, hail, and freezing rain are various forms of precipitation. These different forms of precipitation are influenced by several climatic factors, such as temperature, humidity, atmospheric pressure, and air currents, leading to diverse weather conditions in different regions. These forms of precipitation are described briefly in the following sections.

a. **Rain**: Rain is one of the most common forms of precipitation, occurring when water droplets in the atmosphere coalesce and become heavy enough to fall to the ground. It is typically associated with nimbostratus or cumulonimbus clouds, although it can occur with other cloud types as well. Rainfall is a crucial component of the Earth's water cycle, replenishing freshwater sources and sustaining various ecosystems. Rainfall can vary in intensity, duration, and geographic distribution, influenced by factors such as atmospheric moisture content, temperature, air pressure, and wind patterns. It can occur as light, moderate, or heavy rain and may be accompanied by thunderstorms, lightning, or gusty winds in some cases. Rain plays a significant role in shaping weather patterns and climatic conditions around the world. Different regions experience varying amounts and frequencies of rainfall, leading to diverse ecosystems and agricultural practices. Excessive rainfall can also lead to flooding, while prolonged periods of drought can have severe impacts on water resources, agriculture, and ecosystems. On the basis of intensity, rainfall is classified in Table 2.1 as follows:

Table 2.1: Intensity of different types of rain

Type of rain	Intensity mm/hour
Light rain	< 2.5 mm/hour
Moderate rain	2.5-7.5 mm/hour
Heavy rain	>7.5 mm/hour

b. **Snow**: Snow is a type of precipitation that falls in the form of ice crystals or snowflakes. It occurs when water vapour in the atmosphere condenses directly into the ice crystals at temperatures below freezing (0°C or 32°F). These ice crystals join together to form snowflakes, which accumulate on the ground as snowfall. Snowfall can vary in intensity and accumulation, ranging from light flurries to heavy snowstorms. The characteristics of snow, such as its density and moisture content, can influence how much snow accumulates on the ground. Factors such as temperature, humidity, and atmospheric pressure also play a role in determining the type of snowfall. Snow is common in regions with cold climates, particularly during the winter months. It is an important source of freshwater, contributing to the replenishment of rivers, lakes, and groundwater when it melts. Snow cover also helps to regulate temperatures by reflecting sunlight, which can have significant effects on local climate patterns. Snowfall can impact transportation, infrastructure, and daily activities, sometimes leading to disruptions and hazards such as slippery roads, reduced visibility, and power outages. However, snow also provides opportunities for winter sports and recreation, and it contributes to the scenic beauty of many landscapes.

c. **Sleet**: Sleet is a form of precipitation that consists of small ice pellets or granules. It occurs when raindrops pass through a layer of cold air near the ground and freeze before reaching the surface. Sleet typically forms in winter conditions when there is a shallow layer of warm air above the ground and a deeper layer of cold air near the surface. The process of sleet formation begins when raindrops fall from higher layers of the atmosphere where the temperature is above freezing point. As these raindrops enter the colder layer of air near the ground, they supercool, meaning they remain in a liquid state despite being at temperatures below freezing. Upon contact with surfaces such as the ground, trees, or vehicles, the supercooled raindrops freeze into small ice pellets, resulting in sleet. Sleet is often associated with wintry weather conditions and can create hazardous driving and walking conditions, as the frozen pellets can accumulate on roadways and sidewalks, causing them to become slippery. While sleet may resemble hail, which forms within thunderstorms, it is generally smaller in size and occurs in different atmospheric conditions.

d. **Hail**: Hail is a form of precipitation that consists of solid ice pellets or balls known as hailstones. It forms within thunderstorms that have strong updrafts, which carry raindrops high into the atmosphere where temperatures are below freezing point. As the raindrops are lifted to colder altitudes, they freeze into small ice pellets. These pellets can then collide with supercooled water droplets or other hailstones within the storm cloud, causing them to grow in size. Hailstones can vary in size from small pea-sized pellets to large golf ball-sized or even larger hailstones. The size of hailstones is influenced by factors such as the strength of the updrafts, the duration of time spent in the storm cloud, and the presence of supercooled water droplets. Hailstorms can be destructive, causing damage to crops, vehicles, buildings, and other properties. In severe cases, large hailstones can pose a threat to life and safety. Hailstorms are most common in regions with strong thunderstorm activity, particularly during the warmer months of the year when atmospheric instability is high. While hail is often associated with thunderstorms, it can occur in other types of convective weather systems as well. Forecasters use radar and other tools to monitor atmospheric conditions and issue warnings for hailstorms when necessary to help protect life and property.

e. **Drizzle**: Drizzle is a type of light precipitation characterized by fine water droplets falling slowly from low-lying clouds. Unlike raindrops, which are larger and fall more rapidly, drizzle droplets are very small and tend to float down almost imperceptibly. Drizzle typically occurs when low-stratus clouds, such as nimbostratus clouds, are present in the atmosphere. It often produces misty or hazy conditions and can persist for extended periods, though it usually does not result in heavy accumulation. Drizzle is common in regions with maritime climates, or it occurs during transitional weather patterns, such as frontal passages. While not as intense as heavier precipitation types like rain or snow, drizzle can still impact visibility and create slippery road conditions.

f. **Glaze**: Glaze, also known as freezing rain, in meteorological terms, refers to a thin layer of ice that forms on surfaces such as roads, sidewalks, trees, and power lines when supercooled raindrops freeze upon contact. This phenomenon typically occurs during freezing rain events. Freezing rain occurs when rain falls from the clouds in liquid form and then encounters a layer of air at or near the surface with temperatures below freezing point. As the raindrops come into contact with cold surfaces, they freeze almost instantly, forming a transparent layer of ice known as glaze. This ice glaze can be hazardous, as it creates extremely slippery conditions on roads and walkways, leading to accidents and disruptions in transportation. Glaze can also accumulate on other surfaces, such as tree branches and power lines, adding weight and potentially causing damage or power outages. During severe freezing rain events, thick layers of glaze can build up, resulting in significant impacts on infrastructure and communities. Forecasters closely monitor weather conditions to issue warnings before freezing rain events and advise people to take precautions to stay safe during icy conditions. Measures such as salting roads and sidewalks can help to mitigate the risks associated with glaze formation.

2.3 CHARACTERISTICS OF PRECIPITATION IN INDIA

Precipitation in India is characterized by its variability in both spatial and temporal dimensions due to the country's diverse geographical features, monsoon climate, and regional influences. Understanding the characteristics of precipitation in India is essential for water resource management, disaster preparedness, agriculture, and urban planning to mitigate the impacts of extreme weather events and climate change. Here are some key characteristics of precipitation in India:

a. **Monsoon Dominance**: India's precipitation patterns are heavily influenced by the Indian Ocean monsoon system, which brings the majority of the country's annual rainfall. During the months of June to September, the Southwest Monsoon accounts for about 75-90% of India's annual precipitation. The Retreating Monsoon, which occurs from October to December, brings rainfall to parts of southern India.

b. **Regional Variation**: Precipitation varies significantly across different regions of India. Due to orographic effects and proximity to moisture sources, coastal areas such as the Western Ghats and northeastern states typically receive heavy rainfall. In contrast, arid and semi-arid regions, such as Rajasthan and parts of Gujarat, experience low precipitation.

c. **Seasonal Distribution**: India experiences distinct wet and dry seasons. During the monsoon season, most of the country receives abundant rainfall, while the pre-monsoon and post-monsoon periods are generally drier. However, some regions, particularly in the northeastern part and along the west coast, receive rainfall throughout the year.

d. **Intensity and Frequency**: Precipitation events in India can vary in intensity, ranging from light drizzles to heavy downpours and thunderstorms. These events can occur sporadically or in clusters, leading to localized flooding and waterlogging in urban areas.

e. **Variability and Trends**: Precipitation patterns in India exhibit interannual and intra-seasonal variability, influenced by several climatic factors such as the El Niño-Southern Oscillation (ENSO), Indian Ocean Dipole (IOD), and other climate phenomena. Climate change may also be affecting precipitation trends, with some studies suggesting changes in the frequency and intensity of extreme rainfall events.

f. **Importance for Agriculture**: Precipitation is vital for India's agricultural productivity, providing water for crops, irrigation, and replenishing groundwater resources. During the monsoon season, getting adequate rainfall for food security and the livelihoods of millions of farmers across the country is essential.

2.4 SEASONS IN INDIA

India has four different seasons based on differences in temperature and precipitation.

a. **The Cold Winter Season**: The cold winter season begins in early December and ends at the end of February. The coldest months are January and February. The weather throughout these months is dry and cool. In northern India, the temperature ranges from 10°C to 15°C, while in southern India, it is roughly 25°C. During this time, Tamil Nadu's coast experiences significant rainfall as a result of western disturbances. During this season, there are also occasional snowstorms or rainstorms in the northwest part of India. There is pleasant weather in other regions of the country during the clear and dry months of January and February.

b. **The Hot Summer Season:** March marks the beginning of the hot season, which lasts until mid-June. Because of the vertical sunlight over India at this season, the weather is quite hot. Some locations experience daytime highs that are nearly 50°C. Chottanagpur, Kerala, and the Western Ghats are known to witness pre-monsoon showers due to low-pressure, humid winds originating from the Arabian Sea. During the daytime, loo, or hot winds, blow, and the northern plains continue to be dry. In Punjab, Haryana, and Uttar Pradesh, dust storms can occasionally be followed by a cool breeze and light rain, which significantly reduces the temperature.

During this season, thunderstorms that provide light rain are common. During this season, hail also falls in some regions, which can occasionally harm crops. There are occasionally dust storms across the northern plains that are referred to as *Andhis* locally. These storms are accompanied by strong winds that carry large amounts of dust, which encourage erosion because of the dry weather. Large amounts of loose material, such as papers, dried leaves, polythene bags, etc., are carried by the wind. Only in the event where showers follow the gusts do this material usually settle quickly. Violent thunderstorms with strong gusts and brief rain also occur over the eastern and northeastern parts of the states of Bihar, West Bengal, and Assam.

a. **The Advancing Monsoon Season:** This season starts in mid-June and ends in September. Monsoon winds from the Arabian Sea and Bay of Bengal bring heavy rains to all parts of India. There are occasions observed when rain falls with varying intensities practically in nonstop mode for several days. When the monsoon breaks, the weather turns hot and muggy.
b. **The Retreating Monsoon Season:** The Retreating Monsoon Season spans from October to November. The sky is often clear, and the humidity is low throughout this season. Early October or late September is when the monsoon begins to recede. The hot, wet season gives way to the dry winter season throughout the months of October and early November.

Cyclones arise during the transitional phase when a low-pressure region moves from the northwest of India to the Bay of Bengal. These cyclones can occasionally get so fast that they cause terrible damage. Coastal regions of Bangladesh, Orissa, and Andhra Pradesh are typically impacted by these cyclones. These types of devastations caused significant loss of life and property, like in Bangladesh in 1970, Andhra Pradesh in 1977, and Andhra Pradesh in 1997. Individually, the 1999 Orissa disaster, which claimed over 10,000 lives, will never be forgotten.

2.5 MONSOONS

The Arabic term for "monsoon" refers to the shifting of the winds. Along the Indian Ocean coast, it is often used to show when the wind direction changes seasonally. On the Indian subcontinent, winds typically originate from the southwest half of the year and the northeast the other. The air rises and is replaced by warm, humid air from over the Indian Ocean when the Himalayan plateau warms up in the summer. This reversal of direction is caused by differential heating. At the end of May, when the temperature in India reaches its peak, the southeast trade winds originate from the south Indian Ocean and across the equator. The earth's rotation then swiftly deflects them, causing them to spread out as southwesters over the Arabian Sea, the Bay of Bengal, and the north Indian Ocean. These redirected trade winds arrive in the south Bay of Bengal early in May and then move over both marine areas. The term "Indian southwest monsoon" refers to this westerly stream that crosses India from the coast of Arabia to the China Sea.

The most notable aspect of the Indian climate is the summer monsoon or SW. This monsoon originates in the equatorial belt and travels in two separate currents before reaching the Indian subcontinent. These are known as the Bay of Bengal branch, which starts in the northeast part of the country, and the Arabian Sea branch, which affects the southern part of the peninsula. Together, the first and second branches travel the entire nation, with the first going west and the second north. Rainfall falls on the Andaman and Nicobar Islands from the Bay of Bengal branch and usually begins around May 20. Early June sees the monsoon move along the Konkan coast, arriving in Kerala by June 1. The monsoon normally reaches every part of the country by mid-July. During the monsoon season, the sky is usually dark, and there are frequent bouts of intense rain. The dates on which the nationwide monsoon begins are shown in Figure 2.1.

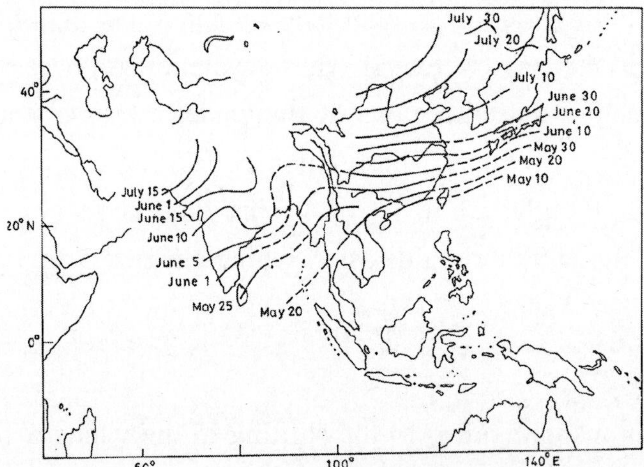

Figure 2.1: Dates on which the nationwide monsoon begins

All facets of Indian society, including politicians, planners, decision-makers, farmers, dealers, and so on, eagerly await the coming of the major rainy season, which is heralded by the summer monsoon. The monsoon season begins with the arrival of cool rains that relieve the stifling heat and signal the beginning of extensive plantation.

"Monsoon depressions" are low-pressure cyclonic systems that start in the Bay of Bengal around this time of year. In the northern part of the Bay of Bengal, they occur two or three times a month on average. They continue to move northward or northwestward, bringing evenly distributed rainfall to the northern and central regions of the nation. These depressions' route has a significant impact on how much rainfall falls in northern and central India. The SW monsoon circulation weakens and starts to retreat from the northwest regions of India in the second half of September. By the end of September, it virtually leaves the nation completely and is gradually replaced by a continental airflow that points towards the north. Along Tamil Nadu's east coast, the withdrawing monsoon winds occasionally produce showers; however, these are less frequent in the interior.

It does not rain every day throughout the monsoon season. There are times throughout the monsoon season known as "breaks in the monsoon" when there is no rainfall, and the temperature is unbearably high. Breaks often last three to seven days, and they occur frequently in August. Extended pauses could seriously harm the crops. The variability of monsoon rains from year to year is influenced by global factors such as El Nino, temperatures in the northern hemisphere, and snow cover, etc. During the monsoon, torrential downpours in the hilly catchments produce flooding throughout the lowlands. Normally, in the month of September, the monsoon withdraws. The typical monsoon withdrawal dates are displayed in Figure 2.2.

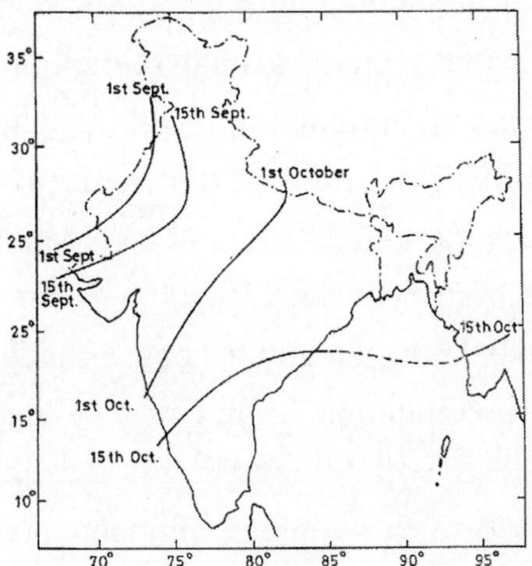

Figure 2.2: Normal dates of withdrawal of monsoon

2.6 ANNUAL PRECIPITATION IN INDIA

Including snowfall, India receives about 4,000 cubic kilometers or 120 cm, of precipitation annually. For a country of similar size, this sum is the biggest in the world. However, the yearly precipitation varies widely. The country receives 1,100 cm of rainfalls above the

Khasi and Jaintia Hills, whereas 200 cm falls in the Brahmaputra Valley to the north. At a height of 1,330 meters, Cherrapunji records up to 104 cm of rainfall per day and about 1,142 cm of annual precipitation. Cyclonic depressions, western disturbances, and the North and South East monsoons are the ways that India receives its rainfall. Most rainfall occurs between June and September under the influence of the Southwest monsoon, with the exception of Tamil Nadu, where it rains between October and November under the influence of the northeast monsoon. The rainfall in India varies greatly, with irregularities in its seasonal and geographical distribution as well as frequent departures from the normal.

A day is classified as a rainy day by the India Meteorological Department (IMD) if there is at least 2.5 mm of precipitation. There are less than 20 mean yearly rainy days in the northwest (West Rajasthan and the Kutchh region of Gujarat), but more than 180 days in the northeast. The southern sections of the West Coast also see about 140 days of precipitation annually. About 40 to 60 rainy days fall in India's central regions each year. Based on the reported spatial pattern, the mean intensity of rainfall varies every rainy day between 10 and 40 mm. In the far north, the lowest amount is less than 10 mm/day. Every rainy day in northwest India has an intensity of over 10 mm. The maximum value is seen in several regions of northeastern India and along the West Coast, where it is about 40 mm/day.

Figure 2.3 displays the average annual rainfall for the entire nation. The graphic illustrates the significant regional variation in India's annual precipitation. The annual rainfall ranges from over 1,100 mm in the northeastern regions to about 100 mm in the western deserts of the country. The wettest states in India include Arunachal Pradesh and Meghalaya. Other states that receive rainfall between 250 and 400 cm annually include Assam, the sub-Himalayas, West Bengal, the Andaman and Nicobar Islands, Konkan, Coastal Karnataka, and Kerala. Over 50% of the precipitation occurs in less than 15 days or roughly 100 hours annually. In the northeast, there are 150 rainy days on average, compared to roughly 5 in the western deserts. The work of water resources engineers is extremely difficult because of this huge variation.

India's west coast experiences heavy rainfall on the windward side of the Western Ghats slopes, which rapidly diminishes on the leeward side. On the east coast of India, the amount of rainfall increases near the sea and decreases in land. In the northern plains, the amount of precipitation decreases from 150 cm in West Bengal to roughly 10 cm in Rajasthan's extreme west.

The eastern Himalayas receive more annual precipitation than the western ones, with the former receiving over 200 cm while the later receives just over 70 cm. Additionally, it rapidly drops in the upper mountains and is higher in the lower foothills.

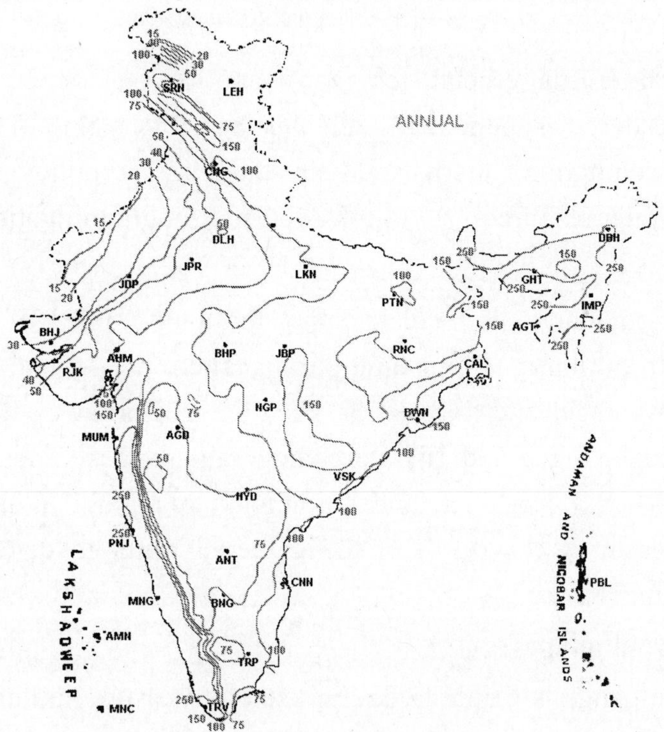

Figure 2.3: Average annual rainfall over India

2.7 MEASUREMENT OF PRECIPITATION

Precipitation is the process by which water falls from the atmosphere to the surface of the ground can be quantified using various tools and techniques. Overall, the measurement of precipitation is crucial for understanding and predicting weather patterns, assessing water resources, and managing various environmental and societal impacts associated with precipitation, such as flooding, droughts, and agriculture. The most common unit for measuring precipitation is depth, typically expressed in millimeters (mm) or inches (in). Here are some methods used to measure precipitation:

2.7.1 Measurement of rainfall

A rain gauge, a basic meteorological device intended to gather and quantify the quantity of precipitation that falls at a certain area during a predetermined amount of time, is commonly used to measure rainfall. Although there are many different designs of rain gauges, the most popular one is a cylindrical container with a funnel at the top to collect rainwater and route it into a graded measurement tube or cylinder. The collected water is measured in millimeters or inches.

Measuring rainfall is crucial for many applications, such as climate research, agriculture, water resource management, and weather forecasting. They provide valuable information about precipitation patterns and help scientists and policymakers make informed decisions related to water resources, flood control, and drought mitigation.

Rain Gauges

Rain gauges are the popular tools for calculating the total amount of precipitation(rainfall) that falls on the earth during a certain time period. There are several types of rain gauges, each with its own advantages and limitations. The rain gauges can be divided into two types: (i) Non-recording rain gauges and (ii) Recording rain gauges. These instruments collect precipitation (rainfall) in a container, and the amount of rainfall is measured manually or automatically by measuring the depth of the gathered water to determine the amount of precipitation.

a. Non-recording rain gauges:

A non-recording rain gauge is a simple device used to measure rainfall without providing a continuous or automatic recording of the data. It is similar to the Syphon rain gauge but without automated recording. Unlike recording rain gauges, which provide a continuous record of rainfall over time, non-recording rain gauges typically require manual observation and measurement. Rainwater is collected in a container equipped with a syphon tube, and the user manually measures the water passing through the syphon to determine rainfall depth.

Non-recording rain gauges (Figure 2.4) are often used in situations where continuous monitoring of rainfall is not necessary or in locations where automated recording systems are not feasible due to cost or maintenance considerations. These are commonly used in manual weather observation networks, educational settings, and for occasional or spot measurements of rainfall.

A standard non-recording rain gauge is a simple device used to measure rainfall without providing a continuous or automatic recording of the data. It is a basic instrument typically employed in manual weather observation networks, educational settings, or where automated systems are not feasible due to cost or maintenance considerations. A standard non-recording rain gauge consists of a cylindrical container with straight sides and an open top. Rainfall collection in most rain gauges is through a funnel-shaped aperture on the top. Rainwater gathered by the funnel collects at the bottom of the cylindrical container after falling into it. To measure rainfall using a standard non-recording rain gauge, an observer needs to manually check the gauge after a rainfall event. The collected rainwater depth in the container is measured using a ruler or graduated cylinder. The depth of the collected water is then recorded as the amount of rainfall that occurred during the observed period.

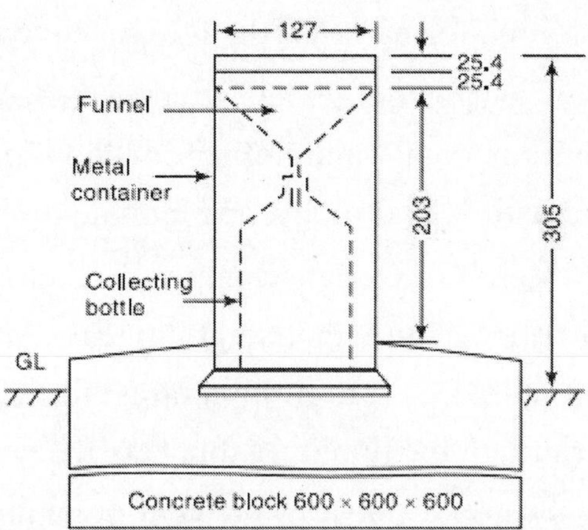

Figure 2.4: Non-recording rain gauge (syphon's rain gauge)

The accuracy of measurements from a standard non-recording rain gauge depends on various factors, including the gauge's design, placement, and the skill of the observer. These gauges provide a single-point measurement for a specific location and time, so they might not capture localized variations in rainfall. Standard non-recording rain gauges require regular maintenance, including cleaning and calibration, to ensure accurate measurements over time.

Overall, while standard non-recording rain gauges are simple and cost-effective tools for measuring rainfall, they rely on manual observation and are suitable for applications where continuous monitoring is not necessary.

b. Recording Rain gauges

i. **Tipping Bucket Rain Gauge:** One kind of rain gauge that is frequently used to measure precipitation is the tipping bucket rain gauge. It operates on a simple principle: as rain falls into the funnel-shaped top of the gauge, it is collected in a small container or bucket. When a particular volume of water accumulates in it, it tips or empties, allowing the next increment of rainfall to be collected. Each tip of the bucket corresponds to a known volume of rainfall, making it relatively easy to measure precipitation accurately.

Tipping bucket rain gauges are popular because they offer several advantages. These rain gauges provide accurate and reliable measurements of precipitation. They are relatively simple to operate and maintain and can measure both the intensity and total accumulation of rainfall. Tipping bucket rain gauges can be automated and integrated into weather monitoring networks for real-time data collection.

Tipping bucket rain gauges are widely used in meteorological stations, research institutions, and environmental monitoring networks for various applications, including weather forecasting, hydrology, and climate research.

ii. **Weighing Bucket Rain Gauge:** A weighing bucket rain gauge, also known as a weighing precipitation gauge or weighing rain gauge, is another type of instrument used to measure rainfall. In contrast to the tipping bucket rain gauge, which measures rainfall by tipping a bucket, the weighing bucket rain gauge directly measures the weight of the collected precipitation. This gauge measures rainfall by the increase in weight of the collected water. It consists of a container placed on a sensitive balance. As rain falls into the container, the weight of the water increases and this change in weight is recorded as rainfall.

Weighing bucket rain gauges offer several advantages. These rain gauges provide highly accurate measurements of precipitation, including rainfall intensity and total accumulation. They are not affected by wind or other environmental factors that may influence the accuracy of measurements in other types of rain gauges. They can be used in remote or harsh environments where other types of rain gauges may be less practical. However, weighing bucket rain gauges also have some limitations, including their relatively high cost and the need for periodic maintenance and calibration to ensure accuracy.

Overall, weighing bucket rain gauges are valuable tools for meteorological stations, research institutions, and environmental monitoring networks, particularly in applications requiring precise measurements of precipitation.

iii. **Optical Rain Gauge:** This type of gauge uses optical sensors to detect raindrops. It typically emits light or laser beams across a known distance and measures disruptions in the beams caused by raindrops passing through. This data is then used to calculate rainfall intensity.

iv. **Acoustic Rain Gauge:** This gauge measures rainfall by analyzing the sound of rain hitting a surface. It typically uses sensors to detect the sound waves generated by raindrops and converts this information into measurements of rainfall intensity.

v. **Radar Rain Gauge:** Radar technology can be used to estimate rainfall by measuring the reflectivity of precipitation particles in the atmosphere. This method provides spatial and temporal information about rainfall over a large area.

vi. **Satellite-based Rain Gauge:** Satellites equipped with remote sensing instruments can provide estimates of rainfall over large geographical areas by analyzing cloud patterns and other atmospheric parameters.

2.7.2 Measurement of snowfall

Measuring snowfall presents unique challenges compared to measuring rainfall due to the variability in snow density and accumulation patterns. Several methods are commonly used to measure snowfall. It is notable that each method has its advantages and limitations, and the choice of measurement technique depends on factors such as accuracy requirements, available resources, and environmental conditions. Combining multiple measurement techniques can provide a more comprehensive understanding of snowfall patterns and their impacts. The methods commonly used to measure snowfall are described in brief in the following sections.

a. **Snow Gauges:** In regions where snowfall is common, snow gauges are used to measure the depth of snow accumulation. These gauges typically consist of a flat surface where snow accumulates, and a ruler or measuring stick are used to manually measure the depth of the deposited snow.

b. **Snowboard:** A snowboard is a flat, elevated platform placed on the ground to collect snow. Snow accumulates on the board, and its depth is measured at regular intervals using a ruler or measuring stick. This method provides a simple and inexpensive way to measure snow accumulation but may underestimate the true amount of snowfall due to compaction.

c. **Snow Stake:** Similar to a snowboard, a snow stake is a vertical pole with markings to measure snow accumulation. The stake is installed in an open area, and snow depth is measured by observing the height of the snow relative to the markings on the stake. This method is also susceptible to compaction but can provide useful data when combined with other measurement techniques.

d. **Snow Pillow**: A snow pillow is a large, inflatable bladder placed on the ground to measure the weight of the snow above it. As snow accumulates, it compresses the pillow, and the change in pressure is recorded electronically. This method is commonly used in remote areas and automated weather stations.

e. **Snow Depth Sensor:** These sensors use various technologies, such as ultrasonic pulses or lasers, to measure the distance between the sensor and the ground surface. By continuously monitoring this distance, snow depth can be determined accurately.

f. **Snow Course**: A snow course involves measuring snow depth at multiple locations along a transect using manual or automated techniques. This method provides a spatially distributed view of snow accumulation and is commonly used for hydrological and water resource management purposes.

2.8 MEASUREMENT OF RAINFALL

2.8.1 Adequacy of Rain Gauge Stations

If there are already a few rain gauge stations in a catchment, statistical analysis can establish the optimal number of stations that should exist to have a given percentage of error to determine the estimate of mean rainfall as Equation (2.2):

$$N = \left(\frac{C_v}{\varepsilon}\right)^2 \qquad \text{.............(2.2)}$$

Where,

N = optimal number of stations,

Σ = allowable degree of error to determine the estimate of mean rainfall and

C_v = coefficient of variation of the rainfall values (in %) at the existing m stations.

If the catchment has m stations, each of which records the amount of rainfall values X_1, X_2, ...,X_i, ...,X_m in a given time, the coefficient of variation C_v can be obtained as Equation (2.2a):

$$C_v = \frac{100 * \sigma_{m-1}}{\bar{X}} \qquad \text{.............(2.2a)}$$

where,

$$\sigma_{m-1} = \sqrt{\left[\frac{\sum_1^m (X_i - \bar{X})^2}{m-1}\right]} = \text{standard deviation}$$

X_i = magnitude of precipitation in the ith station

$\bar{X} = \frac{1}{m}(\sum_1^m X_i)$ = average precipitation

Consider the existing m rain gauges, having C_v as their coefficient of variation and $\bar{X}$ as their mean rainfall. To find the percentage of inaccuracy ($\in_{ex}$) in the mean estimation in the existing m rain gauge system, then Equation (2.2) can be rewritten (by replacing N by m) as Equation (2.3):

$$m = \left(\frac{C_v}{\in_{ex}}\right)^2 \qquad \textbf{.............(2.3)}$$

where,

$$\in_{ex} = \frac{C_v}{\sqrt{m}} \qquad \textbf{..........(2.3a)}$$

The term "$\in_{ex}$" in the aforementioned Equation (2.3a) denotes the predicted error (in percentage) in the mean $\bar{X}$, estimation. Standard error in the estimation of mean precipitation is used to assess the precision of mean precipitation estimation in the current system.

Typically, one uses Equation (2.3) to calculate the number of rain gauges N for a particular amount of error, taking $\in_{ex}$ to be 10%. It may be observed that a smaller value of $\in_{ex}$ will necessitate a larger number of rain gauges. WMO guidelines state that self-recording rain gauges should make up at least 10% of all rain gauges.

Example 2.1: There are seven rain gauge stations inside the catchment of a watershed. The annual rainfall in a given year that the rain gauges recorded is as follows:

Station	A	B	C	D	E	F	G
Rainfall (cm)	82.8	105.9	170.3	116.5	99.9	130.7	115.8

Obtain the standard error in the average rainfall estimation for the current set of rain gauges. Determine the ideal number of stations in the catchment, accounting for a 9% error in the mean rainfall estimate.

Solution

For the given data,

No of rain gauges M = 7;

Mean annual rainfall $\bar{P} = \frac{1}{m}(\sum_1^m P_i)$

$$= \frac{82.8+105.9+170.3+116.5+99.9+130.7+115.8}{7}$$

$$= 117.41 \text{ cm}$$

Standard deviation: $\sigma_{m\text{-}1} = \sqrt{\left[\frac{\sum_1^m (P_i-\bar{P})^2}{m-1}\right]}$

$$= 27.73 \text{ cm}$$

Coefficient of variation $C_v = \frac{100*\sigma_{m-1}}{\bar{P}}$

$$= \frac{100*27.73}{117.41}$$

$$= 23.62$$

Standard error in the mean rainfall Calculation:

$$\epsilon_{ex} = \frac{C_v}{\sqrt{m}} = \frac{23.62}{\sqrt{7}} = 8.9\%$$

When the error is limited to 9%, $\epsilon = 9$; optimum number of raingauges in the watershed can be determined by:

$$N = \left(\frac{C_v}{\varepsilon}\right)^2 \left(\frac{23.62}{9}\right)^2 = 6.88 \text{ say } = 7 \text{ stations}$$

For the watershed, the optimal number of stations is seven, which seems that the number of rain gauges available in the catchment are optimum.

2.8.2 Preparation of data

Prior to using a station's rainfall records, the data must be verified for consistency and continuity. A rain gauge malfunction or other damage can cause missing data, disrupting the continuity of a record for a variety of causes. By utilizing the data from nearby stations, estimation of the missing data is possible. The typical rainfall is utilized as a benchmark in these computations. The average amount of rainfall at a specific date, month, or year for a given 30-year period is known as the normal rainfall. Every ten years, the 30-year average is recalculated. Accordingly, the average annual precipitation at station A based on a given 30-year record is referred to as "normal annual precipitation at station A."

a. Estimation of missing data

At neighboring M stations 1, 2, 3....*M*, respectively, annual precipitation values P_1, P_2, $P_3 ... P_m$ is known. The yearly precipitation P_x that is missing at station X, which is not one of the M stations listed above, is to be obtained. Additionally, each of the aforementioned (M + 1) stations, including station X, has known typical annual precipitation values N_1, N_2 ...N_i .... If the normal annual precipitation at several stations is within 10% of the normal annual precipitation at station X, then P_x can be estimated using a simple arithmetic average method as Equation (2.4). Consequently,

$$P_x = \frac{1}{M}[P_1 + P_2 + \ldots + P_m] \qquad \text{.........(2.4)}$$

By measuring the precipitation at different stations based on the ratios of normal yearly precipitations, the value of P_x can be determined as Equation (2.5), if the normal precipitations vary significantly. This method is known as the *normal ratio method.* According to the *normal ratio method:*

$$P_x = \frac{N_x}{M}\left[\frac{P_1}{N_1} + \frac{P_2}{N_2} + \ldots + \frac{P_m}{N_m}\right] \qquad \text{.......(2.5)}$$

Example 2.2: In a particular catchment, at stations A, B, C, and D, the average yearly rainfall is 85.90, 63.15, 75.18, and 89.91 cm, respectively. Station D was not operational in 1996; annual precipitation at stations A, B, and C was 91.71, 69.72, and 82.98 cm, respectively. Calculate the amount of rain that year at station D.

Solution:

Maximum variation in normal rainfall at station D = 89.91-63.15 = 26.76 cm

Maximum variation with respect to normal at station D in percentage

$$= \frac{26.76}{89.91} x100 = 29.76\%$$

As the variation in normal rainfall values is more than 10%, the normal ratio method is used.

In the normal ratio method M=3

Estimated rainfall at station D= $P_D = \frac{N_D}{M}\left[\frac{P_1}{N_1} + \frac{P_2}{N_2} + \ldots + \frac{P_C}{N_C}\right]$

$$P_D = \frac{89.91}{3}\left[\frac{91.71}{85.90} + \frac{69.72}{63.15} + \frac{82.98}{75.18}\right]$$

=29.97(1.06+1.10+1.10)

= 97.70 cm

Example 2.3: Four rain gauge stations, namely Jhabua, Alirajpur, Dhar and Sardarpur, are located in a typical region, as shown in the Fig below. The rainfall observed during a month at Jhabua, Dhar and Alirajpur is 132.2 mm, 113.3 mm, and 150.5 mm, respectively, while the corresponding rainfall value at Sardarpur is missing. If the normal monthly rainfall at Sardarpur, Jhabua, Dhar and Alirajpur are 246.2 mm, 203.2 mm, 233.4 mm, and 220.8 mm, respectively, obtain the value of missing rainfall data at Sardarpur rain gauge station using the normal ratio method.

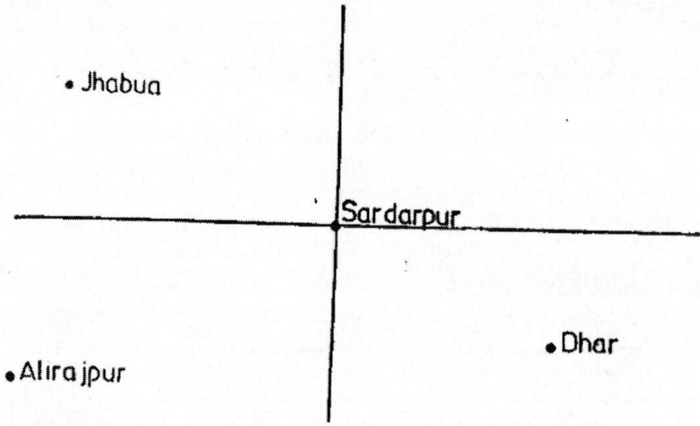

Figure 2.5: Estimation of missing rainfall data

Solution

Maximum variation in normal rainfall at Sardarpur = 246.2 - 203.2 = 43.0 cm

Maximum variation with respect to normal at station Sardarpur in percentage

$$=\frac{43}{246.2} x100 = 17.46\%$$

As the variation in normal rainfall values is more than 10%, the normal ratio method is used.

In the normal ratio method M=3

Estimated rainfall at station Sardarpur

$$=P_D=\frac{N_D}{M}\left[\frac{P_1}{N_1}+\frac{P_2}{N_2}+\cdots+\frac{P_C}{N_C}\right]$$

$$P_D=\frac{246.2}{3}\left[\frac{132.2}{203.2}+\frac{113.3}{233.4}+\frac{150.5}{220.8}\right]$$

=82.07(0.65+0.49+0.68)

= 149.36 cm

2.8.3 Test for Consistency of Record

The rainfall data from a rain gauge station will become inconsistent if there has been a substantial change in the relevant conditions during the recording period. This discrepancy would be apparent as soon as the big shift occurred. Several common reasons for inconsistent records include (i) shifting the site of a rain gauge station, (ii) noticeable changes occurring in the station's neighborhood, (iii) changes in the ecosystem as a result of natural calamities such as forest fires and landslides; and (iv) observational errors occurring after a specific date. The double-mass curve technique can be used to analyze a record for discrepancies. The backbone of this approach is the concept that reported data are reliable if they come from the same parent population.

A selection is made of five to ten base stations that are located in the vicinity of the problematic station X. The annual (or monthly) mean rainfall at station X and the rainfall average of base stations over an extended period of time are presented in reverse chronological order, with the most recent record displaying first and the oldest record displaying last in the list. The cumulative precipitation of station X (i.e., $\sum P_x$) and the accumulated value of the average of the base station group (i.e., $\sum P_{av}$) are computed commencing with the most recent record.

Values of $\sum P_x$ versus $\sum P_{av}$ are plotted for different successive time intervals, as shown in Figure 2.6. The relation is used to rectify a clear break in the slope of the derived values at station X following the regime change phase (see Figure 2.6, point 63). The corrected precipitation at station X for any time period t_1 can be obtained using Equation (2.6).

$$P_{cx} = P_x \frac{M_c}{M_a} \qquad \text{……….(2.6)}$$

where,

P_{cx} = corrected precipitation at station X for any time period t_1

P_x = original corrected precipitation at any time period t_1 at station X

M_c = corrected slope of the double-mass curve

M_a = original slope of the mass curve

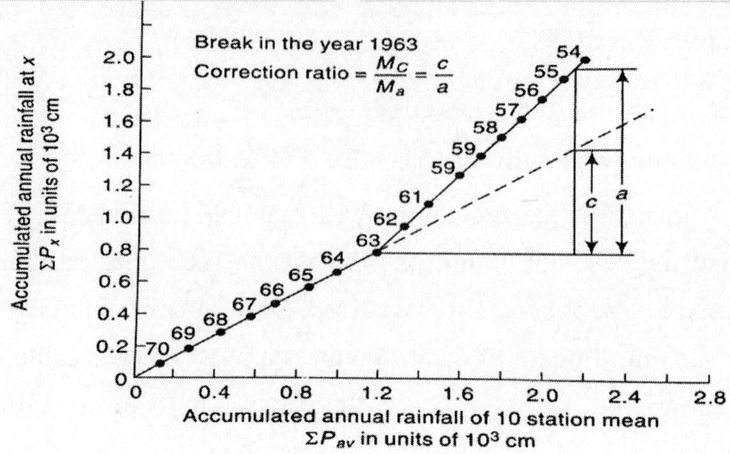

Figure 2.6: Double-mass curve

Example 2.4: The annual rainfall data for station M and the average annual rainfall values for ten neighboring stations situated in a region with a uniform climate are provided below.

Year	Rainfall of station M (mm)	Average annual rainfall of the group (mm)	Year	Rainfall of station M (mm)	Average annual rainfall of the group (mm)
1950	676	780	1965	1244	1400
1951	578	660	1966	999	1140
1952	95	110	1967	573	650
1953	462	520	1968	596	646
1954	472	540	1969	375	350
1955	699	800	1970	635	590
1956	479	540	1971	497	490
1957	431	490	1972	386	400
1958	493	560	1973	438	390
1959	503	575	1974	568	570
1960	415	480	1975	356	377
1961	531	600	1976	685	653
1962	504	580	1977	825	787

Year	Rainfall of station M (mm)	Average annual rainfall of the group (mm)	Year	Rainfall of station M (mm)	Average annual rainfall of the group (mm)
1963	828	950	1978	426	410
1964	679	770	1979	612	588

Evaluate consistency in available yearly rainfall data for station M and modify the record if any discrepancies are found. Determine the average annual precipitation at station M.

Solution:

The data is sorted by year in descending order, commencing from 1979, which is the most recent year. Table 2.2 displays the computation of the ten station average rainfall values ($\sum P_{av}$) and the cumulative rainfall value of station M ($\sum P_m$). Plotting the data using ($\sum P_m$) on the Y-axis and ($\sum P_{av}$) on the X-axis results in a double-mass curve plot (Figure 2.7). Each plotted point's corresponding year value is also displayed on the plot. Two straight lines are used to represent the data, with a grade break occurring in 1969. This represents a change in Station M management after 1968. The best straight line for the years 1979–1969 has a slope of M_c = 1.0295, while the slope for the years 1968–1950 is M_a = 0.8779. The ratio of modification to update the previous records, which span the years 1950–1968, is the 1968 record

$$= \frac{M_c}{M_a} = \frac{1.0295}{0.8779} = 1.173.$$

Table 2.2: Evaluation of Double-Mass curve of Example 2.4.

Year	P_m (mm)	ΣP_m (mm)	P_{av} (mm)	ΣP_{av} (mm)	Adjusted values of P_m(mm)	Finalized values of P_m (mm)
1979	612	612	588	588		612
1978	426	1038	410	998		426
1977	825	1863	787	1785		825
1976	685	2548	653	2438		685
1975	356	2904	377	2815		356
1974	568	3472	570	3385		568

Year	P_m (mm)	ΣP_m (mm)	P_{av} (mm)	ΣP_{av} (mm)	Adjusted values of P_m(mm)	Finalized values of P_m (mm)
1973	438	3910	390	3775		438
1972	386	4296	400	4175		386
1971	497	4793	490	4665		497
1970	635	5428	590	5255		635
1969	375	5803	350	5605		375
1968	596	6399	646	6251	698.92	699
1967	573	6972	650	6901	671.95	672
1966	999	7971	1140	8041	1171.51	1172
1965	1244	9215	1400	9441	1458.82	1459
1964	679	9894	770	10211	796.25	796
1963	828	10722	950	11161	970.98	971
1962	504	11226	580	11741	591.03	591
1961	531	11757	600	12341	622.70	623
1960	415	12172	480	12821	486.66	487
1959	503	12675	575	13396	589.86	590
1958	493	13168	560	13956	578.13	578
1957	431	13599	490	14446	505.43	505
1956	479	14078	540	14986	561.72	562
1955	699	14777	800	15786	819.71	820
1954	472	15249	540	16326	553.51	554
1953	462	15711	520	16846	541.78	542
1952	95	15806	110	16956	111.41	111
1951	578	16384	660	17616	677.81	678
1950	676	17060	780	18396	792.73	793

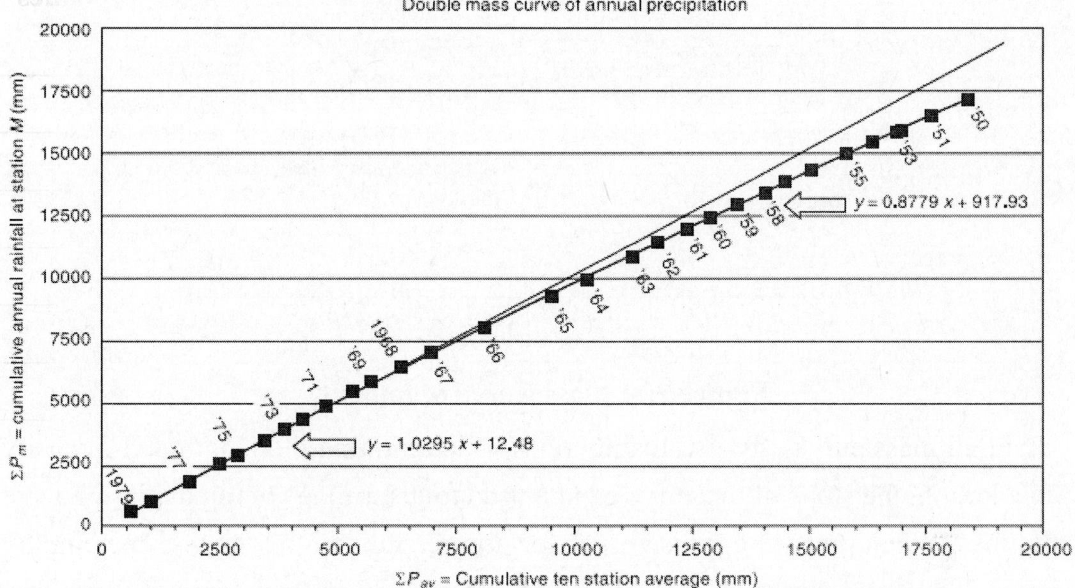

Figure 2.7: Double mass curve of yearly rainfall at station M

Multiply each annual rainfall value before 1969 by the adjustment ratio of 1.173 to obtain the corrected value. Column 6 of Table 2.2 shows the updated outcomes at station M. The finalized P_m values (rounded to the nearest millimeter) for each of the 30 years of data are shown in Column 7.

The mean annual precipitation at station M (based on the corrected time series)

= 19004/30 = 633.5 mm

2.9 PRESENTATION OF RAINFALL DATA

The following list includes often used techniques for calculating rainfall and precipitation data that have been shown to be beneficial in the processing and interpretation of such data.

2.9.1 Mass Curve of Rainfall

The mass curve plot of rainfall is a representation of the total precipitation against time, organized chronologically. This kind of record is applicable to float-type and weighing bucket-type gauges. A typical mass curve of rainfall at a place during a storm is shown in Figure 2.8.

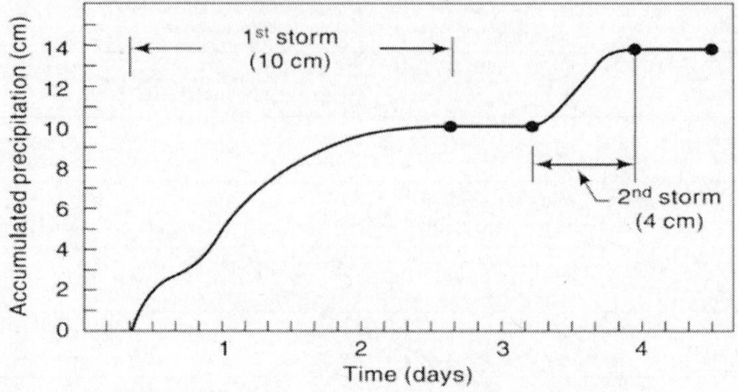

Figure 2.8: Mass Curve of rainfall

Rainfall mass curves are a valuable tool for determining the length and intensity of a storm. In addition, the slope of the curve can be used to determine the intensities of a storm at different times. When preparing mass curves for non-recording rain gauges, one employs the mass curves of adjacent recording gauge stations as a gauge and approximates storm start and end times.

2.9.2 Hyetograph

A hyetograph can be defined as plotting the amount of precipitation against a time duration. The mass curve is the source of the hyetograph, which is typically shown as a bar chart (Figure 2.9). It is a very important way to show the characteristics of a storm and can be extremely beneficial when creating design storms to forecast catastrophic flooding. The entire amount of precipitation received during the period is shown by the area under a hyetograph. The duration employed varies depending on the goal; for example, brief durations are used in urban drainage problems, whereas intervals of roughly six hours are used in larger watershed flood-flow estimates.

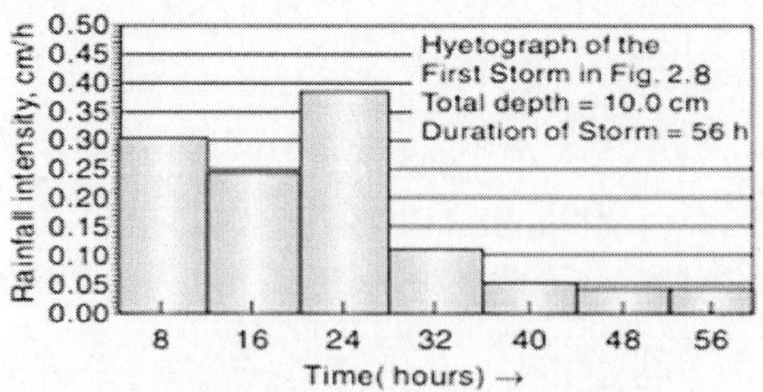

Figure 2.9: Hyetograph of a storm

2.9.3 Point Rainfall

Point rainfall, also known as station rainfall, is the term used to describe the rainfall data obtained from a station. Depending on the demands, data may be presented as daily, weekly, monthly, seasonal, or annual values for various time periods. The magnitude vs time graphs for these data are displayed graphically as bar diagrams. However, such a plot is not valuable for recognizing a trend in the rainfall because there will be huge variances in the rainfall values that cause the plot to shift quickly. A moving average plot, which shows the average precipitation value over three or five consecutive time intervals at the middle of the period, can be used to smooth out differences and highlight patterns.

2.10 MEAN PRECIPITATION OVER AN AREA

Only the location of the rain gauge's installation area is represented by its representation of rainfall. Nevertheless, in practical hydrological analysis, the estimate of point rainfall is transformed into the average rainfall throughout a catchment using the following techniques: (a) Arithmetic mean technique, (b) Thiessen Polygon technique and (c) The isohyet technique.

2.10.1 Arithmetical-Mean technique

When there is little variation in the rainfall recorded at various stations within the watershed, the mean precipitation over the watershed is calculated by taking the arithmetic mean of the station data. Thus, using the arithmetic-mean technique the mean precipitation $\bar{P}$ over the catchment area Equation (2.7) with the following values $P_1, P_2, ..., P_i, ..., P_n$ of measured precipitation at n stations rainfall during a specific period can be expressed as follows:

$$\bar{P} = \left[\frac{P_1+P_2+\cdots P_i+\cdots+P_n}{N}\right] = \frac{1}{N}\sum_{i=1}^{N} P_i \qquad ....(2.7)$$

The Arithmetic method is suitable for basins with a large number of rain gauge stations spaced uniformly and if the variation of individual gauge reading over the mean is not large. However, it may lead to considerable error, if the rain gauge network over the catchment is not uniform and sparse and when the variability of rainfall distribution over a given catchment is large.

2.10.2 Thiessen Polygon Method

In this method, every rain gauge is allocated a certain area of the catchment which is closest to that rain gauge. The allocation of area, A_i to the ith rain gauge is done using Thiessen polygon method (Figure 2.10). Because different stations are assigned weights on a rational basis, the Thiessen polygon approach is better than the arithmetic mean technique.

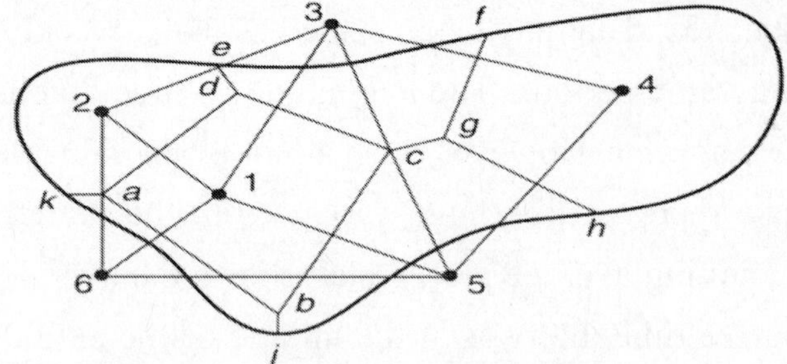

A = total catchment area

Station	Bounded by	Area	Weightage
1	abcd	A_1	A_1/A
2	kade	A_2	A_2/A
3	edcgf	A_3	A_3/A
4	fgh	A_4	A_4/A
5	hgcbj	A_5	A_5/A
6	jbak	A_6	A_6/A

Figure 2.10: Thiessen polygon

Three stations exist in the neighborhood, although they are not inside the catchment. The scale-drawn catchment region shows the exact locations of the total six rain gauge stations. A network of triangles is created by connecting stations 1 through 6. For every side of the triangle, the perpendicular bisectors are drawn. Around every station, these bisectors create a polygon. If the catchment boundary intersects the bisectors, it is considered the polygon's outer boundary. Therefore, the bounding polygon for station 1 is abcd. kade is used to enclose the polygon for station 2. Like this, all the names have been given based on the Thiessen polygons. These six Thiessen polygons had their areas measured using an overlay grid or a planimeter. If $A_1, A_2, \ldots, A_6$ are the respective areas of the Thiessen polygons and $P_1, P_2, \ldots, P_6$ are the magnitudes of rainfall obtained by the stations 1, 2,..., 6 accordingly, then the mean rainfall over the watershed can be denoted by Equation (2.8 and 2.9). The weighting factor for each station is defined as the ratio A_i/A.

$$\bar{P} = \frac{P_1A_1+P_2A_2+\cdots+P_6A_6}{A_1+A_2+\cdots+A_6} \qquad \text{...........(2.8)}$$

Thus, in general, for M rain gauge stations,

$$\bar{P} = \frac{\sum_{i=1}^{M} P_iA_i}{A} = \sum_{i=1}^{M} P_i\frac{A_i}{A} \qquad \text{...........(2.9)}$$

The Thiessen-polygon method of calculating the mean precipitation over a region is better than the arithmetic-average method because each station is given an appropriate amount of weight.

In the case of a fixed station network, the computation is rather simple once the weightage factors are established. This approach has an advantage over the arithmetic mean method in that it considers the catchment's non-uniform rain gauge distribution by assigning distinct areas to the appropriate rain gauges. This is merely a geometrical exercise that ignores the catchment's true variability in rainfall frequency.

Example 2.5: Four rainfall stations are located inside a catchment region, roughly represented by a circle with a diameter of 100 km, and one station is located nearby. The five stations' coordinates as well as the catchment centers are listed below. The yearly precipitation totals from the five sites in 1982 are also provided. Calculate the average annual precipitation using (a) the Thiessen-mean approach (b) the arithmetical mean method.

Centre: (100,100)   Diameter: 100 km   Distance is in km

Station	1	2	3	4	5
Coordinates	(30,80)	(70,100)	(100,140)	(130,100)	(100,70)
Precipitation (cm)	95.0	136.2	95.4	145.9	101.9

***Solution*: Thiessen-mean approach:** The stations are indicated on a scale drawing of the catchment region (Figure 2.11). After joining the stations to create a set of triangles, each side's perpendicular bisector is drawn. The stations are then identified by the Thiessen-polygon enclosing each station. It may be noted that station 1 in the given problem has no influence within the catchment. One can use a planimeter or overlay a grid to calculate the areas of different Thiessen polygons. The calculation of weighted rainfall by the Thiessen polygon method is presented in Table 2.3 below.

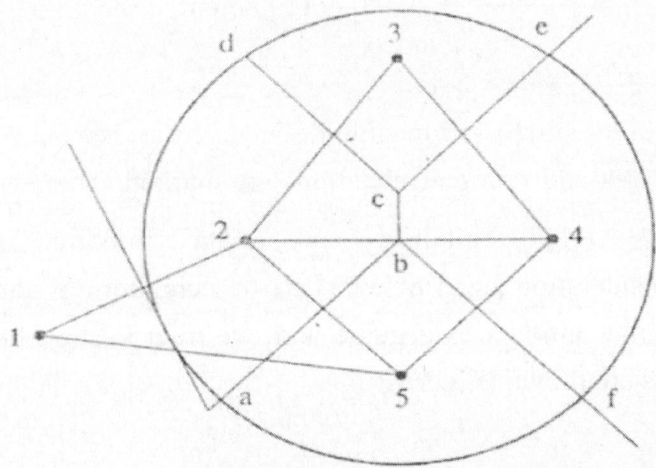

Figure 2.11: Thiessen polygons

Table 2.3: Calculation of weighted rainfall by Thiessen polygon method

Station	Boundary of area	Area (km^2)	Fraction of total area	Rainfall	Weighted P(cm) (col. 4 x col. 5)
1	-	-	-	85.0	-
2	abcd	2140.9	0.2726	136.2	37.13
3	dce	1608.8	0.2048	95.4	19.54
4	ecbf	2141.1	0.2726	145.9	39.77
5	fba	1962.7	0.2500	101.9	25.48
Total		7853.5	1.0000		121.92

Mean precipitation = 121.92 cm.

Arithmetical mean method: Out of five rain gauge stations, only four stations 2,3,4, and 5, fall within the basin. Station no 1 fall outside the basin. Hence, only four stations 2,3,4, and 5 will be considered for obtaining the arithmetic mean. Hence, the mean annual precipitation for the basin for 1982:

$$\bar{P} = \left[\frac{P_1 + P_2 + \cdots P_i + \cdots + P_n}{N}\right]$$

$$\bar{P} = \left[\frac{136.2 + 95.4 + 145.9 + 101.9}{4}\right]$$

$$\bar{P} = \left[\frac{P_1+P_2+\cdots P_i+\cdots+P_n}{N}\right] = 119.85 \text{ cm}$$

2.10.3 Isohyet Method

A line connecting points of equal rainfall magnitude is called an isohyet. Using the isohyet technique, the catchment area and rain gauge stations are marked. The recorded values that need to be obtained areal average rainfall $\overline{P}$ and are then indicated at the appropriate stations on the plot. Consideration is also extended to neighboring stations outside the catchment. Next, using point rainfalls as a guide and eye interpolation between them, the isohyets of different values are drawn (Figure 2.12).

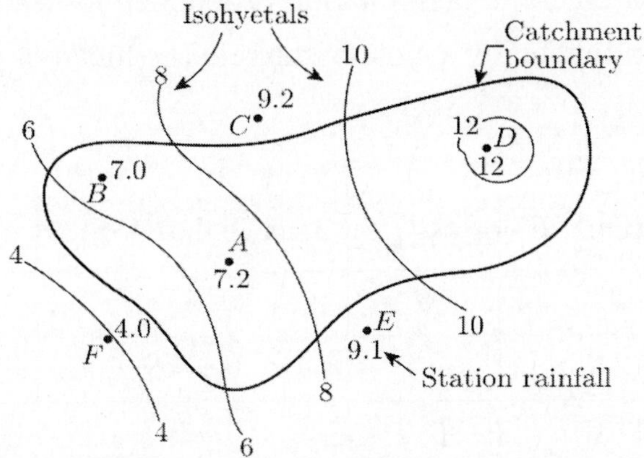

Figure 2.12: Isohyets of a storm

The process is like sketching contours of height using spot levels. The next step is to calculate the distance between two nearby isohyets using a planimeter. If the isohyets escape the catchment, the catchment boundary acts as the bounding line. It is assumed that the average rainfall determined by two isohyets applies to the inter-isohyet region. Thus, if P_1, $P_2 \ldots P_n$ are the isohyet values, and a_1, $a_2 \ldots$ a_{n-1} are the inter-isohyet regions, then the mean precipitation over area A catchment is given by Equation 2.10.

$$\bar{P} = \frac{a_1\left(\frac{P_1+P_2}{2}\right)+a_2\left(\frac{P_2+P_3}{2}\right)+\cdots+a_{n-1}\left(\frac{P_{n-1}+P_n}{2}\right)}{A} \quad \text{.......(2.10)}$$

When there are a lot of stations, the isohyet technique performs better than the other two methods.

Example 2.6: A watershed area surrounded by isohyets was estimated as follows, and the isohyets caused by a storm in the catchment were illustrated (Figure 2.12).

Isohyets(cm)	Area (km^2)
Station – 12.0	35
12.0 – 10.0	150
10.0 – 8.0	75
8.0 – 6.0	170
6.0 – 4.0	25

Determine the storm's mean precipitation.

Solution: An estimate of 12.0 cm of precipitation is obtained for the first area, which consists of a rain gauge station surrounded by a closed isohyet. The mean of two bounding isohyets is used for all other areas.

Isohyets	Mean value of P (cm)	Area (km^2)	Fraction of total area (col. 3/455)	Weighted P (cm) (col. 2 x col. 4)
1	2	3	4	5
12.0	12.0	35	0.0769	0.923
12.0 – 10.0	11.0	150	0.3297	3.627
10.0 – 8.0	9.0	75	0.1648	1.483
8.0 – 6.0	7.0	170	0.3737	2.616
6.0 – 4.0	5.0	25	0.0549	0.275
Total		455	1.0000	8.924

Mean precipitation $\overline{P}$ =8.924 cm.

2.11 STORM ANALYSIS

While estimating the mean rainfall for a given duration in a catchment, the volume of water as rainfall can be obtained as $\overline{P} \times A$. For instance, if a 150 km^2 catchment area experiences mean rainfall of 5 cm, the amount of water that falls as rainfall ($\overline{P}$):

$$\overline{P} = \frac{5}{100} \times 150 \times 10^6 = 7.5 \times 10^6 \text{ m}^3 \qquad \text{......(2.11)}$$

Hydrological studies sometimes entail estimating the likelihood of major floods brought on by heavy rains. Details about the highest quantity of precipitation in a variety of durations that trigger different types of catchment areas to respond are therefore essential. Depth-Area-Duration (DAD) analysis and maximum depth-duration analysis are the terms used to describe these relationships. Below is a quick summary of the DAD analysis:

2.11.1 Depth Area Relation

The "depth-area relation" typically refers to a relationship between the depth of water and the area it covers. This concept is often encountered in hydrology, geography, and engineering, particularly in studies related to rivers, lakes, and other bodies of water. Understanding the depth-area relation is crucial in various fields, including hydrology, hydraulic engineering, flood risk assessment, and environmental management. It helps predict flood extents, design hydraulic structures, manage water resources, and assess the ecological health of aquatic ecosystems.

In general, the depth-area relation can be summarized as follows:

a. **Direct Relationship:** For rainfall of a given duration, as the depth of water increases, the area it covers typically increases as well. This is an initiative, as deeper water will spread out over a larger surface area. This method can obtain mean depth (cm) over an area by Equation 2.12.

$$\bar{P} = P_o \exp(-KA^n) \qquad \text{.......(2.12)}$$

where,

$\bar{P}$= Mean depth (cm) over an area A sq. km

P_o =Maximum value of rainfall in cm at the center of a storm

K = Coefficient for a given region

n = Coefficient for a given region

Dhar and Bhattacharya (1975) obtained the following values for K and n for storms of varying duration based on 42 of the most severe storms in north India:

Duration	**K**	**n**
1 day	0.0008526	0.6614
2 days	0.0009877	0.6306
3 days	0.001745	0.5961

b. **Non-linear Relationship:** The relationship between depth and area is often non-linear. In other words, a slight increase in depth may result in a larger increase in the covered area, especially in shallow water bodies.

c. **Topographic Influence:** The shape and characteristics of the basin or water body, including its topography, play a significant role in determining the depth-area relation. For example, a flat, wide basin may have a different depth-area relation than a narrow, steep-sided canyon.

d. **Hydraulic Geometry:** This concept refers to how the shape and size of a river channel or water body change in response to the variations in flow and sediment transport. The depth-area relation is a key component of hydraulic geometry studies.

e. **Cross-Sectional Analysis:** The depth-area relation is often analyzed in conjunction with cross-sectional data of rivers or other water bodies. This involves measuring the depth and width of the water at various points along its course to understand how its geometry changes.

2.11.2 Depth Area Duration Analysis

The preparation of maximum DAD curves for a given region is proceeded as follows:

i. The severe rainstorms that have occurred in the region are first noted.

ii. Isohyet maps are drawn for a given duration largest amount of areal rainfall occurs in a small area of the region. As the area of the region increases, the areal rainfall also decreases. For example, it is observed that 15 km^2 of the area received a maximum of 9 cm of average rainfall and also 150 km^2 of the area received a maximum of 7 cm of average rainfall in 1 hour. A similar analysis can be carried out to estimate the maximum amount of the average rainfall occurring with respect to the areas of the region with duration as the third parameter.

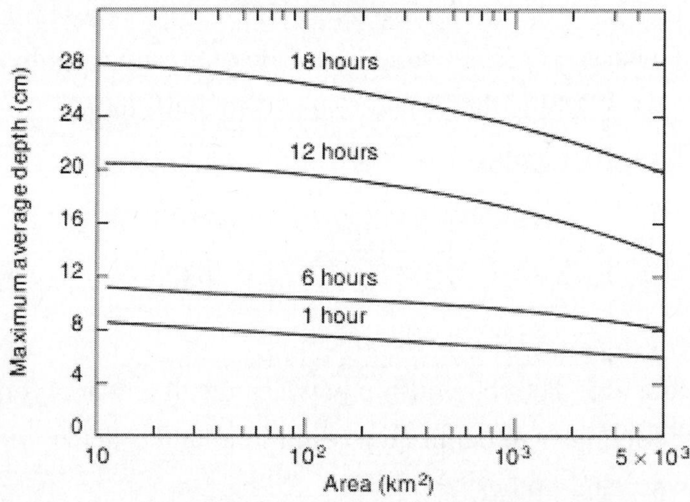

Figure 2.13: Typical DAD curve

DAD curves help create design storms that may be used to calculate design floods for major structures such as dams etc. These curves, which show the relationship between storm depth and area for different durations, are essential in hydrological analysis. They help to ensure that structures are able to handle the maximum potential rainfall that could occur over their area. We can plot the DAD curve for that region (Figure 2.13) from the previous rainfall record (In case records at the specific site are unavailable, even the data of neighboring sites, which are hydrometeorological homogeneous, can be used).

The maximum storm duration can be 24 hours with a maximum rainfall depth of 20 cm. Again, one follows the depth duration analysis and can say that 20cm of average rainfall in 24 hours will be distributed. Knowing this rainfall time distribution, one can do the hydrograph analysis (which we will study later) to estimate not only peak flood but also the time distribution of stream flow caused by this storm. A typical streamflow duration is shown in Figure 2.14.

However, one can notice that the use of 'maximum' has certain ambiguity. Maximum is related to the observed set of records of the second variable i.e., for the10 years; the maximum depth of rainfall during the 10 years will have one value (let us say 5 cm), but if the record is for 100 years, then the maximum observed value is likely to be larger than 5 cm (let us assume it is 9 cm). In other words, the word maximum is ill-defined; instead, we should bring the concept of return period or probability of occurrence. In the above example, 5 cm or more depth of rainfall has 10% probability of occurrence, similarly 9 cm or more depth has a probability of occurrence of only 1 percent. In other words, instead of maximum DAD curves we should Depth-Area-Duration frequency curves (Figure 2.15).

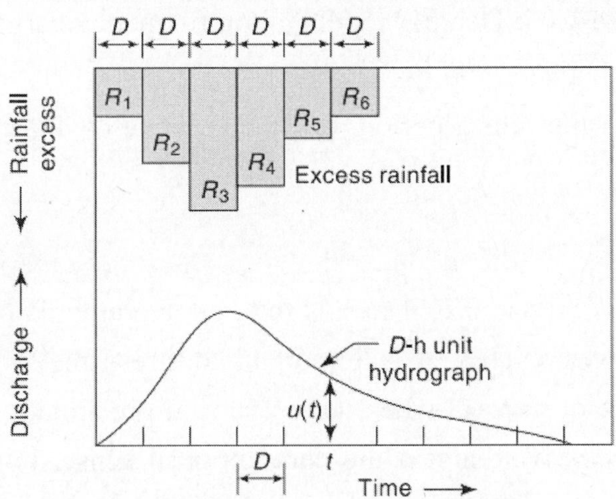

Figure 2.14: Streamflow distribution

The probability or the frequency of an event, be it rainfall or storm flow, is an important concept in hydrology, and we will deal with this concept in detail in flood frequency analysis. However, it will suffice to introduce the terms 'probability of occurrence' and 'the return period' at this stage.

The probability of exceedance, $p(X \geq x)$ (X is the name of the process, be it max daily rainfall in a year, and x is the value of rainfall), is defined as the chance that the rainfall will exceed the amount x. $p(X \geq 12)$ is equal to 0.1, meaning there is only a one percent chance that x will be ≥ 12 cm. Obviously $p(X<12)$ will be equal to 0.9. It means that the sum of $p(X<x)$ and $p(X \geq x)$ will be equal to 1.0.

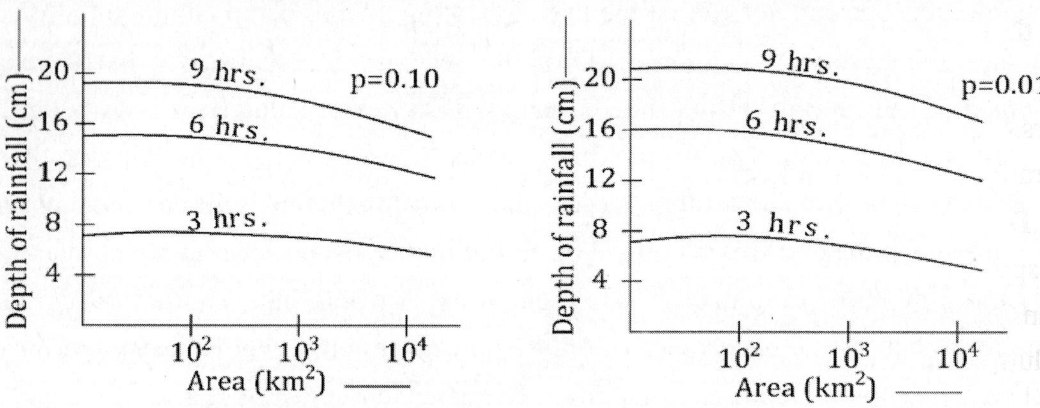

Figure 2.15: Depth-Area-Duration frequency curves

The average time between the occurrence of an event with a magnitude equal to or higher than x is known as the return period, or T. The value of T can be obtained through Equation 2.13.

$$T = 1/p\,(X \geq x) = 1/(1-p\,(X < x)) \qquad \text{...(2.13)}$$

For example, if $p(X \geq 12)$ is 0.1, it means that, on average, daily rainfall of 12 cm or more occurs once in 10 years. Here it is important to stress the word 'average' since it indicates that the sequence of interval when the amount of precipitation per day is 12 cm or more is not necessary every 10 years in the chronological sense, but that 12 cm or more rainfall may occur 100 times in 1000 years or on an average once in every 10 years.

How is probability assigned to a given value? Let us assume that one has n years record of daily rainfall from 1949 (inclusive) to 1997. Now arrange the series in descending order of magnitude m. The largest value x in 49 years of record has an order 1, 2nd largest value has an order 2, and so on. Then the probability is defined as: $p(X>x)$ =m/(n+1) and Return Period T = 1/$p(X>x)$.

2.11.3 Depth-Duration Curve

The rainfall is defined by three variables: - Rainfall depth, Duration & Frequency.

- **Rainfall Depth (D):** It is the magnitude of rainfall expressed in terms of total depth (usually in cm or in mm) in the duration of rainfall.
- **Duration:** the period in hours in which the rainfall event occurred.
- **The frequency** of a rainfall event is represented by its return period (T) within which the magnitude of the event will be exceeded or equaled once.

For example, we can say that a basin x can have 10 cm of rainfall in 3 hours once in 50 years. The Depth-Duration Curve represents the relationship between the maximum depth and duration of rainfall.

Use of Depth Duration Curve: The maximum quantity of rainfall that is expected to fall over a specific period of time is required for flood estimation. It is possible to determine the duration of rainfall from a hypothetical storm using India's Depth Duration Curve. Typical depth duration data with a curve is given in example 2.6 as Figure 2.16.

Example 2.7: The maximum intensity of rainfall was noted from a precipitation gauge during a period of 50 years. Draw the max depth-duration curve.

Time	Rainfall (Intensities/hr)
1 hr	1.84
2 hrs	1.74
3 hrs	0.84
4 hrs	0.83
5 hrs	0.78
6 hrs	0.37
7 hrs	0.12
8 hrs	0.05

Solution: Cumulative Depth of rainfall

Time	Rainfall (Intensities/hr)
1 hr	1.84
2 hrs	3.58
3 hrs	4.42
4 hrs	5.25
5 hrs	6.03
6 hrs	6.40
7 hrs	6.52
8 hrs	6.57

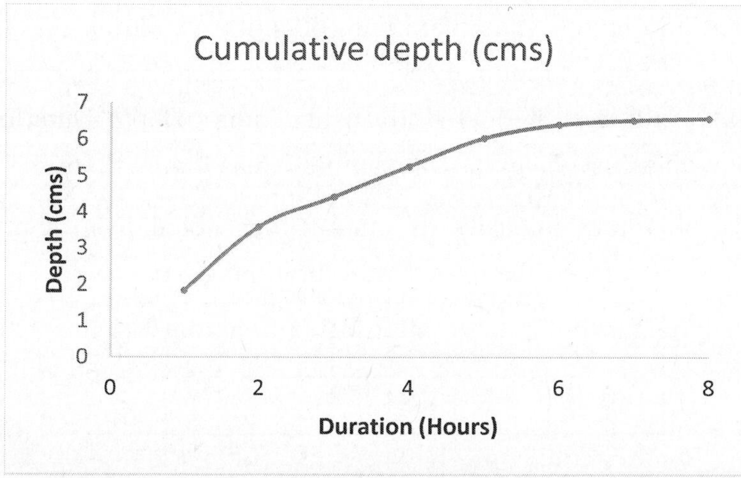

Figure 2.16: Maximum Depth-Duration Curve

2.11.4 Rainfall - Depth - Duration Frequency Curve

While estimating floods, we may not be interested only in the maximum possible flood but also the floods that may occur 1 in 50 years, 1 in 100 years, 1 in 500 years, etc. For example, if an important structure is built on a given stream, it is to be designed for a larger flood than if the structure is of comparatively less importance.

The depth-duration frequency curve refers to a curve that illustrates the relationship between rainfall depth and duration for a specific frequency. By keeping the frequency as a parameter, we can draw curves. For drawing such curves, maximum rainfall intensities for short-term durations such as 5, 10, 15, 30 minutes, 1, and 2 hours are needed. In a developing country like India, where there are only a few recording rain gauges, it will take many years before maps can be prepared from actual rain gauge duration frequency data for short-term rainfall depth based on the rainfall relationship of the U.S.A. A typical rainfall depth-duration curve for a frequency of occurrence of 1 in 50 years for India is as shown in Figure 2.17.

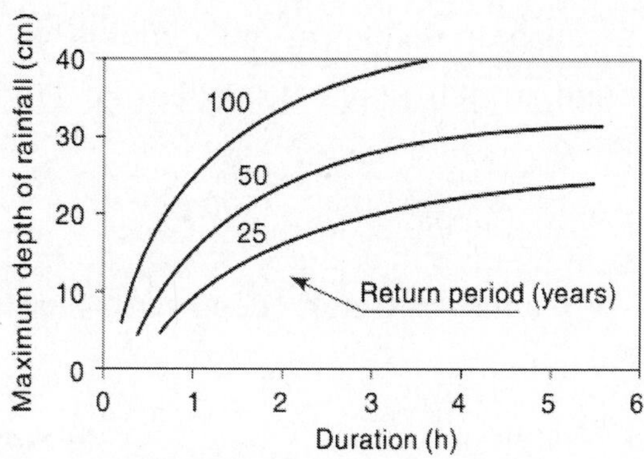

Figure 2.17: Maximum Depth-Duration-Frequency Curves

2.11.5 Depth-Area-Duration Curve

The largest amount of precipitation for a range of sizes and durations over various locations are found using the Depth-Area-Duration curve, or DAD curve. For example, in a catchment area of 650 km^2 in Rajasthan, the maximum amount of rainfall in 6 hrs is 4.00 cm. The preparation of such curves requires elaborate data.

For a given rainfall duration, the maximum average depth of rainfall will decrease as the area increases. This is because the space distribution of rainfall is not uniform. The maximum rainfall will occur in a localized zone called the Storm centre. In addition, as the duration of rainfall increases in a particular area, the rainfall depth also rises.

2.12 PROBABLE MAXIMUM PRECIPITATION (PMP)

In order to design a large dam's spillway, it is necessary to use a much higher precipitation estimate than in the case of almost any other structure. The reasons are: (i) the structure being very costly, its loss would involve a large economic expense, and (ii) failure of such structure carries a significant danger of downstream property and life losses that would not occur otherwise. Meteorologists are consulted by design engineers to estimate the probable maximum precipitation for such structures.

The estimates represent the best judgment of the meteorologists of the realistic upper limit of precipitation that can occur. Many meteorologists have thought there is no upper limit on precipitation amount - that any given amount can conceivably occur. Such a view is not realistic physically. It is, therefore, certainly horrible to put an upper bound on the amount of precipitation that can be possible. In precipitation, to fix an upper bound, there must exist a least upper bound, which might be called the Possible Maximum or (PMP) Probable Max Precipitation.

The procedure involves two steps:

a. The preparation of probable maximum depth-area-duration curves represents the region of which the basin under study is a part.
b. The selection using these curves of a pattern storm for use in the basin.

2.13 RAINFALL DATA IN INDIA

Rainfall data in India varies significantly across different regions and seasons. India has a diverse climate, influenced by the factors such as monsoons, proximity to the sea, topography, and more. The Indian Meteorological Department (IMD) is the primary authority responsible for collecting and disseminating rainfall data in the country. For the most up-to-date and detailed rainfall data for specific regions in India, it is advisable to refer to reports and publications from the Indian Meteorological Department or other relevant governmental agencies such as state water resources departments/state revenue departments. Here is an overview of rainfall patterns in India:

- **Monsoon Season**: For most parts of India, the southwest monsoon, which typically occurs from June to September, is the primary source of precipitation. This seasonal rainfall is crucial for agriculture and water resources. The amount and distribution of rainfall during the monsoon season vary from year to year and from region to region. Coastal regions and areas with orographic features (mountains) tend to receive higher rainfall compared to interior regions.

- **Regions with High Rainfall**: States along the western coast of India, such as Kerala, Karnataka, Maharashtra, and Goa, receive substantial rainfall during the monsoon season due to the Western Ghats. Similarly, northeastern states like Assam, Meghalaya, and West Bengal receive heavy rainfall due to their proximity to the Bay of Bengal and the Himalayas.
- **Regions with Low Rainfall**: Some regions in India experience low rainfall or are classified as arid or semi-arid. Examples include Rajasthan, Gujarat, parts of Punjab, and Haryana. These regions rely heavily on irrigation for agriculture.
- **Rainfall Variability**: India experiences inter-annual and intra-seasonal variability in rainfall. Factors such as El Niño and La Niña events, Indian Ocean Dipole, and other atmospheric phenomena influence the monsoon patterns that can lead to droughts or excess rainfall in different parts of the country.
- **Data Sources**: IMD maintains a network of rain gauge stations across the country to monitor rainfall. Data collected from these stations are used to analyze rainfall patterns, predict weather, and assess the overall climate situation. In addition to this, in some states, water resources departments/ revenue departments also collect rainfall data.
- **Impact on Agriculture**: Rainfall plays a crucial role in Indian agriculture, as a significant portion of the population relies on farming for their livelihood. Adequate and well-distributed rainfall is essential for crop growth, while inadequate rainfall can lead to droughts and crop failures.
- **Climate Change**: Climate change is expected to influence the rainfall patterns in India, leading to more extreme weather events, altered monsoon patterns, and potential shifts in agricultural productivity.

SUMMARY

Precipitation, which might include rain, snow, hail, dew, and more, is the primary source of water supply in engineering hydrology. The primary study subjects are thought to be the varieties of precipitation, their availability, measurement, and use. The various types of precipitation, their intensity, and its effects on ecosystems, agriculture, and water resources are all addressed during this subject. Numerous subjects pertaining to precipitation patterns and India's several seasons have been discussed in this module. Several methods of measuring rainfall and snowfall have been described for India's yearly precipitation. Furthermore, techniques for estimating missing data and testing data consistency have also been covered in the session. This unit presents various techniques for calculating the mean rainfall over a watershed. Finally, the Probable Maximum Precipitation (PMP), the Depth Area Duration (DAD) curve, and the availability of rainfall data in India have all been explained.

EXERCISE

Revision Questions

1. Explain the term precipitation.
2. Describe the different forms of precipitation.
3. What do you understand by the Index of wetness?
4. What is the difference between light rain, moderate rain, and heavy rain?
5. Describe the different methods of recording rainfall.
6. Discuss the current practice and status of rainfall recording in India.
7. Describe the salient characteristics of precipitation in India.
8. Explain the different methods of determining the average rainfall over a catchment due to a storm. Discuss the relative merits and demerits of the various methods.
9. Describe the different seasons in India based on differences in temperature and precipitation.
10. What do you understand by monsoon depression?
11. Kindly explain the methods of measurement of rainfall in India.
12. What is the difference between recording & non-recording rain gauges?
13. Describe the different types of recording rain gauges.
14. Describe the different methods for measurement of snowfall.
15. Explain a procedure for checking rainfall data for consistency.
16. Explain a procedure for supplementing the missing rainfall data.
17. Explain briefly the following relationships relating to the precipitation over a basin:
    a. Depth-Area Relationship;
    b. Maximum Depth-Area-Duration Curves;
    c. Intensity Duration Frequency Relationship.

18. What is meant by Probable Maximum Precipitation *(PMP)* over a basin? Explain how PMP is estimated.
19. Consider the statement: The 50-year-24-hour maximum rainfall at Bangalore is 160 mm. What do you understand by this statement?
20. Describe briefly the procedures adopted to measure the water equivalent of snowfall.
21. Write brief notes on (i) moving average, (ii) Thiessen polygon, (iii) isohyetals maps.
22. What do you understand by Hyetograph?
23. Describe an overview of rainfall patterns in India.

Numerical Problems

1. There are eight rain gauge stations inside the catchment of a watershed. The annual rainfall in a given year that the rain gauges recorded is as follows:

Station	A	B	C	D	E	F	G	H
Rainfall (cm)	73.8	95.9	160.3	117.5	89.9	121.7	123.8	126.5

a. Determine the standard error in the mean rainfall estimation for the current set of rain gauges.

b. Determine the ideal number of stations in the catchment, accounting for an 8% error in the mean rainfall estimate.

2. In a particular catchment, at stations A, B, C, D, and E the average yearly rainfall is 88.90, 73.15, 81.19, 76.78 and 91.91 cm, respectively. Station E was not operational in 1998; annual precipitation at stations A, B, C and D was 89.71, 75.62, 83.88 and 74.39 cm, respectively. Calculate the amount of rain that year at station E.

3. Six rain-gauge stations, namely Roorkee, Bhagwanpur, Manglore, Laksar, Purkaji & Bahadrabad located in a typical region. The rainfall observed during a month at Roorkee, Bhagwanpur, Manglore, Laksar, Purkaji are 162.2 mm, 153.3 mm, 156.5 mm, 145.1 mm, and 151.1 mm, respectively, while the corresponding rainfall value at Bahadrabad is missing. If the normal monthly rainfall at Roorkee, Bhagwanpur, Manglore, Laksar, Purkaji & Bahadrabad is 246.2 mm, 244.2 mm, 240.4 mm and 238.8 mm, 230.2 mm, and 245.2 mm, respectively, Obtain the value of missing rainfall at Bahadrabad rain-gauge station using the Normal ratio method.

4. The normal annual precipitation of five rain gauge stations A, B, C, D and E, respectively 124, 113, 86, 121 and 143 cm. During a particular storm, the precipitation recorded by stations *A,* B, *C,* and D are 14.3, 9.5, 7.8 and 11.3 cm respectively. The instrument at station *E was* inoperative during that storm. Estimate the rainfall at station E during that storm.

5. Test the consistency of the 26 years of data of the annual precipitation measured at station P. Rainfall data for station P, as well as the average annual rainfall measured at a group of ten neighboring stations located in a meteorologically homogeneous region, are given as follows.

Year	Annual rainfall at station P (mm)	average annual rainfall of ten stations group (mm)	Year	Annual rainfall at station P (mm)	average annual rainfall of ten stations group (mm)
1967	165	145	1980	152	183
1968	147	155	1981	143	173
1969	139	197	1982	146	159
1970	148	156	1983	149	158
1971	170	146	1984	148	172
1972	168	176	1985	158	181
1973	159	154	1986	155	158
1974	151	144	1987	165	159
1975	155	168	1988	167	135
1976	169	148	1989	169	189
1977	156	176	1990	171	140
1978	155	173	1991	165	168
1979	157	139	1992	164	149

a. In which year is a change in regime indicated?

b. Adjust the recorded data at station *A* and determine the mean annual precipitation.

6. In a storm of 240 minutes duration, the incremental rainfall at various time intervals is given below.

Time since the start of the storm (minutes)	30	60	90	120	150	180	210	240
Incremental rainfall in the time interval (cm)	1.85	2.27	6.10	4.55	2.60	1.58	0.85	0.75

a. Obtain the ordinates of the hyetograph and represent the hyetograph as a bar chart with time in chronological order in the x-axis.

b. Obtain the ordinates of the mass curve of rainfall for this storm and plot it. What is the average intensity of a storm over the duration of the storm?

7. Represent the annual rainfall data of station *A, which is* given below as a bar chart with time in chronological order. If the annual rainfall less than 75% of the long-term mean is taken to signify meteorological drought, identify the drought years and suitably display the same in the bar chart.

Year	1960	1961	1962	1963	1964	1965	1966	1967	1968	1969
Annual rain (mm)	750	752	437	381	470	627	551	630	614	510
Year	1970	1971	1972	1973	1974	1975	1976	1977	1978	1979
Annual rain (mm)	401	366	710	570	524	112	525	920	610	408

8. The watershed of a stream has five rain gauge stations inside the basin. When Thiessen polygons were constructed, three more stations lying outside the watershed were found to have weights. The details of Thiessen polygons surrounding each rain gauge and the recordings of the rain gauges in the month of July 2022 are given below:

Rain gauge station	*A*	*B*	*C*	*D*	*E*	*F*	*G*	*H*
Thiessen Polygon area (km^2)	730	1390	1448	1050	908	2230	429	1457
Recorded rainfall in mm during July 2022	145	148	136	129	112	125	100	102

Stations *B*, *D* and *F* are outside the watershed. Determine the average depth of rainfall on the watershed in July 2022 by (i) arithmetic mean method and (ii) Thiessen mean method.

9. For a drainage basin of 600 km^2, isohyets drawn for a storm gave the following data:

Isohyets (interval) (cm)	18–15	15–12	12–9	9–6	6–3
Inter-isohyetal area (km^2)	92	128	120	175	85

Estimate the average depth of precipitation over the catchment.

10. There are 10 rain gauge stations available to calculate the rainfall characteristics of a catchment whose shape can be approximately described by straight lines joining the following coordinates (distances in kilometers):

(30, 0), (80, 10), (110, 30), (140, 90), (130, 115), (40, 110), (15, 60). Coordinates of the rain gauge stations and the annual rainfall recorded in them in the year 2011 are given below.

Station	1	2	3	4	5
Co-ordinates	(0, 40)	(50, 0)	(140, 30)	(140, 80)	(90, 140)
Annual Rainfall (cm)	132	136	93	81	85
Station	6	7	8	9	10
Co-ordinates	(0, 80)	(40, 50)	(90, 30)	(90, 90)	(40, 80)
Annual Rainfall (cm)	124	156	128	102	128

Determine the average annual rainfall over the catchment by using the isohyetal method.

11. Following the figure shows a catchment with seven rain gauge stations inside it and three stations outside. The rainfall recorded by each of these stations are indicated in the figure. Draw the figure to an enlarged scale and calculate the mean precipitation by (a) Thiessen- mean method, (b) isohyetal method, and by (c) arithmetic-mean method.

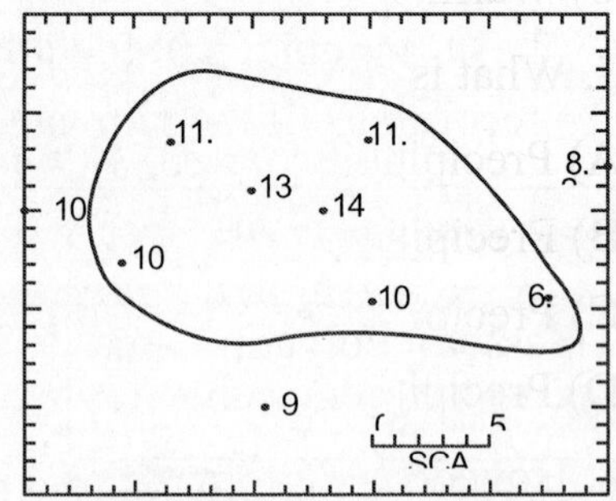

Multiple Choice Questions

1. What is precipitation?

A) Water vapor in the atmosphere

B) Any form of water, liquid or solid, that falls from the sky

C) Water that evaporates from the surface

D) Water stored in lakes and rivers

2. Which of the following is NOT a type of precipitation?

A) Rain B) Snow

C) Sleet D) Humidity

3. What is the primary factor that influences the amount and type of precipitation a region receives?

A) Soil type B) Temperature

C) Wind patterns D) Elevation

4. Which type of precipitation occurs when raindrops freeze before hitting the ground?

A) Rain B) Snow

C) Sleet D) Hail

5. What is the term for precipitation that falls in the form of ice pellets?

A) Snow B) Sleet

C) Hail D) Freezing rain

6. Which of the following conditions is most likely to produce heavy rainfall?

A) High pressure systems B) Cold fronts

C) Warm, moist air rising D) Clear skies

7. What is "orographic precipitation"?

A) Precipitation caused by temperature changes

B) Precipitation resulting from the cooling of air over oceans

C) Precipitation that occurs when moist air rises over mountains

D) Precipitation associated with thunderstorms

8. How does climate change affect precipitation patterns?

A) It has no effect

B) It can lead to increased variability and extreme events

C) It uniformly increases precipitation everywhere

D) It decreases precipitation in all regions

9. What is the most common form of precipitation?

A) Snow B) Rain

C) Hail D) Sleet

10. Which form of precipitation consists of ice pellets?

A) Rain B) Snow

C) Sleet D) Drizzle

11. What type of precipitation falls as crystalline ice and typically occurs in winter?

A) Rain B) Snow

C) Hail D) Fog

12. Which form of precipitation is characterized by larger ice balls and usually occurs during thunderstorms?

A) Drizzle B) Sleet

C) Hail D) Snow

13. What is drizzle?

A) Light rain with very small droplets B) Heavy rain with large droplets

C) Ice pellets that bounce on impact D) Snowflakes falling gently

14. What is the primary difference between sleet and freezing rain?

A) Sleet is frozen before reaching the ground, while freezing rain forms liquid droplets that freeze upon contact.

B) Sleet is heavier than freezing rain.

C) Sleet falls during summer, while freezing rain occurs in winter.

D) There is no difference; they are the same.

15. In which conditions is snow most likely to form?

A) High humidity and warm temperatures B) Low humidity and high temperatures

C) Low temperatures and moisture in the air D) Warm air with no moisture

16. What type of precipitation typically occurs in tropical regions?

A) Snow B) Hail

C) Rain D) Sleet

17. Which of the following is the primary season for monsoon precipitation in India?

A) Winter
B) Summer
C) Spring
D) Autumn

18. What type of precipitation is most common in the Himalayan region of India during winter?

A) Rain
B) Hail
C) Snow
D) Drizzle

19. Which of the following types of precipitation is primarily associated with the southwest monsoon in India?

A) Orographic precipitation
B) Convective precipitation
C) Cyclonic precipitation
D) All of the above

20. What is the term for the rainfall that occurs due to the lifting of moist air over the Western Ghats?

A) Convectional rainfall
B) Orographic rainfall
C) Cyclonic rainfall
D) Frontal rainfall

21. In India, what is the typical effect of the northeast monsoon?

A) Heavy rainfall in the northern plains
B) Drought conditions in southern India
C) Moderate rainfall along the southeastern coast
D) Snowfall in the northern mountains

22. What kind of precipitation is characterized by short, intense showers often followed by clear skies in India?

A) Continuous rainfall
B) Convectional rainfall
C) Orographic rainfall
D) Cyclonic rainfall

23. Which Indian state receives the highest average annual rainfall?

A) Rajasthan
B) Gujarat
C) Kerala
D) Meghalaya

24. What is "rain shadow" effect, and how does it occur in India?

A) Increased rainfall on the windward side of a mountain range
B) Decreased rainfall on the leeward side of a mountain range
C) Equal distribution of rainfall across a region
D) Rainfall caused by urban heat islands

25. What instrument is commonly used to measure precipitation?

A) Barometer
B) Anemometer
C) Rain gauge
D) Thermometer

26. What unit is typically used to express the amount of precipitation?

A) Meters
B) Liters
C) Millimetres
D) Degrees

27. Which type of rain gauge is designed to automatically record precipitation over time?

A) Standard rain gauge
B) Tipping bucket rain gauge
C) Weighing rain gauge
D) Non-recording rain gauge

28. What does a "pluviometer" measure?

A) Wind speed
B) Atmospheric pressure
C) Humidity
D) Precipitation

29. How is precipitation data typically expressed in meteorological reports?

A) As an average over a week
B) As total precipitation over 24 hours
C) As a percentage of humidity
D) As a temperature range

30. What is the purpose of a "snow gauge"?

A) To measure wind speed
B) To measure the depth of snow
C) To measure rainfall
D) To measure temperature

31. Which method involves estimating precipitation using radar technology?

A) Manual measurement
B) Doppler radar
C) Satellite imagery
D) Standard rain gauge

32. How often should precipitation measurements be taken for accurate data collection?

A) Once a month
B) Once a week
C) Daily
D) Every hour

33. What is the primary season for monsoon rainfall in India?

A) Winter
B) Summer
C) Autumn
D) Spring

34. Which region in India receives the highest annual precipitation?

A) Thar Desert
B) Western Ghats
C) Himalayan region
D) Indo-Gangetic Plains

35. What is the average annual rainfall required for an area to be classified as a "humid" region in India?

A) Less than 500 mm
B) 500 mm to 1000 mm
C) 1000 mm to 2000 mm
D) More than 2000 mm

36. Which phenomenon primarily affects the distribution of rainfall in India?

A) El Niño
B) La Niña
C) Intertropical Convergence Zone (ITCZ)
D) Polar Vortex

37. During which month does the southwest monsoon typically begin in India?

A) May
B) June
C) July
D) August

38. Which of the following states is known for its high levels of orographic rainfall?

A) Rajasthan
B) Punjab
C) Assam
D) Haryana

39. What is the main source of precipitation during the winter months in North India?

A) Southwest Monsoon
B) Western Disturbances
C) Local Convection
D) Tropical Cyclones

40. What is the impact of deforestation on precipitation patterns in India?

A) Increased rainfall
B) Decreased rainfall
C) No impact
D) Unpredictable changes

41. What is point rainfall?

A) Rain measured over a large area
B) Rain measured at a specific location
C) Average rainfall in a region
D) Rainfall measured in a specific time frame

42. Which instrument is primarily used to measure point rainfall?

A) Anemometer
B) Rain gauge
C) Barometer
D) Thermometer

43. What is the purpose of a consistency test in rainfall measurement?

A) To measure rainfall intensity
B) To ensure data reliability over time
C) To predict future rainfall
D) To calculate evaporation rates

44. How can point rainfall be affected by local geography?

A) It is not affected at all
B) By the presence of mountains or valleys
C) Only by temperature variations
D) By wind patterns alone

45. What does a consistency test typically compare?

A) Rainfall amounts over different seasons
B) Data from different rain gauges at the same location
C) Rainfall amounts from multiple regions
D) Rainfall intensity and temperature

46. Which of the following is a common method for performing a consistency test?

A) Statistical analysis of historical data
B) Visual inspection of rainfall data
C) Comparing rainfall to evaporation rates
D) Measuring wind speed

47. Why is it important to conduct a consistency test on point rainfall data?

A) To avoid data collection errors
B) To ensure data fits a theoretical model
C) To improve rainfall prediction accuracy
D) To assess the environmental impact

48. What type of rainfall can a point rainfall measurement capture?

A) Only heavy rainfall
B) All types, including light, moderate, and heavy
C) Only rainfall from storms
D) Rainfall occurring during specific seasons

49. In a consistency test, what is the significance of outliers?

A) They confirm the accuracy of the data
B) They indicate potential measurement errors
C) They are ignored in analysis
D) They are used to calibrate instruments

50. What is a common statistical method used in consistency tests?

A) Regression analysis
B) Mean and standard deviation calculations
C) Time series analysis
D) All of the above

51. What is a hyetograph?

A) A graph showing temperature changes over time

B) A graph that represents the amount of rainfall over a specific time period

C) A chart of wind speed variations

D) A diagram of geographical features

52. Which of the following best describes the Thiessen Polygon method?

A) A method to estimate evaporation rates

B) A technique to analyse stormwater runoff

C) A spatial analysis method to determine rainfall distribution based on gauge locations

D) A way to measure soil moisture content

53. In a hyetograph, what does the area under the curve represent?

A) Temperature

B) Total rainfall over the time period

C) Wind speed

D) Evaporation rate

54. Which of the following is a limitation of the Thiessen Polygon method?

A) It requires extensive data collection

B) It assumes uniform rainfall within each polygon

C) It is complex to construct

D) It is only applicable to urban areas

55. How is the Depth-Area relation typically used in hydrology?

A) To analyse soil properties

B) To estimate the volume of runoff from rainfall

C) To predict drought conditions

D) To calculate evaporation rates

56. What does the term "depth" refer to in the Depth-Area relation?

A) The height of rainfall measured

B) The thickness of snow

C) The depth of water in rivers

D) The total area affected by rainfall

57. Which graphical representation is often used to illustrate the relationship between rainfall depth and area?

A) Line graph

B) Scatter plot

C) Bar chart

D) Cumulative frequency curve

58. What is a common application of hyetographs in hydrology?

A) Determining soil erosion rates
B) Flood forecasting and modelling
C) Assessing water quality
D) Mapping land use changes

59. In a Thiessen Polygon analysis, what determines the boundaries of each polygon?

A) Average rainfall amounts
B) Proximity to the nearest rain gauge
C) Equal distances to surrounding rain gauges
D) Topographical features

60. Why is the Depth-Area relation important in designing stormwater management systems?

A) It helps identify vegetation types
B) It aids in predicting potential flooding
C) It measures soil compaction
D) It assesses groundwater levels

Answer: 1-B; 2-D; 3-C; 4-C; 5-C; 6-C; 7-C; 8-B; 9-B; 10-C; 11-B; 12-C; 13-A; 14-A; 15-C; 16-C; 17-B; 18-C; 19-D;20-B; 21-C; 22-B; 23-D; 24-B; 25-C; 26-C; 27-B; 28-D; 29-B; 30-B; 31-B; 32-C; 33-B; 34-B; 35-C; 36-C; 37-B; 38-C; 39-B; 40-B; 41-B; 42-B; 43-B; 44-B; 45-B; 46-A; 47-A; 48-B; 49-B; 50-B; 51-B; 52-C; 53-B; 54-B; 55-B; 56-A; 57-D; 58-B; 59-C; 60-B.

KNOW MORE

For more information related to this topic scan the QR code.

OR

Type this link in your browser

https://drive.google.com/file/d/1Vf2SWG0WqvROL-wPDlVwEf-eQuoXJFkQ/view?usp=sharing

BIBLIOGRAPHY

1. Asawa, G.L. (2005), "Irrigation and Water Resources Engineering, New Age International (P) Ltd., ISBN-13: 978-8122416732
2. Chow. V.T, (Ed), (1964), "Handbook of Applied Hydrology", McGraw-Hill, New York, NY, 1964.
3. IMD (1971), "Scientific Report No. 136, India Meteorological Department,1971, Government of India.
4. Jain S. K., Agarwal, P. K. And Singh, V. P. (2007), Hydrology and Water Resources of India, Water Science and Technology Library, Vol 57, Springer, The Netherland, ISBN-10 1-4020-5179-4(HB).
5. Punmia, B.C. and Lal, Pande B.B., (1992), "Irrigation and Water Power Engineering", Twelfth Edition. Laxmi Publications (P) Ltd.
6. Singh, V. P. (1992). Elementary hydrology. Prentice Hall, New Jersy, 973.
7. Subramanya, K. (2015), "Engineering Hydrology", McGraw-Hill Education (India) Private Limited, New Delhi, 2015, ISBN(13): 978-1-25902997-4.

3 Abstraction from Precipitation

UNIT SPECIFICS

Once the precipitation falls on the Earth's surface, several abstractions, like evaporation, transpiration, evapotranspiration, infiltration, surface detention, and storage, may be considered losses before taking runoff. This unit describes the evaporation process and its measurement through the physical evaporimeters, empirical equations, and analytical means such as water budget, energy budget, and mass-transfer methods. Evapotranspiration is the term used to describe the combination of transpiration from vegetation and evaporation from soil and water bodies. This unit discusses potential evapotranspiration (PET) and actual evapotranspiration (AET) in detail. The unit also discusses depression storages and interceptions, which are "losses" in the runoff production process. Furthermore, the infiltration process, its modelling, and measurement are discussed. Also, soil infiltration capacity and the infiltration indices (Φ and W) are discussed, which is a significant abstraction from precipitation to enhance soil moisture storage and groundwater recharge.

RATIONALE

To know about the abstractions from precipitation through various processes like evaporation, evapotranspiration, interception, depression storages, and infiltration.

PRE-REQUISITE

Nil

UNIT OUTCOMES

The list of outcomes of this unit is as follows:

U3-O1: Learn about the abstractions from precipitation through various processes like evaporation, evapotranspiration, interception, depression storages, and infiltration

U3-O2: Assessment of evaporation using evaporimeters, empirical equations, and analytical means

U3-O3: Assessment of PET & AET

U3-O4: Assessment of infiltration indices (Φ and W), infiltration, and infiltration capacity, and applications in engineering problems

Unit Outcomes	Expected Mapping with Course Outcomes *(1- Weak Correlation; 2- Medium correlation; 3- Strong Correlation)*							
	CO-1	CO-2	CO-3	CO-4	CO-5	CO-6	CO-7	CO-8
U3-O1	3	3	3	3	3	3	3	3
U3-O2	3	3	3	3	3	3	1	3
U3-O3	1	1	3	3	3	3	3	3
U4-O4	1	3	1	2	2	2	3	3

3.1 EVAPORATION PROCESS

Thermal energy is transferred from a liquid to a gas at the free surface, or below the boiling point, by the process of evaporation. Think about a pond or other body of water. With a vast variety of instantaneous velocities, water molecules are always moving. Both this range and average speed rise with the addition of heat. Certain molecules may be able to pass across the water's surface if they have enough kinetic energy. In a similar manner, some of the water molecules in the moving water vapour that are found in the atmosphere close to the water's surface might make their way. The net passage of water molecules from their liquid state into their gaseous state is known as evaporation. A water body must provide the latent heat of vaporization, which happens at about 585 cal/g of evaporated water, during the cooling process of evaporation.

The rate of evaporation is affected by the following variables: (a) the vapour pressures in the air above the water's surface; (b) air and water temperatures; (c) wind speed; (d) atmospheric pressure; (e) water quality; and (f) size of the water body.

- **Vapour Pressure**

John Dalton (1802) was the first to identify Dalton's law of evaporation, which is represented by equation (3.1) below. According to Dalton's law, the rate of evaporation is determined by the difference between the saturation vapour pressure at the water temperature, e_w, and the actual vapour pressure in the air, e_a. Further, based on the law, evaporation is going until $e_w = e_a$. Condensation occurs when $e_w > e_a$. Consequently,

$$E_L = C(e_w - e_a) \qquad \text{............(3.1)}$$

where,

E_L = rate of evaporation (mm/day),

C = constant,

e_w= saturation vapour pressure at the water temperature (mm of mercury), and

e_a = actual vapour pressure in the air (mm of mercury).

- **Temperature**

Considering other factors, when the water temperature rises, the rate of evaporation also rises. There is no strong association between evaporation rate and air temperature, even though there is a general increase in evaporation rate with rising temperatures. Therefore, evaporation in a lake may occur to varying degrees in different months at the same mean monthly temperature.

- **Wind**

Wind plays a crucial role in weather patterns and can have significant effects on various aspects of life, including agriculture, transportation, and energy generation. Humans have harnessed wind power for centuries, initially for sailing ships and later for windmills and wind turbines to generate electricity. Today, wind energy is an important renewable energy source contributing to global efforts to reduce reliance on fossil fuels and mitigate climate change.

More space for evaporation is created as a result of the wind's assistance in clearing the evaporated water vapour from the evaporation zone. Any additional increase in wind velocity, however, has no effect on evaporation if it is high enough to remove all of the evaporated water vapour. Consequently, the rate of evaporation rises in proportion to wind speed up to a critical speed, after which additional wind speed increases have no effect on the rate of evaporation. The essential wind speed is dependent on the water's surface area. High-speed turbulent winds are required to cause the maximum rate of evaporation in big water bodies.

- **Atmospheric Pressure**

If all other parameters are constant, evaporation will rise at high altitudes where a drop in barometric pressure is observed.

- **Dissolvable Salts**

During the dissolution of a solute in water, the rate of evaporation decreases because the solution's vapour pressure is lower than that of pure water. The percentage rise in specific gravity roughly matches the percentage decrease in evaporation. For instance, evaporation from seawater is roughly 2-3% lower than that from Freshwater under the same conditions.

- **Heat Storage in Water Bodies**

Compared to shallow water bodies, deep water bodies can store more heat. Compared to a shallow lake subjected to the same circumstances, a deep lake may store radiation energy received in the summer and release it in the winter, resulting in less evaporation in the summer and more in the winter. The annual evaporation rate is rarely impacted by heat storage, and its main influence is to alter the seasonal evaporation rates.

3.2 EVAPORIMETERS

In many hydrologic issues related to the design and management of reservoirs and irrigation systems, estimating evaporation is crucial. This assessment is especially crucial in arid zones to preserve the limited water supplies. One of the difficult tasks to do is to precisely quantify the amount of water that evaporates from a big body of water.

The amount of water that evaporates from a water surface can be calculated using three different techniques: (i) empirical evaporation equations, (ii) evaporimeter data, and (iii) analytical procedures.

3.2.1 Types of Evaporimeters

Pans of water that are exposed to the atmosphere and periodically measure water loss from evaporation are called evaporimeters. Meteorological data such as humidity, wind direction, air and water temperatures, and precipitation are also captured in addition to the evaporation measurement. There are numerous varieties of evaporators in use; here is a description of a few popular pans.

- **Class A Evaporation Pan:** The US Weather Bureau uses a standard pan, known as a Class A Land Pan, which is a conventional pan with dimensions of 1210 mm in diameter and 255 mm in depth. The water's depth is consistently between 18 and 20 cm (Figure 3.1). Typically, the pan is normally made from unpainted galvanized iron sheets. When corrosion is an issue, Monel metal is employed. The pan is set up on a wooden platform that is 15 cm above the floor to enable unrestricted airflow beneath the pan. In a stilling well, the depth of water is measured using a hook gauge to determine the amount of evaporation.

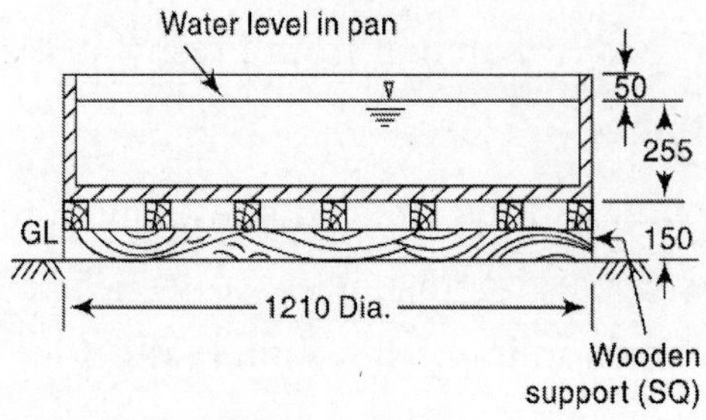

Figure 3.1: U.S. Class A Evaporation Pan

- ***ISI Standard Pan:*** This modified Class A Pan, or pan Evaporimeter, as defined by IS: 5973-1970, has a diameter of 1220 mm and a depth of 255 mm. The pan is composed of 0.9 mm thick copper sheet that is painted white on the exterior and tinned inside (Figure 3.2). The water level is shown using a fixed-point gauge.

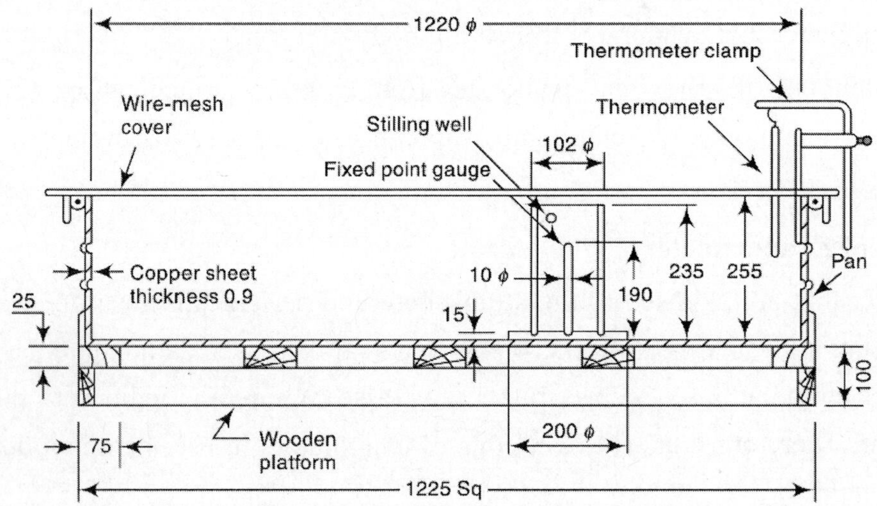

Figure 3.2: ISI Evaporation Pan

Water is added or removed using a calibrated cylinder measure to keep the pan's water level at a set level. To keep birds away from the pan's water, a hexagon-shaped net of galvanized iron covers the whole top of the pan. Furthermore, the water's temperature is more consistent day and night when a wire mesh is present. It is discovered that this pan evaporation is approximately 14% lower than that of an unsealed pan. The pan is set on top of a square hardwood platform that is 100 mm high and 1225 mm wide to allow air to circulate underneath the pan.

- **Colorado Sunken Pan:** The Colorado Sunken Pan Evaporimeter (Figure 3.3) is a specific type of evaporation pan used for this purpose. It is composed of an unpainted galvanized iron sheet i.e., 920 mm square area and 460 mm deep, placed at ground level within 100 mm of the top, filled with water, and exposed to the atmosphere. By measuring the decrease in water level over time, scientists can calculate the rate of evaporation. These Evaporimeters are used in various research and monitoring programs related to agriculture, water management, and climate studies, particularly in arid and semi-arid regions like Colorado, where water resources are particularly important.

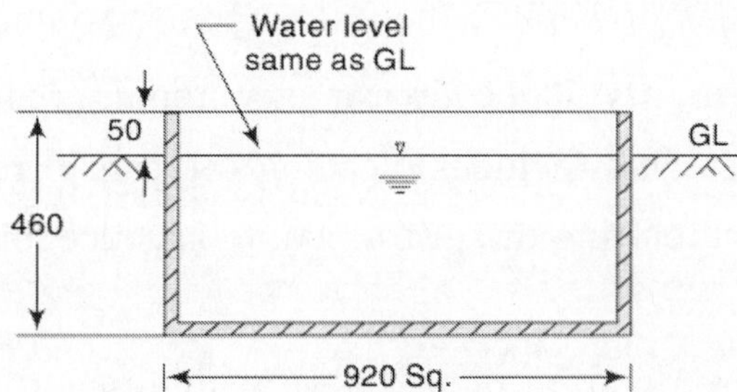

Figure 3.3: Colorado Sunken Pan Evaporimeter

- **United States Geological Survey Floating Pan:** The United States Geological Survey (USGS) uses various tools and instruments for hydrological and meteorological research, including evaporation measurement devices. One such device is the floating pan evaporation pan. The floating pan evaporation pan is a specialized instrument used to measure the rate of evaporation from a water surface. It consists of a shallow square pan 900 mm side and 450 mm deep, filled with water, typically mounted on a floating platform. The pan is placed on the surface of a body of water, such as a lake or reservoir, and exposed to the atmosphere. As water evaporates from the pan, the water level drops. By measuring the decrease in water level over time, scientists can calculate the rate of evaporation. This information is valuable for understanding water balance, estimating water loss from reservoirs, and studying climate patterns.

3.2.2 Limitations and drawbacks

While pan Evaporimeters provide valuable information for estimating evaporation rates, it is essential to consider their limitations and potential sources of error when interpreting the data and making decisions based on the results. Some limitations and drawbacks of pan evaporimeter are described in the following sections:

- **Limited Representativeness:** Evaporation rates measured by pan evaporimeter may not always represent the actual evaporation occurring over a larger area. The conditions within the pan, such as temperature, humidity, and wind speed, may differ from those in the surrounding environment, leading to potential inaccuracies in the estimates.
- **Sensitivity to Environmental Conditions:** Pan evaporation rates are sensitive to environmental factors such as wind speed, air temperature, humidity, and solar radiation. Changes in these conditions can influence evaporation rates measured by the pan, making it challenging to capture the full range of variability in evaporation.
- **Maintenance Requirements:** Pan evaporometers require regular maintenance to ensure accurate measurements. This includes cleaning the pan to remove debris, refilling the water as needed, and calibrating the instrument to account for changes in environmental conditions over time.
- **Spatial Variability:** Evaporation rates can vary significantly across different locations due to variations in climate, land cover, and other factors. Pan Evaporimeters provides point measurements at a single location, which may not capture the spatial variability of evaporation across a larger area.
- **Influence of Pan Design:** The design of the evaporation pan can influence the measured evaporation rates. Different pan designs may yield different results, making it challenging to compare data collected using different types of pans or to extrapolate findings to other locations.
- **Evaporation from Different Surfaces:** Pan Evaporimeters measure evaporation from open water surfaces, such as lakes or reservoirs, which may not accurately represent evaporation from other surface types, such as soil, vegetation, or impervious surfaces.

- **Uncertainty in Estimates:** Despite efforts to account for environmental factors and calibrate the instrument, there is inherent uncertainty associated with evaporation estimates obtained from pan Evaporimeters. This uncertainty can affect the reliability of the data for water resource management and other applications.

3.2.3 Pan coefficient C_p

The "pan coefficient" (C_p) is a term used in evaporation measurement, specifically with reference to Class A evaporation pans, which are commonly used in hydrological and meteorological studies to estimate potential evaporation rates from the earth's surface.

The pan coefficient represents a correction factor applied to the measured water loss from the Class A evaporation pan to estimate actual evaporation rates. This correction factor accounts for various factors that may affect evaporation, such as wind speed, humidity, solar radiation, and pan design. The formula for estimating evaporation using a Class an evaporation pan typically involves multiplying the observed water loss by the pan coefficient. Therefore, a coefficient is introduced in Equation (3.2) as follows:

Lake evaporation = Pan Coefficient C_p x Pan evaporation **.............(3.2)**

The pan coefficient is determined based on calibration studies conducted under local conditions. Different regions and climates may have different pan coefficients due to variations in environmental factors. Therefore, it is crucial to calibrate the Class A evaporation pan and determine an appropriate pan coefficient specific to the location where it will be used. Table 3.1 shows the values of Pan Coefficient C_p for different pans.

Table 3.1: Value of pan coefficients C_p in use for different pans

S.N.	Types of pans	Mean value	Range
1.	Class A Land pan	0.70	0.60-0.80
2.	ISI pan (modified class A)	0.80	0.65-1.10
3.	Colorado sunken pan	0.78	0.75-0.86
4.	USGS floating pan	0.80	0.00-0.82

3.3 EMPIRICAL EVAPORATION EQUATIONS

3.3.1 Dalton type equation

There are numerous empirical equations for estimating lake evaporation that use widely available meteorological data. Most of the formulas are based on the Dalton type equation, which is a simple empirical equation used to estimate potential evaporation rates from open water surfaces. It is named after John Dalton, an English scientist who made significant contributions to meteorology and chemistry. The Dalton type equation provides a relatively straightforward way to estimate potential evaporation rates based on meteorological data, such as air temperature, humidity, wind speed, etc. The Dalton type equation (Equation 3.3) relates potential evaporation to available energy at the earth's surface and the vapour pressure deficit, which can be represented as the difference between the saturation vapour pressure (e_w) and the actual vapour pressure (e_a) of the air. The equation is typically expressed as:

$$E_L = Kf(u)[e_w - e_a] \qquad \text{.............(3.3)}$$

where,

E_L = evaporation rate(mm/day),

K = coefficient,

e_w= saturation vapour pressure at the water surface temperature (mm of H_g),

e_a =actual vapour pressure in overlying air at a specific height (mm of H_g), and

f(u)= correction function for wind speed

(*In Table 3.2: A is the slope of the saturation vapour pressure versus temperature curve at the mean air temperature, in mm of* H_g/ *°C; A will be described in the panman equation in subsequent sections*).

Two evaporation formulas known as Meyer's formula (Equation 3.4) and Rohwer's formula (Equation 3.5) are given in sections 3.2.2 and 3.2.3, respectively.

3.3.2 Meyer's formula

$$E_L = K_M[e_w - e_a]\left(1 + \frac{u_9}{16}\right) \qquad \text{.............(3.4)}$$

where,

E_L = evaporation rate (mm/day),

K_M = Meyer's coefficient (0.36 for large deep water and 0.50 for small shallow water,

Table 3.2: Saturation vapour pressure of water

Temperature (°C)	Saturation vapour pressure e_w (mm of Hg)	A(mm/°C)
0	4.58	0.30
5.0	6.54	0.45
7.5	7.78	0.54
10.0	9.21	0.60
12.5	10.87	0.71
15.0	12.79	0.80
17.5	15.00	0.95
20.0	17.54	1.05
22.5	20.44	1.24
25.0	23.76	1.40
27.5	27.54	1.61
30.0	31.82	1.85
32.5	36.68	2.07
35.0	42.81	2.35
37.5	48.36	2.62
40.0	55.32	2.95
45.0	71.20	3.66

e_w= saturation vapour pressure at the water surface temperature (mm of H_g),

e_a =actual vapour pressure in overlying air at a specific height (mm of H_g), and

u_9= monthly average velocity in km/hr at about 9 m above the earth.

3.3.3 Rohwer's formula

This formula (Equation 3.5) is used to adjust for the influence of pressure in addition to the effect of wind speed, and it may be stated as:

$$E_L = 0.77(1.465 - 0.000732p_a)(0.44 - 0.0733u_0)[e_w - e_a] \quad \text{.............(3.5)}$$

where,

E_L = rate of evaporation (mm/day) and

e_w= saturation vapour pressure at the water's surface temperature (mm of H_g),

e_a = the real vapour pressure in the air above at a given altitude (mm of H_g),

u_0= average wind velocity (km/hr) at earth, represents the velocity at a height of 0.6 meters above the ground and

p_a= average barometric pressure (mm of H_g).

Example 3.1: During a week, the average value of climate parameters of a lake with a surface area of 300 hectares are as follows: Water temperature = 25°C; Relative humidity = 50%; Wind velocity at 1.0 m above ground surface =18 km/hour. Using Meyer's formula, calculate the average daily evaporation from the lake.

Solution:

e_w= saturation vapour pressure at the water surface temperature (mm of H_g)

= From Table 3.2= 23.76 mm of Hg

e_a =actual vapour pressure in overlying air at a specific height (mm of H_g).

= e_w x relative humidity = 23.76x0.50 = 11.88 mm of Hg

u_9 = monthly mean velocity in km/hr at about 9 m above ground

= $u_1 x(9)^{1/7}$= 18 x 1.368 = 24.64 km/h

Average daily evaporation by Meyer's formula:= $K_M[e_w - e_a]\left(1 + \frac{u_9}{16}\right)$

= 0.36 (23.76-11.88) (1+24.64/16)

=0.36 x11.88x2.54

= 10.86 mm/day

3.4 ANALYTICAL METHODS FOR EVAPORATION ESTIMATION

There are three basic groups of analytical approaches for determining lake evaporation.:

a. Water-budget method,

b. Energy-balance method, and

c. Mass-transfer method.

3.4.1 Water-Budget Method

In terms of analytical methods, the water-budget method is the most straightforward and least trustworthy. It entails calculating the evaporation based on estimates or knowledge of other factors and creating the lake's hydrological continuity equation. Thus, while considering a lake's daily average values, the continuity equation (Equation 3.6) can be expressed as follows:

$$P + V_{is} + V_{ig} = V_{os} + V_{og} + E_L + \Delta S + T_L \qquad \text{.............(3.6)}$$

where,

P = daily precipitation,

V_{is} = daily surface inflow into the lake,

V_{ig} = daily groundwater inflow,

V_{os} = daily surface outflow from the lake,

V_{og} = daily seepage outflow,

E_L= daily lake evaporation,

ΔS = increase in lake storage in a day,

T_L= daily transpiration loss.

All the values in equation 3.6 are expressed in terms of depth (mm) or volume (m^3) over a reference region. It is possible to rewrite the equation (3.6) as equation (3.7).

$$E_L = P + (V_{is} - V_{os}) + \left(V_{ig} - V_{og}\right) + T_L - \Delta S \qquad \text{.............(3.7)}$$

The terms P, V_{is}, V_{os} and ΔS are measurable in equation 3.7. However, variables like V_{ig}, V_{og} and T_L can only be estimated because they cannot be measured. In certain reservoirs, transpiration losses can be deemed negligible. It is possible to estimate E_L with greater precision if the time unit is kept big, such as weeks or months. It is unrealistic to expect extremely accurate results from the water-budget technique because of the many uncertainties in the predicted values and the potential for inaccuracies in the measured variables. Controlled trials, like the one conducted in 1952 in Lake Hefner in the United States, have produced results using this method that are reasonably accurate.

3.4.2 Energy-Budget Method

The law of energy conservation is applied through the energy-budget approach. When estimating the amount of energy, all incoming, departing, and stored energy in the water body during a predetermined period is taken into consideration. Considering the specified water body seen in Figure 3.4, the daily energy balance to the evaporating surface is provided by equation 3.8 as follows:

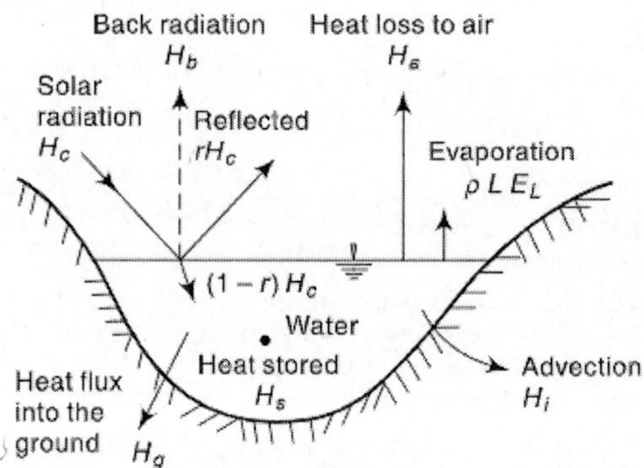

Figure 3.4: Energy Balance in a Water Body

$$H_n = H_a + H_e + H_g + H_s + H_i \qquad \text{.............(3.8)}$$

where,

H_n = net thermal energy derived from the surface of the water

$H_n = H_c(1\text{-}r) - H_b$

in which,

$H_c(1 - r)$ = sun radiation entering a surface with an albedo (reflection coefficient) r,

H_b = long-wave back radiation from a body of water,

H_a = sensible heat transfer from the water's surface to the atmosphere,

H_e = heat energy lost during evaporation= ρLE_L,

where,

ρ= water density,

L = latent heat of evaporation and

E_L = amount of evaporation (mm)

H_g= heat flux into the earth

H_s= heat retained in a body of water

H_i = net heat that the water flow conducts out of the system (advected energy)

Calories per square millimeter per day are used for all energy expressions. The terms H_s and H_i are negligible and can be ignored when time intervals are short. Every term, apart from H_a, is measurable or evaluable indirectly. Using Bowen's ratio β, the sensible heat term H_a, which is challenging to measure, is estimated as Equation 3.9.

$$\beta = \frac{H_a}{\rho L E_L} = 6.1 \times 10^{-4} \times P_a \frac{T_w - T_a}{e_w - e_a} \quad \text{.............(3.9)}$$

where,

P_a = atmospheric pressure in mm of Hg,

e_w = saturated vapour pressure in mm of Hg,

e_a = real vapour pressure of air in mm of Hg,

T_w = temperature of water's surface in degrees Celsius, and

T_a = temperature of air in degrees Celsius.

Based on equations (3.8) and (3.9) E_L can be assessed as equation (3.10).

$$E_L = \frac{H_n - H_e - H_g - H_s - H_i}{\rho L(1+\beta)} \quad \text{.............(3.10)}$$

The energy balance method of calculating evaporation in a lake has been found to produce good results, with errors in the order of 5%, when applied to durations shorter than a week.

3.4.3 Mass-Transfer Method

The mass transfer of water vapour from the surface to the surrounding atmosphere is calculated using this method, which is based on theories of turbulent mass transfer in boundary layers. This method can produce satisfactory results if it uses quantities measured by sophisticated (and expensive) instrumentation.

Example 3.2: During a period of a month of 31 days, following observations were obtained to conduct a lake's water budget.

Mean Surface area= 20 km^2; Rainfall= 15 cm; Mean surface inflow rate = 12 m^3/s; Mean surface outflow rate= 20 m^3/s; Fall in lake level = 1.30 m; Pan evaporation = 22 cm; Assuming Pan evaporation coefficient =0.70, calculate the average seepage discharge during that month.

Solution

Area of lake = 20 x 10^6 $\mathbf{m^2}$

Surface inflow = 12 x 31 x 24 x 3600 = 32,140,800$\mathbf{m^3}$

Surface outflow = 20 x 31 x 24 x 3600 = 53,568,000$\mathbf{m^3}$

Evaporation = 0.7 x 22/100 x 20 x 10^6 = 3,080,000$\mathbf{m^3}$

Rainfall = 15 x 20 x 10^6/100 = 3,000,000$\mathbf{m^3}$

Reduction in storage = 20 x 10^6 x 1.3 =26,000,000$\mathbf{m^3}$

Inflow		**Outflow**	
Item	**Volume (m^3)**	**Item**	**Volume (m^3)**
Surface inflow	32,140,800	Surface outflow	53,568,000
Rainfall	3,000,000	Evaporation	3,080,000
Total inflow volume	35,140,800	Seepage Volume S_e	S_e
Reduction in storage	26,000,000	Total Outflow	56,648,000+ S_e

Total volume of Outflow – Total volume of inflow = change in storage

(56,648,000 + S_e) – (35,140,800) = 26,000,000

S_e= 4,492,800 = 4.4928 Mm^3

3.5 RESERVOIR EVAPORATION AND METHODS FOR ITS REDUCTION

One can estimate reservoir evaporation using any of the previously described techniques. While analytical approaches yield superior outcomes, they involve parameters that are expensive to obtain or difficult to assess. Empirical equations can, at most, be used to obtain estimated values of the appropriate order of magnitude. Thus, the pan measures are generally acknowledged for practical application. The mean monthly and annual evaporation statistics that IMD gathers are very useful for field estimations. The monthly amount of water lost from a reservoir as a result of evaporation is computed as equation 3.11 as follows:

$$\mathbf{V_E = AE_{pm}\,C_p} \qquad \text{.............(3.11)}$$

where,

V_E = volume of water lost in evaporation in a month (m^3),

A= average reservoir area during the month (m^2),

E_{pm} = pan evaporation loss in meters in a month (m),

= E_L in mm/day× No. of days in the month× 10^{-3} , and

C_p= relevant pan coefficient

Evaporation is a continuous process by which water evaporates from a water surface continuously. In India, the average annual evaporation loss from a water body is approximately 160 cm, with higher values in arid regions. Given the size of many of the nation's lakes, both natural and artificial, the amount of water that evaporates each year is in fact significant. A tiny tank (lake) may only have a surface area of about 20 hectares, but large reservoirs like Narmada Sagar have a surface area of about 90,000 ha. Table 3.3 lists the surface areas and capacities of some of the largest reservoirs in India.

3.5.1 Methods to Reduce Evaporation Losses

Reducing evaporation losses can be crucial in various industries, especially in sectors like agriculture, water management, and chemical processing. Here are several methods to mitigate evaporation losses:

(i) **Covering reservoirs or water bodies**: One of the most effective methods is to cover open water bodies such as reservoirs, ponds, or tanks with impermeable materials like plastic sheets, geomembranes, or floating covers. This prevents direct exposure to air and significantly reduces evaporation.

(ii) **Using floating covers**: Floating covers made of materials like foam or plastic can be placed directly on the surface of water bodies. These covers reduce the surface area exposed to air and can also provide insulation, thereby reducing evaporation losses.

(iii) **Implementing windbreaks**: Wind can accelerate evaporation by increasing the rate of water-air exchange. Planting windbreaks such as trees, shrubs, or constructing barriers can reduce wind speed over water bodies, thereby decreasing evaporation.

(iv) **Using mulch**: In agriculture, applying mulch to the soil surface helps to retain moisture and reduces evaporation from the soil. Mulch can be made of materials like straw, wood chips, or plastic film.

(v) **Optimizing irrigation methods**: Compared to conventional overhead irrigation techniques, the use of efficient irrigation systems like drip irrigation or micro-sprinklers minimizes surface evaporation by delivering water directly to the root zone of plants.

(vi) **Regulating water levels**: Maintaining optimal water levels in reservoirs or tanks can minimize the exposed surface area, thereby reducing evaporation losses.

(vii) **Applying chemical monolayers**: Chemical monolayers, such as biodegradable organic compounds or synthetic substances, can be applied to the surface of water bodies to form a thin film. These films inhibit evaporation by reducing surface tension and suppressing vaporization.

(viii) **Increasing humidity**: In certain controlled environments, such as greenhouses or industrial settings, increasing humidity can reduce the vapour pressure gradient between the water surface and the surrounding air, thereby decreasing evaporation rates.

(ix) **Optimizing temperature control**: Controlling the temperature of water bodies or substances prone to evaporation can help reduce evaporation losses. This may involve shading, insulation, or actively cooling the environment.

(x) **Implementing technological solutions**: Advanced technologies such as evaporation suppressants or evaporation ponds with recirculation systems can be employed in industrial processes to minimize evaporation losses effectively.

Implementing a combination of these methods tailored to specific environments and requirements can significantly reduce evaporation losses and improve water conservation efforts.

Table 3.3: Surface Areas and Capacities of Some Indian Reservoirs

SI. No	Reservoir	River	State	Surface area at MRL (km^2)	Gross reservoir capacity (Mm^3)
1.	Narmada Sagar	Narmada	Madhya Pradesh	914	12,230
2.	Nagarjuna Sagar	Krishna	Andhra Pradesh	285	11,315
3.	Sardar Sarovar	Narmada	Gujarat	370	9510
4.	Bhakra	Sutlej	Punjab	169	9868
5.	Hirakud	Mahanadi	Orissa	725	8141
6.	Gandhi Sagar	Chambal	Madhya Pradesh	660	7746
7.	Tungabhadra	Tungabhadra	Karnataka	378	4040
8.	Shivaji Sagar	Koyna	Maharashtra	115	2780
9.	Kadana	Mahi	Gujarat	172	1714
10	Panchet	Damodar	Jharkhand	153	1497

3.6 TRANSPIRATION

The process by which moisture moves from roots to tiny pores on the underside of leaves, where it transforms into vapour and is expelled into the atmosphere, is called transpiration. It is one of the major components of the water cycle and plays a crucial role in plant physiology, water movement within plants, and the overall moisture balance in the environment. Transpired water contributes to the moisture content of the atmosphere, ultimately influencing weather patterns and precipitation. Understanding transpiration is crucial for various fields, including agriculture, ecology, and hydrology, as it influences water availability, ecosystem dynamics, and climate patterns. Transpiration is a significant component of the water cycle, along with processes like evaporation, condensation, and precipitation.

The rate of transpiration can be affected by a number of environmental conditions, including temperature, humidity, wind speed, and soil moisture content. For example, high temperatures and low humidity typically increase transpiration rates, while factors that reduce water availability, such as drought stress, can lead to stomatal closure and decreased transpiration.

While transpiration is essential for plant growth and function, excessive transpiration can lead to water stress in plants, particularly in arid or drought-prone regions. Therefore, strategies to manage transpiration, such as optimizing irrigation practices and selecting drought-tolerant plant species, are important for sustainable water management in agriculture and natural ecosystems.

3.7 EVAPOTRANSPIRATION

The combined processes of transpiration from plants and evaporation from the Earth's surface are known as evapotranspiration (ET). It represents the total amount of water lost by vegetation and soil to the atmosphere. Evapotranspiration is a critical component of the hydrological cycle and plays a significant role in the water balance of ecosystems, agricultural systems, and the atmosphere. Changes in land use and land cover, such as deforestation, urbanization, and agricultural expansion, can alter evapotranspiration rates and patterns. These changes can have significant implications for local and regional hydrology, ecosystem function, water availability, and climate dynamics. Understanding evapotranspiration is essential for various applications, including water resource management, agriculture, ecology, climate modelling, and weather forecasting. Overall, evapotranspiration is a complex process that involves interactions between the atmosphere, vegetation, and soil. It is a key driver of Earth's water cycle and has far-reaching effects on both natural and human-dominated landscapes. Some of the important terms related to evapotranspiration are given below:

- **Consumptive use:** The loss by evapotranspiration is often referred to as consumptive use. Consumptive use refers to the portion of water withdrawn from a water source that is consumed or not returned to its original source. It is a measure of the amount of water that is removed from a water body or aquifer and is no longer available for immediate reuse.

  The amount of water that crops require throughout the growing season is referred as consumptive use in agriculture. This includes water absorbed by plant roots, transpired through plant leaves, and evaporated from the soil surface. Agricultural consumptive use is influenced by factors such as crop type, climate, soil properties, irrigation practices, and water management techniques.

- **Potential evapotranspiration (PET):** Naturally, for a particular set of atmospheric conditions, evapotranspiration obviously depends on the availability of water. Potential evapotranspiration (PET) is the amount of evapotranspiration that results when there is always enough moisture present to fully supply the needs of vegetation covering the entire region. Potential evapotranspiration no longer critically depends on the soil and plant variables but mostly depends on climatic factors.
- **Actual evapotranspiration (AET):** The real evapotranspiration occurring in a specific situation is called actual evapotranspiration (AET). Actual evapotranspiration refers to the total amount of water evaporated from the soil and transpired by plants into the atmosphere over a specific area during a given period. It is influenced by various factors such as solar radiation, temperature, wind speed, humidity, soil moisture content, and vegetation type and density.

  Estimating actual evapotranspiration is crucial in various fields, such as agriculture, hydrology, and environmental science, as it helps in understanding water availability, managing water resources, and predicting drought conditions. There are several methods to estimate actual evapotranspiration, including empirical equations, physical models, and remote sensing techniques. These approaches frequently use weather data, soil moisture measurements, and vegetation features to quantify water loss from the ground surface. It is required to introduce two words at this stage: field capacity and permanent wilting point. Field capacity and permanent wilting point are important soil moisture concepts that play a crucial role in understanding plant-water relationships and soil characteristics in agriculture and ecology.
- **Field Capacity:** Field capacity refers to the maximum quantity of water that the soil can retain against gravity after surplus water has drained away, usually within 1-3 days following a rain or irrigation event. At field capacity, the soil is fully saturated, and any excess water has drained off. It represents the maximum amount of soil moisture available to plants. However, plants cannot extract all the water held at field capacity due to gravitational drainage. Field capacity is affected by soil texture, structure, organic matter content, and compaction.

- **Permanent Wilting Point:** The soil moisture level at which plants wilt irreversibly and are unable to recover, regardless of water additions, is known as the permanent wilting point (PWP). It represents the lower limit of soil moisture available to plants for uptake. Plants are unable to draw soil water at this stage because the soil particles are holding it too firmly. The soil's organic matter content, structure, and texture all affect the permanent wilting point, which varies depending on the kind of soil. The range between field capacity and permanent wilting point is often referred to as the plant-available water or available water capacity. This range represents the soil moisture that is accessible to plants for uptake and is crucial for understanding irrigation scheduling, crop water requirements, and soil fertility management. Soil moisture sensors and measurements are often used to determine field capacity and permanent wilting point for specific soil types and conditions.

The moisture that is available for plant growth is known as available water, and it is the difference between field capacity and permanent wilting point. During the adequate water supply to the plant, soil moisture will be at the field capacity, resulting in AET and PET being equal. If the water supply is less than PET, the soil dries out, and the ratio between AET and PET is less than 1.0. The kind of soil and the rate at which it dries out determine how much the ratio of AET to PET decreases with available moisture. The AET goes to zero when soil moisture hits the permanent wilting point. The variation of AET is represented in figure 3.5. The hydrologic budget equation for a watershed in a given time period can be expressed as follows:

$$P - R_s - G_o - E_{act} = \Delta S \qquad \ldots\ldots\ldots\ldots (3.12)$$

where,

P = precipitation,

R_s = surface runoff,

G_o = sub-surface outflow,

E_{act} = actual evapotranspiration (AET) and

ΔS = change in the moisture storage.

E_{act} can be estimated using this water budgeting by computing the remaining elements of Eq. (3.12). The stream flow at the basin exit can be roughly determined by adding the values of G_o and R_s.

All applied hydrology studies, except for a few specialized ones, use PET (instead of AET) as a fundamental parameter in a range of estimations connected to water utilizations associated to the evapotranspiration process. The reliability of PET as a lake evaporation approximation is widely recognized. As a result, PET can be used to predict lake evaporation in situations where pan evaporation data is not available.

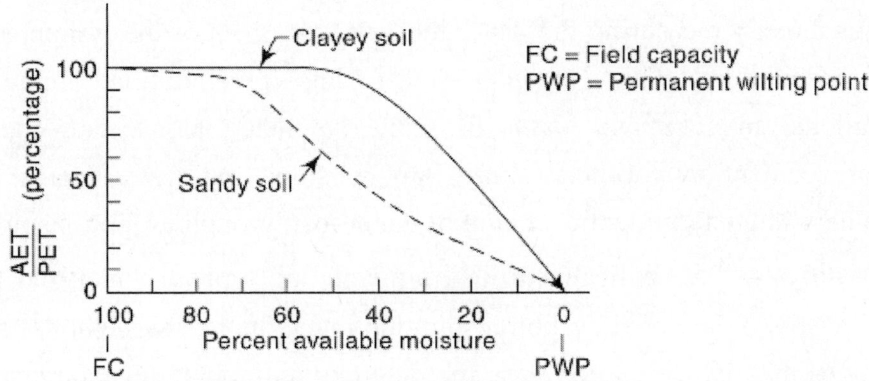

Figure 3.5: Variation of AET

3.7.1 Measurement of Evapotranspiration

There are two methods for measuring evapotranspiration for a particular type of vegetation: (a) using lysimeters and (b) using field plots.

a. **Using Lysimeters:** Lysimeters are specialized instruments used to measure evapotranspiration (ET) directly by monitoring the water balance within a controlled soil column or plot. They offer a very precise means of measuring the volume of water lost from the soil as a result of plant transpiration and evaporation. Here's how lysimeters are typically used to measure evapotranspiration:

- **Installation**: Lysimeters are usually installed in the ground, either at the surface or at various depths, depending on the research objectives. They consist of a soil column or container equipped with sensors to measure changes in soil moisture, as well as instruments to monitor weather parameters such as temperature, humidity, wind speed, and solar radiation.
- **Weighing Method**: The most common technique used with lysimeters is the weighing method. In this method, the lysimeter is periodically weighed to determine

changes in its weight, which directly corresponds to changes in soil water content due to evapotranspiration. By measuring the weight of the lysimeter and accounting for any precipitation inputs, researchers can determine the volume of water lost to evapotranspiration from the soil.

- **Water Balance Approach**: The water balance approach is another approach that involves directly measuring the water inputs and outputs of the lysimeter to calculate evapotranspiration. This includes measuring precipitation inputs, irrigation applications, and drainage losses from the lysimeter, along with changes in soil moisture content over time. Taking into consideration water inputs and outputs, researchers can determine the amount of water lost through evapotranspiration.
- **Data Analysis**: Data collected from lysimeters are typically analyzed to determine daily, weekly, or seasonal evapotranspiration rates. Statistical methods and modelling techniques may be used to analyze the data and extrapolate evapotranspiration rates for different conditions or time periods.

Lysimeters provide highly accurate and site-specific measurements of evapotranspiration, making them valuable tools for research, water management, and agricultural applications. However, they require careful installation, maintenance, and calibration to ensure accurate measurements.

b. **Using Field Plots**: The field plot method for measuring evapotranspiration (ET) involves setting up small plots of vegetation and measuring water use through a combination of direct and indirect methods. Here is a general outline of how the field plot method works:
   - **Plot Setup**: Select an area representative of the vegetation or crop of interest. Divide this area into smaller plots, typically ranging from a few square meters to a hectare, depending on the scale of the study and the heterogeneity of the vegetation.
   - **Vegetation Selection**: Choose the vegetation or crop species for the plots. It is important that the vegetation in each plot is homogeneous in terms of species, age, and canopy cover to minimize variability.
   - **Instrumentation**: Install instruments to measure the key parameters necessary for calculating evapotranspiration. This may include weather stations to measure meteorological parameters like temperature, humidity, wind speed, and solar

radiation; Soil moisture sensors to track the amount of water in the soil at various depths; and devices to measure water inputs (e.g., rainfall) and outputs (e.g., runoff).

- **Measurement of ET**: Evapotranspiration is typically calculated using a water balance approach (Equation 3.13), where inputs (precipitation, irrigation) are measured, and outputs (runoff, deep percolation) are monitored. The difference between these inputs and outputs represents the water loss due to evapotranspiration.

Evapotranspiration = [precipitation + irrigation input – runoff- increase in soil storage -groundwater loss] **.............(3.13)**

Validation: Validation of the evapotranspiration estimation obtained from the field plots against other independent methods or models to ensure accuracy and reliability.

By following these steps, the field plot method provides valuable insights into the water use of different vegetation types or crops under varying environmental conditions, helping researchers and resource managers make informed decisions regarding water management and agricultural practices.

3.7.2 Evapotranspiration Equations

A multitude of techniques have been developed to estimate PET from climatological data due to the scarcity of trustworthy field data and the challenges associated with acquiring trustworthy evapotranspiration data. Several equations are commonly used to estimate evapotranspiration (ET), depending on the available data, the complexity required, and the accuracy desired. Here are some of the most commonly used equations:

a. Penman's Equation

Penman's equation, which combines the mass-transfer and energy-balance approaches, is derived from good theoretical reasoning. After considering some of the changes other researchers have proposed, Penman's equation (Equation 3.14) can be written as:

$$PET = \frac{AH_n + E_a\gamma}{A+\gamma} \qquad \text{.............(3.14)}$$

where,

PET = daily potential evapotranspiration in mm/ day;

A = slope of the saturation vapour pressure versus temperature curve at the mean air temperature, in mm of mercury per °C (Table 3.2)

H_n = net radiation of evaporable water/ day in mm

E_a= parameter comprising wind velocity and saturation deficit

γ = psychrometric constant = 0.49 mm of mercury/°C

The following equation (Equation 3.15) is used to calculate the same net radiation in the energy budget [Equation (3.8)].

$$H_n = H_a(1-r)\left(a + b\frac{n}{N}\right) - \sigma T_a^4(0.56 - 0.092\sqrt{e_a})(0.10 + 0.90\frac{n}{N}) \quad \text{.............(3.15)}$$

where,

H_a = incident solar radiation outside the atmosphere on a horizontal surface, expressed in mm of evaporable water/per day (H_a is the function of latitude and time period as given in Table 3.4)

a = a constant depending upon the latitude φ and is given by a = 0.29 cosφ

b = a constant with an average value of 0.52

n = actual duration of bright sunshine in hours

N= maximum possible bright sunshine hours (function of latitude as indicated in Table 3.5)

r = reflection coefficient (albedo). Table 3.6 displays the typical ranges of r..

σ = Stefan-Boltzman coefficient = 2.01 × 10^9 mm/day

T_a = mean air temperature in degrees kelvin = 273 + °C

e_a = actual mean vapour pressure in the air in mm of H_g

The parameter E_a is estimated as equation 3.16:

$$E_a = 0.35\left(1 + \frac{u_2}{160}\right)(e_w - e_a) \quad \text{.............(3.16)}$$

in which,

u_2 = average wind speed at 2 m above the earth in km/day

e_w = saturation vapour pressure at mean air temperature in mm of H_g (Table 3.2)

e_a = real vapour pressure, previously described

Table 3.4: Mean monthly solar radiation at the top of the atmosphere (Ha), measured in millimeters of evaporable water

North Latitude	Jan	Feb	Mar	Apr	May	June	July	Aug	Sep	Oct	Nov	Dec
0°	14.5	15.0	15.2	14.7	13.9	13.4	13.5	14.2	14.9	15.0	14.6	14.3
10°	12.8	13.9	14.8	15.2	15.0	14.8	14.8	15.0	14.9	14.1	13.1	12.4
20°	10.8	12.3	13.9	15.2	15.7	15.8	15.7	15.3	14.4	12.9	11.2	10.3
30°	8.5	10.5	12.7	14.8	16.0	16.5	16.2	15.3	13.5	11.3	9.1	7.9
40°	6.0	8.3	11.0	13.9	15.9	16.7	16.3	14.8	12.2	9.3	6.7	5.4
50°	3.6	5.9	9.1	12.7	15.4	16.7	16.1	13.9	0.5	7.1	4.3	3.0

Information on n, e_a, u_2, mean air temperature, and nature of the surface (i.e., the value of r) is required for the computation of PET. These can be found by actual observations or by looking up the region's meteorological data. It should be noted that r = 0.05 can be utilized to calculate the quantity of evaporation from a water surface using Penman's equation. Many countries, including Australia, the UK, India, and some areas of the USA, adopt Penman's equation.

Table 3.5: Mean monthly values of possible sunshine hours (N)

North Latitude	Jan	Feb	Mar	Apr	May	June	July	Aug	Sep	Oct	Nov	Dec
0°	12.1	12.1	12.1	12.1	12.1	12.1	12.1	12.1	12.1	12.1	12.1	12.1
10°	11.6	11.8	12.1	12.4	12.6	12.7	12.6	12.4	12.9	11.9	11.7	11.5
20°	11.1	11.5	12.0	12.6	13.1	13.3	13.2	12.8	12.3	11.7	11.2	10.9
30°	10.4	11.1	12.0	12.9	13.7	14.1	13.9	13.2	12.4	11.5	10.6	10.2
40°	9.6	10.7	11.9	13.2	14.4	15.0	14.7	13.8	12.5	11.2	10.0	9.4
50°	8.6	10.1	11.8	13.8	15.4	16.4	16.0	14.5	12.7	10.8	9.1	8.1

Table 3.6: Values of reflection coefficient (r)

Surface	**Range of r values**
Close ground crops	0.15-0.25
Bare lands	0.05-0.45
Water Surface	0.05
Snow	0.45-0.95

Example 3.3: Using Penman's method, determine the probable evapotranspiration from a region close to New Delhi in October. The information listed below is accessible:

Latitude	:	30°0'N
Elevation	:	230m (above sea level)
Mean monthly temperature	:	20°C
Mean relative humidity	:	65%
Mean observed sunshine hours	:	9 h
Wind velocity at the height of 2 m	:	75*km/day*
Type of surface cover	:	Close-ground green crop

Solution:

From Table 3.2,

e_w= 17.54 mm of Hg   A= 1.05 mm/°C

From Table 3.4,

H_a= 11.3 mm of water/day

From Table 3.5,

N = 11.5 h n/N = 9/11.5 = 0.78

From given data

e_a= 17.54 x 0.65 = 11.40 mm of Hg

a = 0.29 cos 30° = 0.2511

b = 0.52

σ= 2.01 x 10^{-9} mm/day

$T_a = 273 + 20 = 293$ K

$\sigma T_a^4 = 14.613$

r = albedo for close-ground green crop is taken as 0.25

Using Equation (3.15),

$$H_n = H_a(1-r)\left(a + b\frac{n}{N}\right) - \sigma T_a^4(0.56 - 0.092\sqrt{e_a})(0.10 + 0.90\frac{n}{N})$$

H_n = 11.3 x (1 - 0.25) x (0.2511 + (0.52 x 0.78)) - 14.813 x (0.56 - 0.092 x(11.40)$^{0.5}$) x (0.10 + (0.9 x 0.78))

H_n = 5.5655 - 2.97 = 2.60 mm of water/day

Using Equation (3.16),

$$E_a = 0.35\left(1 + \frac{u_2}{160}\right)(e_w - e_a)$$

$$E_a = 0.35\left(1 + \frac{75}{160}\right)(17.54 - 11.40)$$

E_a =3.517 mm/day

Using γ = 0.49 mm of mercury/°C in equation (3.14),

$$PET = \frac{AH_n + E_a\gamma}{A + \gamma}$$

$$PET = \frac{1.05x2.60 + 3.517x0.49}{1.05 + 0.49}$$

PET = 2.89 mm/day

b. Reference Crop Evapotranspiration (ET_0)

Reference Crop Evapotranspiration (ET_o), is the rate of evapotranspiration from a hypothetical reference crop, typically grass, under standard meteorological conditions. It serves as a benchmark for estimating the water needs of various crops and landscapes. ET_o represents the evaporative demand of the atmosphere when a crop is not water-limited, meaning it has ample water supply and is actively growing without stress.

The most commonly used method for calculating reference crop evapotranspiration is the Penman-Monteith equation, which considers various meteorological parameters such as temperature, humidity, wind speed, solar radiation, and atmospheric pressure. The equation provides a comprehensive estimation of ET_o based on energy balance and aerodynamic principles.

Reference evapotranspiration is expressed in units of length per unit of time (e.g., millimeters per day or inches per day) and is used as a standard measure for irrigation scheduling, water resource management, and agricultural planning. It allows for the comparison of water requirements among different crops and landscapes under similar climatic conditions.

Any other crop's potential evapotranspiration (ET) can be computed by multiplying the Reference evapotranspiration by a coefficient K (Equation 3.17), whose value varies depending on the crop's stage. The value of K varies from 0.5 to 1.3 depending on the type of crop. Thus

ET = K (ET$_o$) **.............(3.17)**

c. Penman-Monteith Equation

This is one of the most widely used and comprehensive equations (Equation 3.18) for estimating evapotranspiration. It considers both the energy balance and aerodynamic components of ET and requires meteorological data, including temperature, humidity, wind speed, solar radiation, and atmospheric pressure.

$$ET_0 = \frac{0.408\Delta(R_n - G) + \frac{900}{T}\gamma u_2 \delta_e}{\Delta + \gamma(1 + 0.34u_2)} \quad ......(3.18)$$

where,

ET_O = Reference Evapotranspiration rate (mm/day)

Δ = saturation specific humidity change rate in relation to air temperature. (Pa K^{-1})

R_n = Net Radiance flux (M m^{-2} day^{-1}); (MJ= million joule)

G = Ground heat flux (MJ m^{-2} day^{-1}), which is typically equal to zero on a day

T = Air temperature at 2m (K)

u_2 = Wind speed at a height of 2m (m/s)

δe = vapour pressure deficit (kPa)

γ = Psychrometric constant (γ H66 Pa K^{-1})

d. Blaney-Criddle Formula

The Blaney-Criddle formula is an empirical equation used to estimate potential evapotranspiration (PET) based on temperature and latitude data. It is a simplified method and widely used, especially in regions where more complex methods or extensive meteorological data are not available. This formula estimates potential evapotranspiration based solely on temperature, making it a straightforward and easy-to-use method. However, it does not account for other factors, such as humidity, wind speed, and solar radiation, which are important for more accurate estimations of evapotranspiration. Therefore, while the Blaney-Criddle formula can provide a rough estimate of potential evapotranspiration, it may not be suitable for all applications, especially in areas with significant variations in climate and vegetation. Hence, the Blaney-Criddle formula may not be as accurate as more comprehensive methods like the Penman-Monteith equation. The Blaney-Criddle formula (Equation 3.19 and 3.20) is expressed as follows:

$$E_T = 2.54\ KF \quad \text{.............(3.19)}$$

and

$$F = \frac{\Sigma P_h \overline{T}_f}{100} \quad \text{.............(3.20)}$$

where,

E_T = PET in a crop season in cm

K = a coefficient depends on the type of the crop

F = Sum of monthly consumption uses factors

P_h = monthly percent of annual day-time hours depending on the latitude of the place.

$\overline{T}_f$ = Mean monthly temperature °F

3.8 POTENTIAL EVAPOTRANSPIRATION OVER INDIA

Utilizing Penman's formula and pertinent meteorological information, the nation's PET estimate has been produced. Isopleths, or the lines on a map that cross locations with identical evapotranspiration depths, are used to display the mean annual PET (in cm) at different regions of the nation [Figure 3.6 (a)]. The annual PET is observed to vary between

140 and 180 cm across most of the nation. Rajkot, Gujarat, has the greatest annual PET, measuring 214.5 cm. Tamil Nadu's extreme southeast likewise exhibits high average values above 180 cm. At Tiruchirappalli, Tamil Nadu, the PET for the southern peninsula is the highest, measuring 209 cm. Figure 3.6 (b) shows the monthly PET variance at selected locations across the nation's various climate zones.

Example 3.4: Calculate an area's PET during the wheat-growing season, November through February, using the Blaney-Criddle formula. The region is in North India, at latitude of 25°N, and experiences the following mean monthly temperatures:

Month	November	December	January	February
Temp °C	18	15	10	13.5
Value of P_h at 25°C	7.40	7.42	7.53	7.14

Value of K for wheat = 0.65

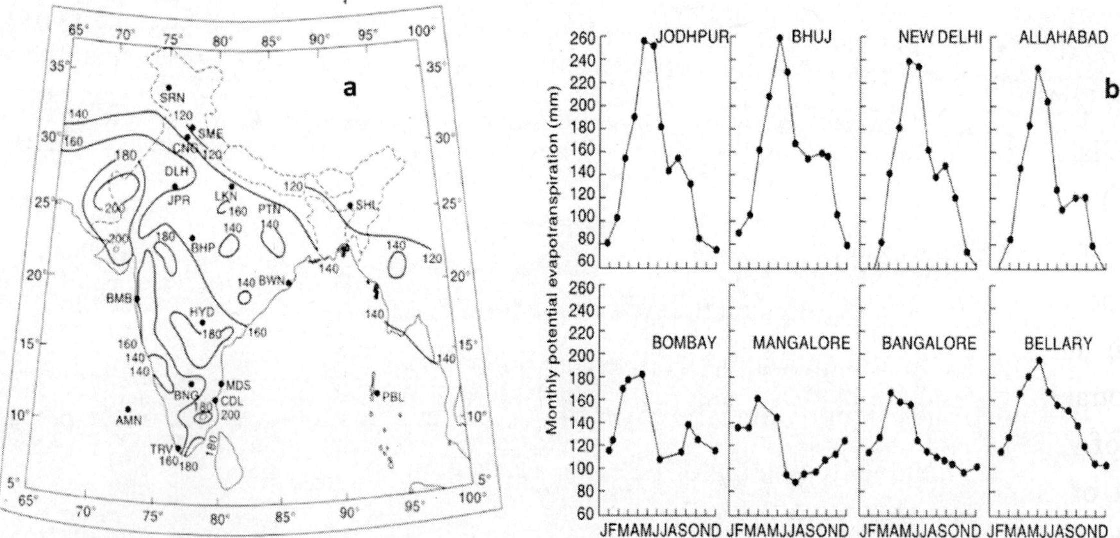

Figure 3.6: (a): Annual PET (cm) over India, (b): Monthly Variation of PET (mm) (Source: Scientific Report No. 136, India Meteorological Department,1971, © GoI)

Solution

The following table computes the temperature to Fahrenheit.

Month	$\underline{T_f}$	P_h	$P_h/100x\overline{T_f}$
November	64.4	7.40	4.77
December	59.0	7.42	4.38
January	50	7.53	3.77
February	56.3	7.14	4.02
		$\frac{\Sigma P_h \overline{T_f}}{100} =$	16.94

From Blaney-Criddle formula: $E_T = 2.54\ KF = 2.54 \times 0.65 \times 16.94 = 27.97$ cm

3.9 ACTUAL EVAPOTRANSPIRATION (AET)

The combined process of evaporation from soil and plant surfaces and transpiration, which moves water from the Earth's surface into the atmosphere, is referred to as actual evapotranspiration. It represents the actual amount of water that is lost from a given area due to these processes, considering factors such as temperature, humidity, wind speed, and the type of vegetation present. The water cycle depends heavily on actual evapotranspiration, which is also a critical component of hydrological research, agricultural water management, and climate modelling. For hydrological and irrigation reasons, actual evapotranspiration (AET) can be calculated by accounting for soil, plant, and atmospheric interactions, as well as water budgeting. Below is a straight forward process according to Doorenbos and Pruit:

The combined process of evaporation from soil and plant surfaces and transpiration, which moves water from the Earth's surface into the atmosphere, is referred to as actual evapotranspiration. It considers variables like temperature, humidity, wind speed, and the type of vegetation present to indicate the real amount of water lost from a given region as a result of these activities. The water cycle depends heavily on actual evapotranspiration, which is also a critical component of hydrological research, agricultural water management, and climate modelling. For hydrological and irrigation reasons, actual evapotranspiration (AET) can be calculated by accounting for soil, plant, and atmospheric interactions, as well as water budgeting. The simple procedure described below is based on Doorenbos and Pruit:

a. The reference crop evapotranspiration (ETo) is determined using the meteorological data that is currently available.

b. Published tables are used to determine the crop coefficient K for the specified crop (and growth stage). The potential crop evapotranspiration ET_c is estimated using the following equation :

$$ET_c = K(ET_o).$$

c. The following formula (Equation 3.21a & 3.21b) is used to get the actual evapotranspiration (ET_a) at a given time t at the field growing the specified crop:

- If AASW ≥ (1- p) MASW

$$ET_a = ET_o \text{ (known as a potential condition)} \quad \text{..............(3.21a)}$$

- If AASW < (1 - p) MASW

$$ET_a = \left[\frac{AASW}{(1-p)\,MASW}\right] ET_c \quad \text{..............(3.21b)}$$

where,

MASW = total available soil water content above the root depth

AASW = actual available soil-water at time t over the root depth

p = soil-water depletion factor for a given crop and soil complex.

(P values vary from roughly 0.1 in sandy soils to roughly 0.5 in clayey soils)

[Observe that the terms used previously, PET =ET_o and AET = ET_a, are equivalent]

Example 3.5: On Day 1, the total accessible soil moisture in a recently irrigated field plot is at its maximum value of 15 cm. Determine the actual evapotranspiration on Days 1, 8, and 9 if the reference crop's evapotranspiration is 7.0 mm/day. Assume that crop factor K = 0.90 and soil-water depletion factor p = 0.20.

Solution

Here ET_o= 7.0 mm and MASW= 150 mm

(1- p) MASW = (1- 0.2) x 150 = 120.0 and

ET_c = K x ET_o = 0.9 x 5.0 = 4.5 mm/day

Day I: AASW = 150 mm> (1 - p) MASW

Hence, a potential condition exists and ET_a = ET_c = 4.5 mm/day. This rate will continue till a depletion of (150 - 120) = 30 mm takes place in the soil. This will take 30/4.5 = 6.66 days.

Thus,

Day 7: will also have $ET_a = ET_c = 4.5$ mm/day

Day 8: At the beginning of Day 8, AASW = (150 - 4.5 x 7) = 118.5 mm

Since AASW < (1 - p) MASW

$$ET_a = \left[\frac{AASW}{(1-p)\ MASW}\right] ET_c = \left[\frac{118.5}{120}\right] 4.5 = 4.44$$

Day 9: At the beginning of Day 9, AASW = (118.5 - 4.44) = 114.06 mm

Since AASW < (1 - p) MASW

$$ET_a = \left[\frac{AASW}{(1-p)MASW}\right] ET_c = \left[\frac{1114.06}{120}\right] 4.5 = 4.28\ mm$$

AASW *at* the end of Day 9 = 114.06- 4.28 = 109.78mm.

3.10 INITIAL LOSS

The "Initial loss" refers to the amount of water that does not contribute directly to runoff or infiltration during the early stages of a rainfall event. When rain falls onto the ground, some of it may initially be absorbed by surface materials such as soil, vegetation, or pavement before it begins to runoff or infiltrate into the soil. This initial loss represents the water that is temporarily held on the surface or within the surface materials before it starts to move elsewhere.

Initial loss can vary depending on several factors, including the type of surface (e.g., soil, vegetation, and pavement), rainfall intensity and its duration, the slope of the terrain, and the antecedent moisture conditions of the surface. In hydrological modelling and engineering, it is important to account for this initial loss when estimating the amount of rainfall that will contribute to runoff or infiltration, as it affects the timing and magnitude of surface water flow and groundwater recharge.

The primary abstraction in precipitation that reaches a catchment's surface comes from the infiltration process. Two other processes, however minor in scope, function as abstractions by decreasing the amount of water available for runoff. These processes are the interception process and the depression storage process. This concept represents the amount of storage that needs to be fulfilled before overland runoff starts.

3.10.1 Interception

Interception refers to the process by which precipitation (such as rain, snow, or hail) is caught and retained by vegetation, such as trees, shrubs, grass, and other forms of vegetation, before it reaches to the ground. This interception process is a crucial component of the hydrological cycle and has significant effects on the movement of water within ecosystems. Interception plays a significant role in hydrological processes, affecting the amount of water that reaches the soil, the timing and magnitude of runoff, soil moisture dynamics, and overall water balance within a watershed or ecosystem. It also influences various ecological processes and ecosystem functions, such as nutrient cycling, microclimate regulation, and habitat availability for organisms. Therefore, understanding interception processes is important for water resource management, ecological studies, and land-use planning.

When precipitation falls onto a vegetated surface, some of it is intercepted by the leaves, branches stems, and other parts of the vegetation. This intercepted water may then undergo various processes:

- **Storage:** The intercepted water is temporarily held on the surface of the vegetation, where it can be stored until it evaporates back into the atmosphere or is eventually released to the ground or other surfaces.
- **Evaporation:** Interception can lead to evaporation directly from the vegetation surface. The intercepted water absorbs solar energy and undergoes an evaporation process, returning moisture to the atmosphere.
- **Drip or Through fall:** Some intercepted water may eventually drip off the vegetation and reach the ground directly (known as through fall), bypassing interception, while some may be transferred along the surface of the vegetation and eventually reach the ground (known as stem flow).

Transpiration, through fall, and stem flow are not included in the definition of intercepted loss; it is only caused by evaporation. Measuring the amount of water intercepted in a particular region is very challenging. It is dependent upon the species composition, density, and storm characteristics of the vegetation. The interception loss is thought to be between 10% and 20% of the total amount of rain that falls in an area while plants are growing. If a region receives a lot of minor storms each year, the annual interception loss from forests will be significant and will exceed 25% of the total precipitation. Figure 3.7 presents the quantitative change of interception loss with storm rainfall magnitude for small storms.

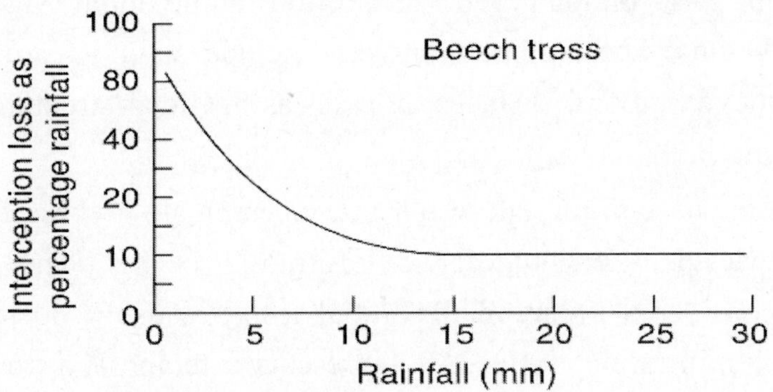

Figure 3.7: Typical Interception Loss Curve

It is observed that for tiny rainfall, the interception loss is high, and for heavier storms, it levels out to a constant amount. The interception loss for a particular storm can be presented as equation 3.22:

$$\mathbf{I_i = S_i + K_i E_t} \qquad \text{.............(3.22)}$$

where,

I_i = interception loss (mm),

S_i = interception storage, which is dependent on the kind of vegetation and ranges in value from 0.25 to 1.25 mm,

K_i = ratio of vegetal surface area to its projected area and,

E_t = evaporation rate in mm/h during the precipitation.

It has been discovered that deciduous trees have less interception loss than conifers. Furthermore, dense grasses can account for nearly 20% of the season's total rainfall and have the same interception losses as fully-grown trees. In their growth season, agricultural crops also account for a large portion of interception losses. Given this, the interception process has a major effect on the region's water balance, in situ water harvesting, and the ecology of the area regarding silviculture issues. Interception loss, however, is rarely considerable and is not separately considered in flood-related hydrological studies. It is customary to let a lump sum amount be subtracted from the storm's starting duration as the initial loss.

3.10.2 Depression storage

Depression storage under initial loss refers to the portion of water that is retained within depressions or low-lying areas on the ground surface during the initial stages of precipitation events before runoff begins. These depressions can include natural features such as small basins, potholes, or uneven ground surfaces, as well as artificial structures like ditches or depressions in pavement.

During the initial phase of rainfall, water may accumulate in these depressions before reaching a threshold level where runoff or infiltration begins. This stored water in depressions is often considered a part of the initial loss because it does not immediately contribute to runoff or infiltration. Instead, it represents a temporary storage reservoir for water until the rate of rainfall exceeds the infiltration capacity of the soil or the capacity of the depression is reached, leading to runoff or further infiltration.

Before rain can flow over the surface, it must first fill in any depressions as it hits the ground. The volume of water that these depressions hold is known as depression storage. This quantity subsequently becomes part of the initial loss due to evaporation and infiltration processes that lead to runoff. Depression storage is influenced by many variables, the main ones being (i) soil type, (ii) surface conditions reflecting depression size and kind, (iii) catchment slope, and (iv) antecedent precipitation as a proxy for soil moisture. It is evident that there are no universal expressions available for the quantitative evaluation of this loss. According to qualitative research, antecedent precipitation significantly reduces the amount of precipitation loss to runoff during a storm because of depression. Values in sand (0.5 cm), loam (0.4 cm), and clay (0.2 cm) can be used as indicators of the loss of depression-storage after strong storms.

In hydrological modelling and engineering, accounting for depression storage under initial loss is important for accurately estimating the timing and magnitude of surface runoff and infiltration, particularly in areas with uneven terrain or significant surface irregularities. It helps in understanding how precipitation interacts with the ground surface and how water moves across the landscape during rainfall events.

3.11 INFILTRATION

Infiltration is a fundamental process in hydrology that refers to the movement of water from the surface into the soil or porous rock layers beneath. When precipitation hits the earth, it can either infiltrate into the soil, runoff over the surface, or evaporate back into the atmosphere. Here's how infiltration typically occurs:

a. **Saturation:** Initially, when precipitation falls onto the soil surface, the soil may absorb water until it reaches its infiltration capacity. This capacity is determined by soil properties such as texture, structure, porosity, and permeability.

b. **Infiltration:** Once the soil reaches its infiltration capacity, excess water starts to infiltrate into the soil pores and move downwards due to gravity. This infiltrating water replenishes soil moisture, which is crucial for plant growth and groundwater recharge.

c. **Percolation:** As water infiltrates deeper into the soil profile, it may continue to move downward, driven by gravity, until it reaches a zone where all soil pores are filled with water (the saturated zone). At this point, further infiltration is restricted, and the infiltrated water may percolate horizontally towards streams, lakes, or other water bodies, or it may contribute to groundwater recharge.

d. **Surface Retention:** Surface retention refers to the mechanism by which water stays on the surface of the land without infiltrating into the soil or running off into streams, rivers, or other water bodies. This water may be retained temporarily on various types of surfaces, including soil, vegetation, pavement, or impermeable surfaces like concrete or asphalt. Understanding surface retention is crucial for effective water management, land use planning, and sustainable development practices, especially in regions prone to water scarcity, flooding, or degradation of natural resources. Efforts to enhance surface retention may involve implementing green infrastructure, such as permeable pavements, green roofs, and rain gardens, and restoring natural hydrological processes in urban and rural landscapes. Surface retention can occur due to various factors, including:

   - **Low Infiltration Rates**: In areas with compacted soils, impermeable surfaces, or saturated soil conditions, water may be unable to infiltrate into the soil, leading to surface retention.
   - **Vegetation and Biomass**: Vegetation, such as grass, shrubs, or trees, can intercept and retain water on their surfaces through processes like interception, evapotranspiration, and storage in biomass.

- **Depression Storage**: Small depressions or low-lying areas on the land surface can act as temporary storage for water, especially during rainfall events, before it evaporates or slowly infiltrates into the soil.
- **Pavement and Impervious Surfaces**: Impermeable surfaces like roads, parking lots, and buildings prevent water from infiltrating into the soil, leading to surface runoff and retention on these surfaces until it evaporates or is drained away through stormwater systems.

Surface retention can have both positive and negative impacts on the environment and human activities:

- **Positive Impacts**: Surface retention can contribute to soil moisture replenishment, support vegetation growth, reduce erosion, and mitigate the effects of drought by providing water for plants and ecosystems.
- **Negative Impacts**: Excessive surface retention, especially on impermeable surfaces in urban areas, can lead to flooding, waterlogging, and increased runoff pollution, as well as reduced groundwater recharge and increased urban heat island effects.

Factors affecting infiltration include soil characteristics (texture, structure, compaction), vegetation cover, slope gradient, land use practices, and antecedent moisture conditions. Infiltration rates can vary widely depending on these factors and can be influenced by human activities such as urbanization, agriculture, and deforestation.

Understanding infiltration is essential for various hydrological applications, including groundwater recharge estimation, flood prediction, watershed management, and soil conservation practices. It plays a crucial role in maintaining water balance within ecosystems and sustaining groundwater resources.

The infiltration rate can be measured from consistent observations of storm rainfall. A specific amount of water is placed on the soil's surface prior to runoff. This surface storage evaporates when the rainfall stops, but some of it keeps seeping into the soil throughout that time. During the process of infiltration, the total volume of precipitation can be defined as equation 3.23:

$$P = Q + D_a + V_d + I_c + F \qquad \textbf{(3.23)}$$

where,

P = Volume of precipitation in cms

Q = Volume of Surface Runoff in cms

$I_c + V_d$ = Volume of initial storage in cms

D_a = Volume of surface retention storage in cms

F = Volume of infiltrated water in cms

After the steady condition is reached

$$P = Q + F \text{ or } F = P - Q$$

3.12 INFILTRATION CAPACITY

The maximum rate at which a soil can absorb water at a particular moment is known as its infiltration capacity. It is denoted as f_p and can be measured in units of cm/h. The actual rate of infiltration f can be expressed as equation 3.24.

$f = f_p$ when $i \geq f_p$

and

$f = i$ when $i < f_p$ ....**(3.24)**

where, i denotes the intensity of the rainfall. When a storm first starts, a soil's infiltration capacity is high and decreases exponentially over time. Figure 3.8 illustrates how a soil's infiltration capacity (f_p) typically changes over time.

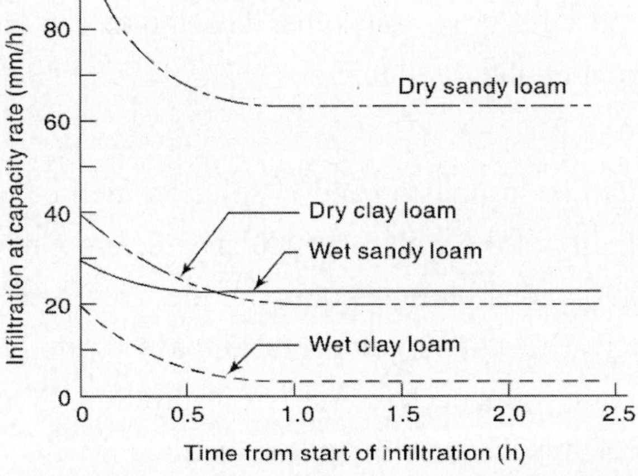

Figure 3.8: Infiltration Capacity Variation

The infiltration capacity of a watershed may be dependent on several factors, out of which some important factors affecting f_p are described below:

- Soil characteristics (Texture, porosity, and hydraulic conductivity)
- Condition of the soil surface
- Current moisture content
- Vegetation cover and temperature of Soil

- **Characteristics of Soil**

The texture, structure, permeability, under drainage, and type of soil: clay, silt, or sand are the key attributes that fall under this category. A sandy soil that is loose and permeable will be able to absorb more water than a soil that is compacted and clayey. It makes sense that a soil with good under drainage, that is, the ability to transfer the infiltrated water downward to groundwater storage would have a larger capacity for infiltration. The total infiltration rate is determined by the transmission capacity of the soil layers when the soils are found in layers. Furthermore, dry soil has a higher capacity to absorb water than soil with fully developed pores (Fig.3.10). Land use has a significant influence on f_p. For instance, the same soil in an urban environment that is subjected to compaction will have a considerably lower value of f_p than the same soil in a forest that is rich in organic matter.

- **Surface of Entry**

Raindrop impact displaces soil particles at the soil's surface, which might block the pore spaces in the soil's higher layers. This has a significant impact on the infiltration capacity. Therefore, a surface covered with grass and other vegetation can slow down this process, which has a significant impact on the f_p value.

- **Fluid Characteristics**

Numerous contaminants, both in suspension and in solution, will be present in the water that seeps into the soil. The turbidity of the water, particularly the amount of clay and colloid, is a significant impact. These suspended particles obstruct the small soil pores and decrease the soil's ability to absorb water. The water's temperature plays a role because it influences the water's viscosity, which in turn influences the rate of infiltration. Water contaminated by the dissolved salts may have an impact on the soil's structure, which may, therefore, have an impact on the rate of infiltration.

3.13 MEASUREMENT OF INFILTRATION

In some problems of operational hydrology, the conduction of controlled experiments on a representative watershed may be desirable. Determining the infiltration capability may be obtained by:

1. Flooding type infiltrometers
2. Simple (Tube type) infiltrometer
3. Double-ring type infiltrometer
4. Rainfall simulator
5. Hydrograph analysis

3.13.1 Simple (Tube type) infiltrometer

A simple (Tube type) infiltrometer is a simple instrument made up of a metal tube that is roughly 60 cm long and 30 cm in diameter and is pushed into the earth, with about 10 cm sticking out above the ground. Depending on the kind of vegetation, water is poured into this ring to a depth of 2.5 to 5 cm. The constant head is maintained by pouring additional water into the ring. Time intervals are used to measure the increased volume. Experiments on infiltration capacity may take two or three hours to reach a uniform rate of infiltration, depending on the average infiltration rate during the measurement period. The measure disadvantage of this type of infiltrometer is that the tube element of the experiment may cause significant disruptions to the field circumstances.

3.13.2 Double-ring infiltrometer

A double-ring infiltrometer is a tool used in hydrology and soil science to measure the rate at which water infiltrates into the soil. It consists of two concentric rings, typically made of metal or plastic, which are pressed into the soil surface. The inner ring is open at the bottom and serves as a reservoir for water, while the outer ring is closed at the bottom and acts as a boundary to control the lateral flow of water.

For installing a double-ring infiltrometer, the soil surface is prepared by removing any debris and levelling the area where the infiltrometer will be installed. The double-ring infiltrometer is pressed into the soil surface using a mallet or other suitable tool. Both rings should be firmly seated in the soil to prevent leakage. A measured volume of water is added to the inner ring of the infiltrometer. This water is allowed to infiltrate into the soil through the bottom of the inner ring. The rate of water infiltration is monitored over time by obtaining the water level in the inner ring at regular intervals. This can be done manually using a graduated ruler or automatically with sensors connected to data loggers. The data collected during the infiltration test can be used to calculate various parameters, such as the infiltration rate, hydraulic conductivity, and soil infiltration capacity. These parameters provide valuable information about the ability of soil to absorb and transmit water, which is essential for various hydrological and engineering applications.

Double-ring infiltrometers are commonly used in field studies to assess soil infiltration properties, evaluate the effectiveness of land management practices, and design drainage systems and erosion control measures. They provide a simple and reliable method for quantifying soil water infiltration under controlled conditions, allowing researchers and practitioners to better understand and manage water resources in agricultural, urban, and natural environments. A section view of the double-ring infiltrometer is given in figure 3.9.

3.13.3 Rainfall Simulators

Rainfall simulators (Figure 3.10) are devices used in hydrology, soil science, and agricultural research to mimic natural rainfall under controlled conditions. These simulators generate artificial rainfall with specific characteristics, such as intensity, duration, and droplet size distribution, allowing researchers to study various aspects of water movement, soil erosion, infiltration, and runoff in a controlled environment. Overall, rainfall simulators provide a valuable tool for researchers to study the complex interactions between rainfall, soil, vegetation, and land use, helping to improve our understanding of water resources management and environmental sustainability.

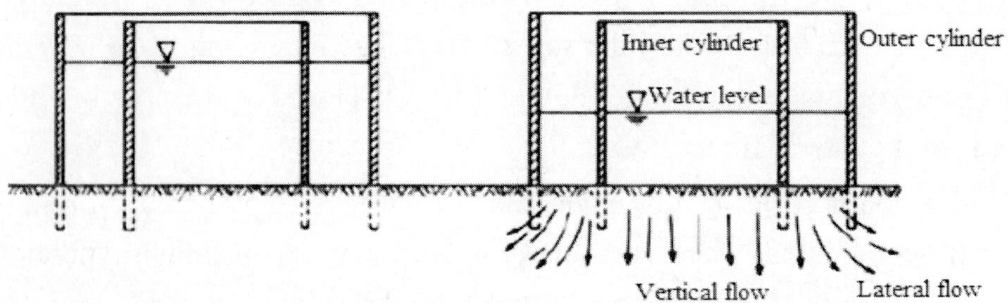

Figure 3.9: Section view of double-ring infiltrometer.

Rainfall simulators come in various designs, ranging from simple portable units to complex systems with multiple components. The design may include a water source, distribution system, and mechanism for generating raindrops. The simulator is connected to a water source, such as a tank, reservoir, or water pump, which supplies water to generate artificial rainfall. The water is distributed over a test plot or soil surface using a distribution system, which may consist of pipes, hoses, or spray nozzles. The distribution system ensures uniform coverage of the test area with artificial rainfall. Rainfall simulators generate raindrops using various methods, such as gravity-fed nozzles, sprinklers, or rotating arms. These devices produce raindrops with controlled characteristics, such as size, velocity, and distribution, to simulate different types of natural rainfall events. Researchers can adjust the intensity, duration, and other parameters of the artificial rainfall using control systems integrated into the simulator. Data loggers and sensors may be used to monitor various parameters, such as rainfall intensity, runoff volume, soil moisture, and erosion rates, during the experiment.

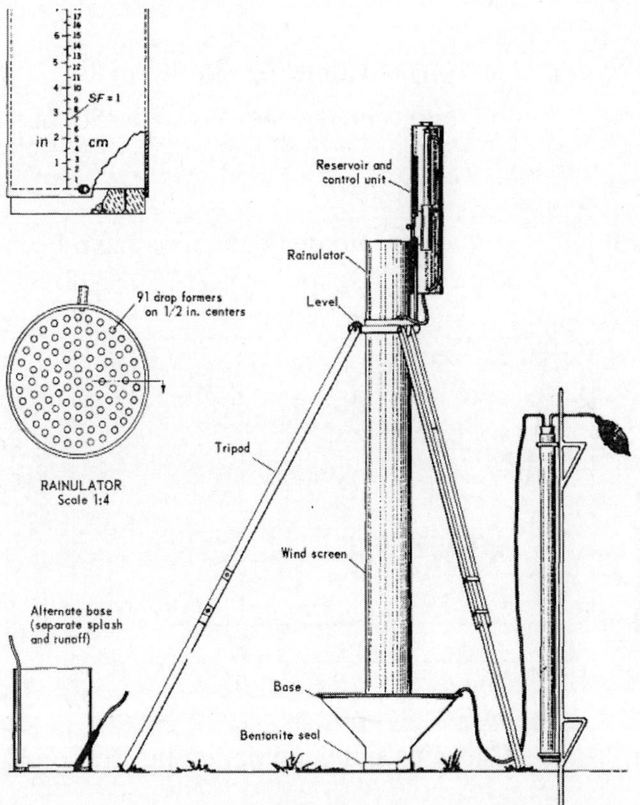

Figure 3.10: Rainfall simulator (Source: I. S. Mcqueen, 1963)

3.13.4 Hydrograph Analysis

An accurate assessment of the infiltration capacity of a small watershed can be obtained by analyzing runoff hydrographs that have been measured and corresponding rainfall records. The water budget equation can be used to estimate the abstraction by infiltration if sufficiently accurate rainfall records and runoff hydrographs corresponding to individual storms in a small watershed with relatively homogeneous soils are available. By understanding the watershed's land cover and use, the evapotranspiration losses are approximated in this case.

3.14 MODELLING INFILTRATION CAPACITY

Typical changes in infiltration capacity f_p over time are shown in Figure 3.11. Cumulative infiltration capacity $F_p(t)$ is the total amount of infiltration volume accumulated over time since the start of the process and is calculated by equation 3.25a.

$$F_p = \int_0^t f_p\ (t)dt \qquad \text{.............(3.25a)}$$

Hence, the curve between $F_p(t)$and time *(t)* shown in Fig. (3.11) is represented as mass curve of infiltration. It can be represented as equation (3.25b) as follows:

$$f_p\ (t) = \frac{dF_P\ (t)}{dt} \qquad \text{.............(3.25b)}$$

For use in hydrological studies, numerous equations have been developed to express the curves $f_p(t)$ or $F_p(t)$. Four such equations will be explained in this section.

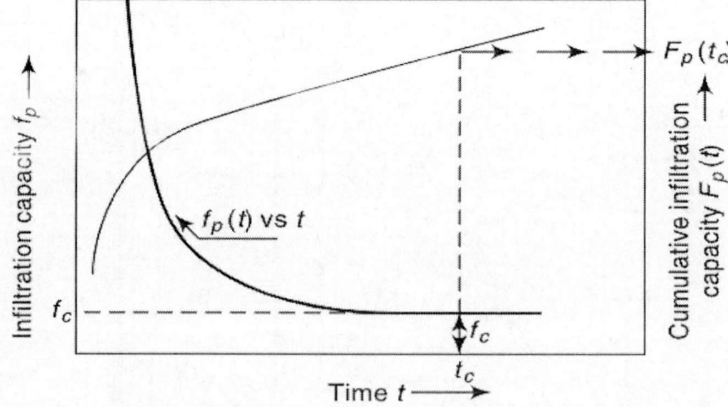

Figure 3.11: Infiltration Capacity and Cumulative Infiltration Capacity Curves

3.14.1 Horton's Equation (1933)

According to Horton, the infiltration capacity decays exponentially over time (equation 3.26) as follows:

$$f_p = f_c + (f_o - f_c)e^{-K_h t} \text{ for } 0 \geq t \leq t_c \qquad \text{.............(3.26)}$$

where,

f_p = infiltration capacity from the beginning of the rainfall at any given time t

f_0= initial infiltration capacity at t = 0

f_c= final steady state infiltration capacity occurring at t = t_c. Furthermore, f_c is also referred to as ultimate infiltration capacity or constant rate on occasion.

K_h=Horton's decay coefficient which is influenced by the vegetation cover and the properties of the soil.

Example 3.6: Using Horton's equation, the infiltration capacity in a watershed is represented as

$$f_p = 3.5 + e^{-2t}$$

where f_p is in cm/h and t is in hours. Estimate the depth of infiltration in (i) the first 30 minutes and (ii) the second 30 minutes of the storm, assuming that infiltration occurs at capacity rates for a storm lasting one hour.

Solution

$$F_p = \int_0^t f_p\ (t)dt$$

and

$f_p = 3.5 + e^{-2t}$

(i) In the first 0.5 hour,

$$F_{p1} = \int_0^{0.5}(3.5 + e^{-2t})dt = \left[(3.5\,t - \tfrac{1}{2}e^{-2t})\right]_0^{0.5}$$

$$= \left(3.5\,x\,0.5 - \tfrac{1}{2}(e^{-2x0.5}\right) - (-1/2)$$

= (1.75-0.184) + 0.5

= 2.066 cm

(ii) In the Second 0.5 hour,

$$F_{p2} = \int_{0.5}^{1.0}(3.5 + e^{-2t})dt = \left[(3.5\,t - \frac{1}{2}e^{-2t})\right]_{0.5}^{1.0}$$

$$=\left(3.5 \, x \, 1.0 - \frac{1}{2}(e^{-2}\right) - \left(3.5 \, x \, 0.5 - \frac{1}{2}(e^{-2x0.5}\right)$$

= (3.5-0.0677) –(1.75-0.184)

= 1.8663 cm

Example 3.7: Before a rainfall event, the infiltration capacity of the soil in a small watershed was found to be 8 cm/h. After ten hours of storming, the same was measured at 1.5 cm/h. Find the value of the decay coefficient K_h in Horton's infiltration capacity equation if the total infiltration over the storm's 10-hour duration was 18 cm.

Solution

Horton's equation:

$$f_p = f_c + (f_o - f_c)e^{-K_h t}$$

and

$$F_p = \int_0^t f_p\ (t)dt$$

$$F_p = \int_0^t f_p\ (t)dt = f_c t + (f_o - f_c)\int_0^t\ \ e^{-K_h t}dt$$

As $t \rightarrow \infty \int_0^{\infty} e^{-K_h t}dt \rightarrow \frac{1}{K_h}$

Hence, for large *t*-values

$$F_p = f_c t + \frac{(f_o - f_c)}{K_h}$$

Here, F_p = 18.0 cm,

f_0 = 8.0 cm/h,

f_c = 1.5 cm/h, and

t = 10 hours.

18.0 = (1.5 x 10) + (8.0 - 1.5)/K_h

K_h= 6.5/3.0 = 2.2 h^{-1}

3.14.2 Philip's Model

Philips represented the expression for F_p as Equation 3.27 as follows:

$$F_p = \frac{1}{2} s t^{\frac{-1}{2}} + Kt \qquad \text{.............(3.27)}$$

where, s = an expression of soil suction potential known as absorptivity

K= Darcy's hydraulic conductivity

Plot the measured f_p values against $t^{-0.5}$ on an arithmetic graph paper. K is the intercept, and (s/2) is the slope of the best-fitting straight line through the depicted points. It is important to remember that K must be positive while fitting Philip's model, and in order to do so, it might be necessary to exclude a few data points from the early phases of the infiltration experiment (viz. at small values of t). K is going to be roughly equal to the asymptotic value of f_p.

3.14.3 Kostiakov Model

In the Kostiakov model, the relation between F_p and t can be given as equation 3.28a as follows:

$$F_p = at^b \qquad \text{.............(3.28a)}$$

where, a and b are known as local parameters; in which $a > 0$ and $0 < b < 1$. Taking logarithms of both sides of equation (3.28a), one gets equation (3.28b) as follows;

$$\ln(F_p) = \ln a + b \ln(t) \qquad \text{.............(3.28b)}$$

On an arithmetic graph paper, the data is shown as *ln(F*$_p$*) vs ln(t)*. The intercept of the best fit straight line across the plotted points is ln a, and the slope is b. Keep in mind that $0 < b < 1$ indicates that b is a positive number.

3.14.4 Green–Ampt Model

Based on Darcy's law, Green Ampt has proposed a model to obtain infiltration capacity, which can be represented as equation (3.29).

$$f_p = m + \frac{n}{F_p} \qquad \text{.............(3.29)}$$

where m and n are Green–Ampt coefficients. In this approach, the values of F_p are plotted against $(1/F_p)$, and the best fit straight line is drawn through the plotted points. The coefficients m and n, respectively, represent the line's intercept and slope. In certain cases, it may be necessary to exclude the values of f_p and the corresponding F_p at very low values of t, in order to obtain the best-fitting straight line with a good correlation coefficient.

3.15 CLASSIFICATION OF INFILTRATION CAPACITIES

Based on the Infiltration capacity, the soil classification is given in Table 3.7.

3.16 INFILTRATION INDICES

In flood-related hydrological computations, it is found to be simple to utilize an infiltration rate that remains constant during the storm. There are two commonly used forms of infiltration indices, which are calculated as the average infiltration rate.

Table 3.7: Classification of soil based on the infiltration capacity

Infiltration Class	Infiltration Capacity (mm/h)	Soil Type
Very low	<2.5	Highly Clayey Soil
Low	2.5 to 25.0	Shallow soils, Clay, and Soils low in organic matter
Medium	12.5 to 25.0	Sandy loam, Silt
High	>25.0	Deep sands, well-drained aggregated soil

3.16.1 Φ-index

The Φ index is the average rainfall above which the volume of runoff and the volume of precipitation are equal. The Φ index is computed using the rainfall hyetograph and the known resultant runoff volume. Infiltration also refers to the initial loss. The Φ value can be determined by treating it as a constant infiltration capacity. However, the infiltration rate is likewise equal to the intensity of the rainfall if the infiltration of the rainfall is more than or equal to the intensity of the rainfall. When the intensity of rainfall and infiltration in a given time frame is more than Φ, the Φ value is discovered, which indicates the runoff volume (Figure 3.12).

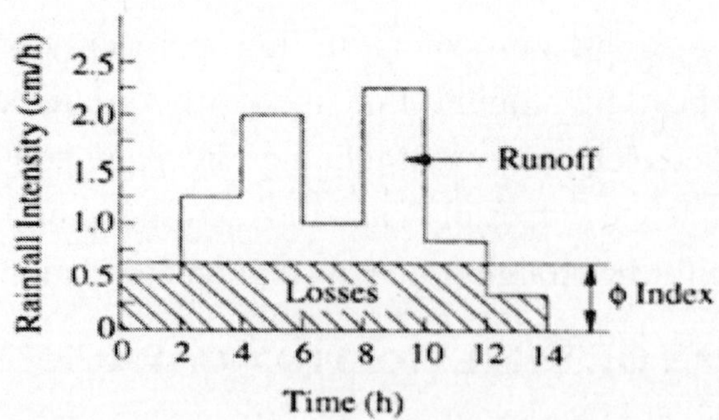

Figure 3.12: Φ-index. (Source: Subramanya, 2006)

Rainfall excess is the quantity of rainfall that exceeds the Φ index. Thus, the Φ -index accounts for the whole abstraction and, for a given rainfall hyetograph, enables the estimation of runoff magnitudes.

Procedure for Calculation of Φ-index

Imagine a rainfall hyetograph where N pulses are spread at Δt time intervals and the event duration is D hours so that:

$N.\Delta t = D$. (In Figure 3.12, $N = 7$)

Let,

I_i = rainfall intensity in i[th] pulse and

R_d = total direct runoff than,

Total rainfall $P = \sum_1^N \; I_i - \Delta t$

If Φ=Φ-index value then: P - Φ. $t_e = R_d$

where,

t_e = duration of rainfall excess.

The Φ -index of the storm can be found using the trial-and-error method described below, provided that the rainfall hyetograph and total runoff depth R_d are known.

- Assume that M pulses out of the given N pulses have excess rainfall. (where $M \leq N$).

  Choose M pulses in decreasing order of rainfall intensity I_i.

- Obtain the value of Φ, which satisfies the relation.

$$R_d = \sum_1^M (I_i - \Phi)\Delta t$$

- Determine the number of pulses (M_c) which give excess rainfall using the value Φ from Step 2. (Therefore, M_c = number of pulses with $1_i \geq$ Φ rainfall intensity).
- If $M_c = M$, then Φ of Step 2 represents the accurate value of the Φ-index. If not, repeat the procedure from Step 1 onwards and use the new value of M in the process. The result of Step 3 can serve as a guide for the subsequent one.

3.16.2 W- Index

To improve the value of the Φ-index, the initial losses are isolated from the overall abstraction, resulting in an average infiltration rate known as the W index obtained, which can be defined as equation 3.30 below.

$$W = \frac{P - R - I_a}{t_e} \quad \text{.............(3.30)}$$

where, P = total precipitation (cm),

R = total storm runoff (cm),

I_a = initial losses (cm),

t_e = Time duration of the rainfall excess, i.e. total time in which the rainfall intensity > W (in hours) and

W = mean infiltration rate (cm/h).

Due to the difficulty of obtaining I_a values, it can be challenging to estimate the W index. W_{min} is the lowest value of the W index that is attained in extremely wet soil conditions and indicates the steady-state minimum rate of catchment infiltration. The values of W-index and the Φ index are not constant among storms and vary from storm to storm.

SUMMARY

Storm event runoff is often regarded as the primary study area in engineering hydrology. A loss in the process of producing runoff could be attributed to any abstraction from precipitation, including precipitation lost through transpiration, infiltration, surface detention, and storage. The primary components of abstractions from precipitation are covered in this lesson and are crucial to comprehend when examining various hydrologic conditions. The combination of transpiration from vegetation and evaporation from soil and water bodies is referred to as evapotranspiration. Several subjects pertaining to evapotranspiration from basins and evaporation from water bodies have been thoroughly covered in this unit. Depression storages and interceptions, which represent "losses" in the runoff production process, are also covered in the module. Additionally, this unit explains the infiltration process. Furthermore, the infiltration process is explained in this module. It is an important process for improving soil moisture storage and groundwater recharge, as well as a considerable drain from precipitation.

EXERCISE

Revision Questions

1. Explain briefly about the several types of abstractions derived from precipitation.
2. Discuss in brief the evaporation process.
3. Explain various factors affecting evaporation from a water body.
4. Define Dalton's law of evaporation,
5. Describe a commonly used evaporimeter. Give an example of a typical evaporimeter.
6. Describe how to estimate evaporation from a lake using the energy budget approach.
7. Explain the importance of evaporation control of reservoir. Give the possible methods for obtaining the same.
8. List the variables influencing the evapotranspiration process.
9. What do you understand about consumptive use?
10. Apply Penman's formula to get potential evapotranspiration from a specific location.
11. Discuss Meyer's and Rohwer's evaporation formulas.
12. Give a brief explanation of (a) the evapotranspiration of reference crops and (b) the actual evapotranspiration.
13. Give a brief explanation of the soil moisture zones that are created because of infiltration.
14. Explain various factors affecting the infiltration capacity of an area.
15. Describe how to fit an experiment's data from a particular plot to Horton's infiltration equation.
16. Describe various methods for measuring the infiltration.
17. Describe the methods that are frequently employed to ascertain a plot of land's infiltration characteristics. Clearly state the respective benefits and drawbacks of each of the listed techniques.
18. Describe the various models that have been used to depict how infiltration capacity changes over time.
19. Differentiate between the following:
    - Infiltration rate and capacity;

- Actual and projected evapotranspiration;
- Field capacity and permanent wilting point;
- Depression storage and interception.

20. Define Φ-index. Explain the procedure for calculating Φ-index.
21. Differentiate between Φ-index and W-index.

Numerical Problems

1. A reservoir had an average surface area of 28 km^2 during June 1983. In that month the mean rate of inflow = 20 m^3/s, outflow = 25 m^3/s, monthly rainfall = 18 cm and change in storage = 18 million m^3. Assuming the seepage losses to be 1.9 cm, estimate the evaporation in that month.
2. Using Penman's method, determine the probable evapotranspiration from a region close to Haridwar in November. The information listed below is accessible:

Latitude	:	30°0' N
Elevation	:	314 m (above sea level)
Mean monthly temperature	:	16°C
Mean relative humidity	:	25%
Mean observed sunshine hours	:	7 h
Wind velocity at the height of 2 m	:	65*km/day*
Type of surface cover	:	Close-ground green crop

3. A class A pan was set up adjacent to a lake. The depth of water in the pan at the beginning of a certain week was 186 mm. That week, there was a rainfall of 38 mm, and 8 mm of water was removed from the pan to keep the water level within the specified depth range. If the depth of the water in the pan at the end of the week was 180 mm calculate the pan evaporation. Using a suitable pan coefficient, estimate the lake evaporation that week.
4. A reservoir has an average area of 60 km^2 over a year. The normal annual rainfall at the place is 110 cm, and the class A pan evaporation is 230 cm. Assuming the land flooded by the reservoir has a runoff coefficient of 0.4, estimate the net annual increase or decrease in the streamflow as a result of the reservoir.

5. Using Horton's equation, the infiltration capacity in a watershed is represented as

$$f_p = 4.5 + e^{-2t}$$

where f_p is in cm/h and t is in hours. Estimate the depth of infiltration in (i) the first 30 minutes and (ii) the second 30 minutes of the storm if infiltration occurs at capacity rates for a storm lasting one hour.

6. The results of an infiltrometer test on soil are given below. Determine the best values of the parameters of Horton's infiltration capacity equation for this soil.

Time since starts in minutes	5	10	15	20	30	40	60	80	100
Cumulative infiltration in mm	21.7	35.7	54.2	68.9	81.4	88.5	99.8	109.6	124.3

7. During a week, the average value of climate parameters of a lake with a surface area of 420 hectares are as follows: Water temperature = 28°C; Relative humidity = 60%; Wind velocity at 1.0 m above ground surface =16 km/hour. Using Meyer's formula, calculate the average daily evaporation from the lake.

8. The following climatic data were observed at a reservoir in the neighborhood of Jaipur. Estimate the mean monthly and annual evaporation from the reservoir using Meyer's formula.

Month	**Temp. (°C)**	**Relative humidity (%)**	**Wind velocity at 2m above GL (km/h)**
Jan	13.5	65	4.6
Feb	14.8	70	5.1
Mar	20.9	35	4.7
Apr	29.0	40	4.9
May	36.5	20	6.9
Jun	33.5	30	9.0

Month	Temp. (°C)	Relative humidity (%)	Wind velocity at 2m above GL (km/h)
Jul	42.6	80	8.2
Aug	34.0	90	5.8
Sep	29.2	67	5.3
Oct	24.9	40	4.6
Nov	15.9	30	4.6
Dec	12.7	26	4.2

9. A wheat field has a maximum available moisture of 14 cm. If the reference evapotranspiration is 7.7 mm/day, estimate the actual evapotranspiration on Day 1, Day 6, and Day 9 after irrigation. Assume soil-water depletion factor $p = 0.25$ and crop factor $K = 0.65$.

10. During a period of a month of 30 days, the following observations were obtained to conduct a lake's water budget.

    Mean Surface area = 25 km^2; Rainfall= 26 cm; Mean surface inflow rate = 14m^3/s; Mean surface outflow rate= 22 m^3/s; Fall in lake level = 1.10 m; Pan evaporation = 23 cm; Assuming Pan evaporation coefficient = 0.75, calculate the average seepage discharge during that month.

11. The results of an infiltrometer test on soil are as follows:

Time since starts in minutes	5	10	15	25	40	60	75	90	110	130
Cumulative infiltration in mm	21.5	37.0	48.2	57.9	65.1	71.8	75.8	81.3	89.0	95.0

Determine the parameters of (i) Kostiakov's equation, (ii) Green–Ampt equation, and Philips equation

12. On Day 1, the total accessible soil moisture in a recently irrigated field plot is at its maximum value of 17 cm. Determine the actual evapotranspiration on Days 1, 8, and 9 if the reference crop's evapotranspiration is 8.2 mm/day. Assume that crop factor K = 0.85 and soil-water depletion factor p = 0.22.

13. The infiltration process at capacity rates in soil is described by Kostiakov's equation as $Fp = 3.5\ t^{0.8}$ where Fp is cumulative infiltration in cm and t is time in hours. Estimate the infiltration capacity at (i) 3.5 h, and (ii) 5.0 h from the start of infiltration.

14. Before a rainfall event, the infiltration capacity of the soil in a small watershed was found to be 10 cm/h. After twelve hours of storming, the same was measured at 1.7 cm/h. Find the value of the decay coefficient K_h in Horton's infiltration capacity equation if the total infiltration over the storm's 12-hour duration was 16 cm.

15. For an area in South India (latitude = 13° N), the mean monthly temperatures are given.

Month	June	July	Aug	Sep	Oct
Temp (°C)	33.4	32.0	28.0	28.5	26.0

Calculate the seasonal consumptive use of water for the rice crop in the season June 16 to October 15, by using the Blaney–Criddle formula.

16. Calculate an area's PET during the wheat-growing season, November through February, using the Blaney-Criddle formula. The region is in North India, at a latitude of 30°N, and experiences the following mean monthly temperatures:

Month	November	December	January	February
Temp °C	16	12	6	14
Value of P_h at 26°C	7.30	7.45	7.50	7.20

Value of K for wheat = 0.65

17. The mass curve of an isolated storm in a 500 ha watershed is as follows:

Time from the start (h)	0	2	4	6	8	10	12	14	16	18
Cumulative rainfall (cm)	0	1.8	3.7	4.8	5.9	9.3	14.8	16.8	19.4	22.6

If the direct runoff produced by the storm is measured at the outlet of the watershed as 0.340 Mm^3, estimate the Φ-index of the storm and duration of rainfall excess.

18. In a 160-min storm, the following rates of rainfall were observed in successive 20-min intervals: 7.0, 7.9, 19.0, 15.0, 2.6, 2.0, 2.0 and 12.0 mm/h. Assuming the Φ-index value as 3.5 mm/h and an initial loss of 0.9 mm, determine the total rainfall, net runoff, and *W*-index for the storm.

Multiple Choice Questions

1. What is evaporation?

A) The process of a liquid turning into a gas at a specific temperature
B) The process of a solid turning into a liquid
C) The process of a gas turning into a liquid
D) The process of a liquid turning into a solid

2. Which factor increases the rate of evaporation?

A) Decreasing temperature
B) Increasing surface area
C) Increasing humidity
D) Reducing air movement

3. At what temperature does evaporation occur?

A) Only at boiling point
B) Only at freezing point
C) At any temperature
D) Only at room temperature

4. Which of the following substances will evaporate fastest at room temperature?

A) Water
B) Olive oil
C) Mercury
D) Alcohol

5. Which of the following scenarios demonstrates evaporation?

A) Ice melting into water
B) Water boiling
C) Sweat evaporating from skin
D) Water freezing into ice

6. How does wind affect the evaporation process?

A) It has no effect
B) It decreases the rate of evaporation
C) It increases the rate of evaporation
D) It changes the state of the liquid

7. What is an evaporimeter?

A) A device used to measure humidity
B) A device that measures the rate of evaporation

C) A device that measures temperature

D) A device used to measure precipitation

8. Which type of evaporimeter uses a pan filled with water to measure evaporation rates?

A) Class A pan evaporimeter
B) Lysimeter
C) Weighing-type evaporimeter
D) Vacuum evaporimeter

9. What is the main principle behind a weighing-type evaporimeter?

A) Measuring temperature changes

B) Measuring water level changes

C) Weighing the amount of water lost over time

D) Measuring humidity levels

10. Which of the following is a disadvantage of using a Class A pan evaporimeter?

A) It is difficult to set up

B) It is expensive

C) It is affected by wind and temperature changes

D) It is not portable

11. What type of evaporimeter is often used in agricultural applications?

A) Class A pan evaporimeter
B) Weighing-type evaporimeter
C) Electronic evaporimeter
D) Both A and B

12. Which of the following types of evaporimeters is known for being highly accurate and typically used in research?

A) Class A pan evaporimeter
B) Weighing-type evaporimeter
C) Simple evaporation gauge
D) Stick-type evaporimeter

13. In which type of evaporimeter would you likely find a built-in sensor to measure evaporation directly?

A) Class A pan evaporimeter
B) Weighing-type evaporimeter
C) Electronic evaporimeter
D) Bucket evaporimeter

14. What is a common application of evaporimeters?

A) Measuring soil moisture
B) Calculating crop water needs
C) Monitoring air pressure
D) Assessing water quality

15. Which of the following methods estimates evaporation using meteorological data?

A) Class A pan method
B) Penman-Monteith equation
C) Lysimeter method
D) Water balance method

16. The Class A pan method primarily measures evaporation by:

A) Using temperature and humidity only
B) Measuring water loss from a standard pan
C) Observing changes in soil moisture
D) Analysing satellite imagery

17. What is a lysimeter?

A) A device that measures rainfall
B) A device that directly measures water loss from soil
C) A container used for Class A pan measurements
D) A method for calculating evaporation using wind speed

18. Which method estimates evaporation based on the energy balance at the surface?

A) Water balance method
B) Penman equation
C) Class A pan method
D) Evapotranspiration method

19. The water balance method for estimating evaporation relies on which key principle?

A) Direct measurement of water temperature
B) Balancing inputs and outputs of water in a system
C) Measuring wind speed and humidity
D) Analysing soil salinity

20. Which method is often used for estimating evaporation in large water bodies?

A) Penman-Monteith equation
B) Energy budget method
C) Class A pan method
D) Bulk transfer method

21. Which method is primarily used for short-term evaporation measurements?

A) Class A pan method
B) Lysimeter method
C) Water balance method
D) Energy balance method

22. What is transpiration?

A) The process of water evaporating from soil
B) The process by which plants lose water vapor through their leaves
C) The process of water moving through the atmosphere
D) The process of liquid water turning into ice

23. Evapotranspiration is defined as:

A) The sum of evaporation and plant transpiration

B) The loss of water only from the soil

C) The process of water freezing in plants

D) The condensation of water vapor in the atmosphere

24. Which of the following factors affects the rate of transpiration?

A) Soil moisture B) Temperature

C) Humidity D) All of the above

25. Which method is commonly used to estimate evapotranspiration?

A) Penman-Monteith equation B) Class A pan method

C) Weighing lysimeter D) Both A and C

26. The primary driver of transpiration in plants is:

A) Nutrient availability B) Water availability

C) Sunlight and temperature D) Soil pH

27. Which plant characteristic can significantly influence transpiration rates?

A) Leaf size and shape B) Root depth

C) Type of soil D) All of the above

28. In what way does humidity affect transpiration?

A) High humidity increases transpiration rates

B) Low humidity decreases transpiration rates

C) Humidity has no effect on transpiration

D) High humidity decreases transpiration rates

29. What role does stomata play in transpiration?

A) They absorb nutrients

B) They facilitate gas exchange and water loss

C) They store water

D) They protect the plant from pests

30. What is meant by "initial loss" in the context of evapotranspiration?

A) The immediate loss of water from precipitation before it enters the soil

B) The gradual loss of water through transpiration

C) The total amount of water lost in a watershed

D) The water retained in the soil after a rainfall event

31. Which of the following factors can influence initial loss?

A) Soil type
B) Vegetation cover
C) Rainfall intensity
D) All of the above

32. What does "depression storage" refer to in hydrology?

A) Water stored in aquifers

B) Water held in surface depressions like ponds and puddles

C) Water lost through evaporation from large water bodies

D) Water captured in the atmosphere

33. Which method is commonly used to estimate initial loss in a watershed?

A) Soil moisture balance
B) Curve Number method
C) Penman-Monteith equation
D) Water balance approach

34. How does depression storage affect the estimation of evapotranspiration?

A) It reduces the overall evapotranspiration rate

B) It increases the amount of water available for plants

C) It does not influence evapotranspiration rates

D) It only affects surface runoff

35. Which equation can be used to account for initial loss and depression storage in hydrological modeling?

A) Water balance equation
B) Continuity equation
C) Darcy's law
D) Penman equation

36. What is the primary impact of initial losses on runoff?

A) They increase runoff
B) They decrease runoff
C) They have no effect on runoff
D) They lead to immediate evaporation

37. In which scenario would depression storage be most significant?

A) A flat, dry landscape

B) An area with heavy rainfall and poor drainage

C) A desert environment

D) A mountainous region

Answer: 1-A; 2-B; 3-C; 4-D; 5-C; 6-C; 7-B; 8-A; 9-C; 10-C; 11-D; 12-B; 13-C; 14-B; 15-B; 16-B; 17-B; 18-B; 19-B;20-D; 21-A; 22-B; 23-A; 24-D; 25-D; 26-C; 27-D; 28-D; 29-B; 30-A; 31-D; 32-B; 33-B; 34-B; 35-A; 36-B; 37-B.

KNOW MORE

For more information related to this topic scan the QR code.

OR

Type this link in your browser

https://drive.google.com/file/d/1tnkQlZLn30RtYEhl-kjlXJXzUdh9Dtbi/view?usp=sharing

BIBLIOGRAPHY

1. Asawa, G.L. (2005), "Irrigation and Water Resources Engineering, New Age International (P) Ltd., ISBN-13: 978-8122416732
2. Chow. V.T, (Ed), (1964), "Handbook of Applied Hydrology", McGraw-Hill, New York, NY, 1964.
3. IMD (1971), "Scientific Report No. 136, India Meteorological Department,1971, Government of India.
4. Jain S. K., Agarwal, P. K. And Singh, V. P. (2007), Hydrology and Water Resources of India, Water Science and Technology Library, Vol 57, Springer, The Netherland, ISBN-10 1-4020-5179-4(HB).
5. McQueen, I.S. (1963), Development of a Hand Portable Rainfall-Simulator Infiltrometer, Geological Survey Circular, 482, Washington, 1963.
6. Punmia, B.C. and Lal, Pande B.B., (1992), "Irrigation and Water Power Engineering", Twelfth Edition. Laxmi Publications (P) Ltd.
7. Singh, V. P. (1992). Elementary hydrology. Prentice Hall, New Jersy, 973.
8. Subramanya, K. (2015), "Engineering Hydrology", McGraw-Hill Education (India) Private Limited, New Delhi, 2015, ISBN (13): 978-1-25902997-4.

4 Runoff

UNIT SPECIFICS

In engineering hydrology, runoff is described as the portion of precipitation that flows through surface drains from a catchment area. It is influenced by various storm and catchment characteristics such as the intensity and duration of rainfall, soil type, slope of the land, vegetation cover, and various anthropogenic activities like urbanization and change in land use and land cover. This unit deals with the runoff process along with the factors affecting the runoff process, as well as methods to estimate runoff volume, including the SCS-CN method. The unit also describes the runoff hydrograph with its components, factors affecting the runoff hydrograph, flow duration curve, flow mass curve, etc. The unit also describes India's effective rainfall and various surface water resources. Finally, environmental flow techniques are described.

RATIONALE

To learn about the Rainfall-runoff process and modelling, flow duration, hydrograph, base flow separation, unit hydrograph, flood hydrograph, and environmental flow.

PRE-REQUISITE

Nil

UNIT OUTCOMES

The list of outcomes of this unit is as follows:

U4-O1: Rainfall-runoff process and modelling, methods of runoff assessment

U4-O2: Excess runoff hydrograph, unit hydrograph, factors affecting unit hydrograph

U4-O3: Flood hydrograph and environmental flow

Unit Outcomes	Expected Mapping with Course Outcomes (1- Weak Correlation; 2- Medium correlation; 3- Strong Correlation)							
	CO-1	CO-2	CO-3	CO-4	CO-5	CO-6	CO-7	CO-8
U4-O1	3	3	3	3	3	3	3	3
U4-O2	3	3	3	3	3	3	1	3
U4-O3	1	1	3	3	3	3	3	3

4.1 INTRODUCTION

Runoff is the term used to describe the portion of precipitation that flows through a surface channel and drains from a catchment area. As a result, it shows the catchment's output for a specific time period. It is the movement of water across the earth's surface, primarily due to rainfall, which has not been absorbed into the soil, evaporated into the atmosphere, or used by plants. This phenomenon is a critical component in the hydrological cycle, acting as a transporter of water from the land surface back into the water bodies like rivers, lakes, and oceans. The runoff process is influenced by various factors such as the intensity and duration of rainfall, soil type, slope of the land, vegetation cover, and human activities like urbanization. Figure 4.1 shows different routes of runoff. A brief overview of the runoff process is as follows:

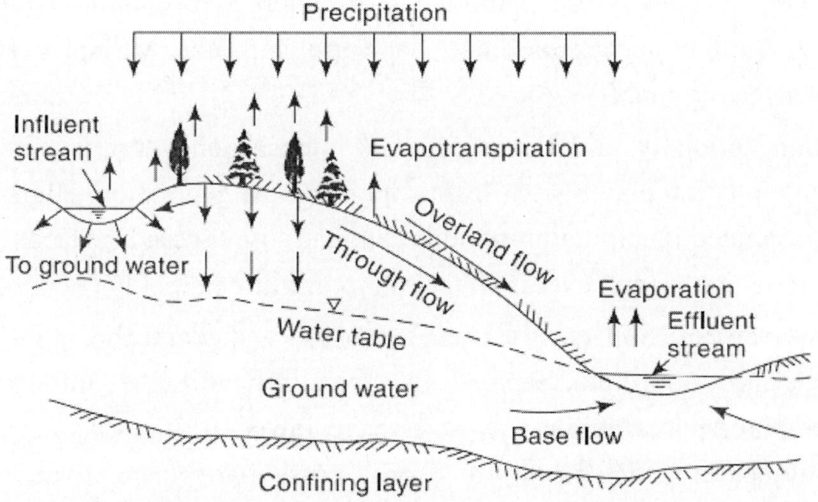

Figure 4.1: Different routes of runoff

a. **Rainfall:** Rainfall, also known as precipitation, is a fundamental component of the Earth's hydrological cycle. Precipitation, in the form of rain, snow, sleet, or hail, falls onto the Earth's surface. It refers to any form of water, in liquid or solid state, falling from the atmosphere to the Earth's surface. Rainfall plays a vital role in shaping climates, sustaining ecosystems, and supporting human activities such as agriculture and water supply.

b. **Infiltration:** Infiltration refers to the process by which some of the precipitation infiltrates into the soil, replenishing groundwater reserves and supporting plant growth. It is a critical component of the hydrological cycle and plays a significant role in groundwater recharge, soil moisture replenishment, and overall water availability. When precipitation, such as rain or snow, falls onto the ground surface, a portion of it infiltrates into the soil. This infiltration occurs primarily through the pores and spaces between soil particles.

c. **Surface Runoff:** If the intensity of rainfall exceeds the rate at which the soil can absorb water (infiltration capacity), excess water starts to flow over the land surface. This is known as surface runoff. Surface runoff collects in low-lying areas, forms streams, and eventually flows into larger bodies of water such as rivers, lakes, or oceans.

4.1.1 Factors Influencing Runoff

Several factors influence the runoff process, determining the amount and speed at which water flows over the land surface. These factors interact in complex ways and can vary depending on geographic location, climate, land use, and other variables. Here are some of the key factors affecting runoff:

a. **Precipitation Intensity and Duration**: The intensity and duration of rainfall or other forms of precipitation play a significant role in runoff generation. High-intensity rainfall events or prolonged precipitation periods can lead to increased surface runoff, as the soil may become saturated, and excess water cannot infiltrate quickly enough.

b. **Soil Characteristics**: Soil type, texture, structure, and permeability influence runoff by affecting the infiltration rate. Sandy soils typically have higher infiltration rates due to their larger pore spaces, while clay soils have lower infiltration rates because of their smaller pore spaces and tendency to form a crust that impedes water movement. Soil compaction, organic matter content, and soil moisture levels also impact runoff.

c. **Vegetation Cover**: Vegetation, including grasses, shrubs, trees, and other vegetation types, can influence runoff by intercepting precipitation, reducing the impact of rainfall on the soil surface, and promoting infiltration. Root systems help to bind soil particles together, enhance soil structure, and increase water absorption capacity. Deforestation, land clearing, and urbanization can reduce vegetation cover and increase runoff rates.

d. **Topography and Slope**: The slope and topography of the land play a crucial role in runoff generation. Steeper slopes generally lead to faster runoff as water flows down slope more rapidly, whereas flatter terrain may allow for more infiltration and slower runoff. Topographic features such as ridges, valleys, and depressions can also affect the distribution and concentration of runoff.

e. **Land Use and Land Cover Changes**: Human activities such as urbanization, agriculture, deforestation, and construction alter the natural landscape and can significantly impact runoff patterns. Impermeable surfaces like roads, parking lots, and buildings increase surface runoff by preventing water from infiltrating into the soil. Changes in land cover, such as the conversion of forests to agricultural fields or urban areas, can exacerbate runoff and lead to downstream flooding, erosion, and sedimentation.

f. **Antecedent Soil Moisture Conditions**: The moisture content of the soil prior to a rainfall event, known as antecedent soil moisture, affects runoff generation. Dry soils have higher infiltration rates and can absorb more water before runoff occurs, whereas wet or saturated soils may produce more runoff because they have less capacity to absorb additional moisture.

g. **Climate and Weather Patterns**: Climate factors such as temperature, humidity, wind, and atmospheric pressure influence precipitation patterns and evapotranspiration rates, which in turn affect runoff. Seasonal variations, weather events like storms or droughts, and long-term climate change can all impact runoff dynamics on regional and global scales.

4.1.2 Need for the study about Runoff

The study of runoff is essential in hydrology for several reasons:

a. **Water Cycle Balance:** Runoff is a key element in maintaining the balance of the water cycle. It ensures the continuous movement of water from the land to water bodies, contributing to the global distribution of water resources.

b. **Flood Prediction and Management:** Understanding the dynamics of runoff enables hydrologists to predict and manage floods. By analyzing runoff patterns and volumes, it is possible to estimate the flood potentiality in certain areas, which is crucial for planning and implementing flood control measures.

c. **Water Resource Planning:** Runoff analysis is vital for water resource planning and management. It aids in the assessment of water availability for various uses such as irrigation, hydroelectric power generation, and domestic consumption.

d. **Environmental Impact:** Studying runoff patterns helps in understanding the environmental impact of various land use changes, such as deforestation, urbanization, and agriculture. Changes in land use can significantly affect runoff characteristics, influencing both the quantity and quality of water in a region.

e. **Hydraulic Structure Design and Construction:** Knowledge about runoff characteristics is essential for the design and construction of various hydraulic structures like dams, reservoirs, canals, and Stormwater drainage systems.

f. **Erosion and Sediment Control:** Runoff can cause soil erosion and transport sediments. Understanding runoff patterns helps in designing effective soil conservation and sediment-controlling strategies.

g. **Water Quality Management:** Runoff can carry pollutants from urban and agricultural areas into water bodies. Studying the runoff process is crucial for implementing effective water quality management practices.

4.1.3 Runoff Classification

Runoff is a fundamental concept in hydrology and water resources engineering with far-reaching implications for environmental management, water resource planning, and infrastructure development, which is altogether meant for water resources development. Its study helps in understanding the dynamics of water movement, predicting, and managing natural events like floods, and ensuring the sustainable management of water resources. The runoff is divided into two groups according to the time delay between the precipitation and the runoff as (a) Direct runoff and (b) Base flow.

a. Direct Runoff

Direct runoff is the amount of runoff that flows into the stream immediately after it rains. Surface runoff, instantaneous interflow, and precipitation on the stream's surface are all included in direct runoff. The flow that results from snowmelt and enters the stream is likewise referred to as direct runoff. Direct runoff is sometimes referred to by expressions like storm runoff and direct storm runoff.

b. **Base Flow**

Base flow is the delayed flow that effectively acts as groundwater flow and reaches a stream. This category frequently includes delayed interflow as well. The base flow of a perennial stream can be easily identified in its yearly hydrograph as the stream's gradually declining flow during dry spells. A typical annual hydrograph has been shown in Figure 4.2 above.

4.1.4 Natural Flow

The response of a watershed to precipitation, or runoff, is an expression of the combined influence of various rainfall, climate, and catchment features. Thus, stream flow in its natural state, that is, without human interference, represents true runoff. A stream's natural flow, also known as its virgin flow, is uninterrupted by human constructions like reservoirs and diversion structures. If a stream has storage or diversion works, then the downstream channel's flow is impacted by the hydraulic and operational properties of these structures and, unless it is adjusted for return and diversion flow, fails to accurately represent the actual runoff.

At a catchment's terminal point, the natural flow (virgin flow) volume in time Δt is expressed by the water balance equation as:

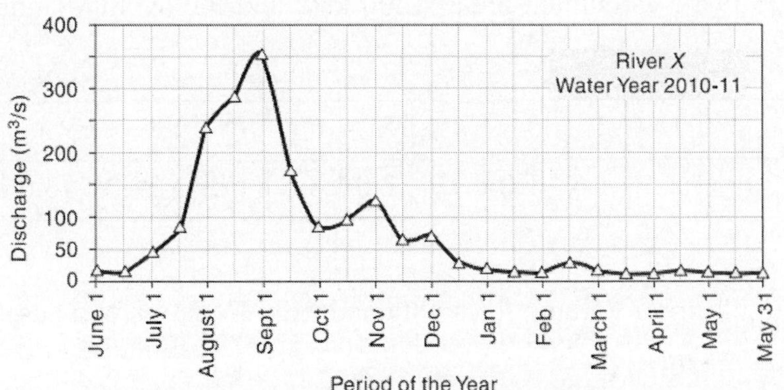

Figure 4.2: Perennial Stream

$R_N = (R_o - V_r) + V_d + E + E_x + \Delta S$ **..............(4.1)**

where,

R_N= Natural runoff volume in time Δt

R_o= observed runoff volume in time Δt at the terminal site

V_r= volume of return flow from irrigation, domestic water supply and industrial use

V_d= volume diverted flow of the stream for irrigation, domestic water supply, and industrial use

E = net evaporation losses from reservoirs and lakes

E_x= net export of water from the basin

ΔS= change in the storage volumes of water storage bodies

Correlation for natural flows is developed in hydrological investigations. On the other hand, natural flows must be estimated from data on abstractions from the stream and observed flows. In real life, however, upstream abstractions have an impact on the observed stream flow at a site and include return flow. Therefore, data on abstractions from the stream and observed flows must be used to determine natural flows.

Example 4.1 The measured discharge values for a given year at a river gauging station are displayed in the following table. A weir constructed across the river, upstream of the gauging site, diverts 4.0 Mm^3 and 1.00 Mm^3 of water each month for irrigation and industrial usage, respectively. Estimated return flows to the river upstream of the gauging location are 0.80 Mm^3 from irrigation and 0.60 Mm^3 from industry. Calculate the natural flow. Find the runoff-rainfall ratio if the catchment area is 200 km^2 and the average annual rainfall is 190 cm.

Month	1	2	3	4	5	6	7	8	9	10	11	12
Gauged Flow (Mm^3)	2.3	1.6	1.0	0.9	2.0	8.6	17.9	24.0	13.6	9.7	7.0	3.5

Solution

In a month the natural runoff volume R_N is obtained from Equation (4.1) as:

$R_N = (R_o - V_r) + V_d + E + E_x + \Delta S$

Here E, E_x and ΔS are assumed to be insignificant and of zero value.

V_r= volume of return flow from irrigation, domestic water supply and industrial use

= 0.80 + 0.60 = 1.40 Mm^3

V_d= volume diverted to the stream for irrigation, domestic water supply and Industrial use

= 4.0 + 1.0 = 5.0 Mm^3

The calculations are shown in the following table:

Month	1	2	3	4	5	6	7	8	9	10	11	12
Gauged Flow (Mm^3)	2.3	1.6	1.0	0.9	2.0	8.6	17.9	24.0	13.6	9.7	7.0	3.5
V_d (Mm^3)	5.0	5.0	5.0	5.0	5.0	5.0	5.0	5.0	5.0	5.0	5.0	5.0
V_r (Mm^3)	1.4	1.4	1.4	1.4	1.4	1.4	1.4	1.4	1.4	1.4	1.4	1.4
R_N (Mm^3)	5.9	5.2	4.6	4.5	5.6	12.2	21.5	27.6	17.2	13.3	10.6	7.1

Total R_N=135.3 Mm^3

Annual natural flow volume = Annual runoff volume = 135.3 Mm^3

Area of catchment = 200 km^2= 2.00 x $10^8 m^2$

Annual runoff depth = $\frac{1.1353x10^8}{2.0x10^8}$ =0.567 m =56.7 cm

Annual Rainfall = 190 cm

runoff-rainfall ratio = 56.7/190 = 0.30

4.1.5 Runoff Volume (Yield)

The Runoff Volume (yield) of the catchment in a given period is the entire amount of surface water that may be anticipated from a river at the outlet of its catchment over that period of time. We have annual yield and seasonal yield, which indicate the catchment's yield each year and season, respectively, depending on the time selected. The word "yield", unless specified otherwise, usually refers to the annual yield. In India, the term "yield" is mostly used by specialists in irrigation engineering. The water balance equation (Equation 4.1) can be used to express the yield of a catchment Y in a period Δt as follows:

$Y = R_N + V_r = R_o + A_b + \Delta S$ **.............(4.1a)**

where,

R_N= Natural runoff volume in time Δt

R_o= observed runoff volume in time Δt at the terminal site

V_r= volume of return flow from irrigation, domestic water supply and industrial use

A_b = abstraction in time, Δt for irrigation, water supply and industrial use and inclusive of evaporation losses in surface water bodies on the stream.

ΔS = change in the storage volumes of water storage bodies

4.2 HYDROGRAPH

Hydrographs are important tools in hydrology for understanding the response of a watershed or river basin to rainfall events or other hydrological inputs. They can provide insights into the timing and magnitude of peak flows, the duration of high-flow events, and the overall behavior of the river system.

A hydrograph is a graphical representation of the response of a given catchment to a rainfall input. It represents the flow rate or discharge of a river or stream over a period of time. It typically shows the variation in water level (or discharge) over time, usually plotted as a line graph with time on the x-axis and discharge (or water level) on the y-axis. Hydrograph comprises flow in all three phases of runoff—surface runoff, interflow, and base flow—and represents the combined effects of numerous complexes interacting with catchment and rainfall characteristics. Let us assume that catchment A, shown in Figure 4.3 has a regular record of rainfall. Also, the discharge of the river AB is measured at the point B.

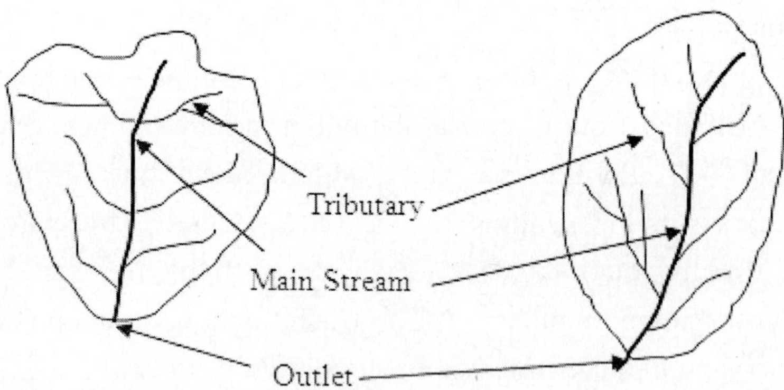

Figure 4.3: (a). Fan-shaped catchment (b) Leaf-shaped catchment

Hydrographs are useful for a variety of purposes, including flood forecasting, water resources management, and understanding the effects of land use changes or climate variability on river systems. They are often generated using stream flow data collected by gauging stations located along rivers and streams, which measure the water level or discharge at regular intervals. These data are then plotted to create the hydrograph. Additionally, hydrographs can be simulated using hydrological models to predict river behavior under different scenarios. A hydrograph has the following parts:

a. **Rising Limb**: The portion of the hydrograph where discharge or water level increases in response to rising rainfall or snowmelt.

b. **Peak Flow**: The highest point on the hydrograph, representing the maximum discharge during the event.

c. **Recession Limb**: The portion of the hydrograph where discharge or water level decreases as the rainfall event subsides and runoff diminishes.

d. **Base flow**: The relatively constant, low-level flow in a river or stream that is sustained between storm events. Base flow contributes to the overall hydrograph shape, especially during dry periods.

4.2.1 Hydrograph generation

It is quite obvious that the rainfall is in isolated spurts, whereas the stream flow is more continuous. Rainfall is generally shown as a hyetograph, whereas the stream flow is represented on a hydrograph. A hydrograph is a graphical representation of discharge measured at a particular point on a stream and is plotted as a function of time. The shape of the hydrograph tells us how the stream flow varies with respect to time. The following elements make up stream flow, as mentioned in the hydrological cycle:

a. **Rain directly falling into the stream:** Streamflow is mostly instantaneously generated when rain falls directly into the stream. However, in a typical catchment, this motion is negligible; however, in a watershed with huge lakes, it could be highly significant.

b. **Surface runoff:** Surface water reaches the stream by the process of overland flow.

c. **Interflow:** The term "interflow" describes the horizontal movement of water through the soil's higher strata. Infiltration produces streamflow much more slowly than direct runoff, but that leads to generally insignificant results compared to surface runoff.

d. **Groundwater flow**: When the surrounding area's water table is higher than the river beds, the water table reaches the stream.

Usually, some water mass continues to flow in the stream even when there is no rainfall in the catchment. This streamflow may be due to the (i) melting of snow in the catchment or (ii) because of the groundwater. In general, surface runoff appears to contribute to the streamflow most significantly and most quickly, which makes it an instrument for generating the peak flow.

4.2.2 Components of a natural hydrograph

A hydrograph represents the response of a watershed or river basin to a hydrological input, typically rainfall. It consists of various components that reflect different aspects of the water flow within the system. By analyzing the different components of a hydrograph, hydrologists can gain insights into the behavior of watersheds and rivers, including their response to rainfall events, the dynamics of flow generation, and the factors influencing flood risk and water availability. A natural hydrograph's components can be broadly categorized as follows:

a. **Direct Runoff**: Direct runoff refers to the portion of the rainfall that quickly runs off the land surface into streams and rivers without infiltrating into the soil. This component contributes to the rising limb of the hydrograph and is often associated with peak flows during storm events.

b. **Base flow**: Base flow represents the portion of streamflow that comes from delayed sources such as groundwater discharge and subsurface flow. It is typically sustained between storm events and contributes to the relatively constant flow in rivers and streams during dry periods. Base flow can influence the recession limb of the hydrograph, helping to sustain flow even after rainfall has ceased.

c. **Quick Flow**: Quick flow refers to the rapid response of a watershed to rainfall, characterized by high flow rates and short response times. It includes both surface runoff and rapid subsurface flow through shallow soil layers or preferential flow paths. Quick flow contributes to the initial rise in streamflow during a storm event.

d. **Slow Flow**: Slow flow represents the delayed response of a watershed to rainfall, characterized by lower flow rates and longer response times compared to quick flow. It includes delayed infiltration into deeper soil layers, groundwater recharge, and the gradual release of stored water within the watershed. Slow flow contributes to the recession limb of the hydrograph, sustaining flow in rivers and streams after the storm event has ended.

e. **Peak Flow**: Peak flow refers to the maximum discharge observed during a storm event, typically occurring during the rising limb of the hydrograph. It reflects the combined effects of direct runoff, quick flow, and any other contributing factors, such as channel routing or reservoir releases.

When there is no rainfall, there may be some flow in the river stream, such as Q_i. This quantity will depend on the groundwater flow. Obviously, if the groundwater continues contributing to the stream, its own water level will decrease very slowly, and its contribution will also decrease. The flow in the river is called base flow, which is constantly decreasing over time. An exponential decay curve represents the base flow hydrograph, and the quantity can be written precisely as equation 4.2 is given below.

$$Q_t = Q_0 \times e^{-kt} \quad \text{............(4.2)}$$

where,

Q_o = discharge at start of time period

Q_t = discharge at end of time t

k = coefficient of aquifer

t = base of natural logarithm

Before any measurable runoff reaches to the stream channels, there is an initial period of interception and infiltration as soon as the rainfall starts. There is not any surface runoff input during this time. Surface runoff starts when rainfall exceeds losses and continues to a peak value, which is determined at a time t_p. The infiltration and percolation that have been continuing during the gross rainfall period result in an elevated groundwater table, which therefore contributes more at the end of the storm than at the beginning, but thereafter, again, it declines along its depletion curve. The streamflow will begin to recede once the peak has been reached, and the rainfall stops until it completely disappears.

However, it is very difficult to differentiate between base flow and surface flow exactly as there are many interrelated factors on which they are dependent. Because of the complexity of the process, several empirical methods have been devised to discriminate between direct runoff.

4.2.3 The contribution to base flow to stream discharge

The base flow represents the discharge from the groundwater. The change in base flow occurs very slowly, and that is a lag between the change in the water table and the contribution from the base flow. This lag depends upon the transmissibility characteristics of the aquifer and may extend up to a period of days and weeks. There are two types of streams, known as influent streams and effluent streams, depending on whether the base flow is positive or negative.

- **Influent stream**: An influent stream is one in which water moves from the stream into the groundwater or when the base flow is negative. This can happen when the Groundwater table is below the water in the stream (Figure 4.4). There are times when an influent stream is entirely dry; these streams are known as ephemeral streams. The hydrograph of such streams is shown in Figure 4.5.
- **Effluent stream:** An effluent stream is fed up by groundwater as shown in Figure 4.6.

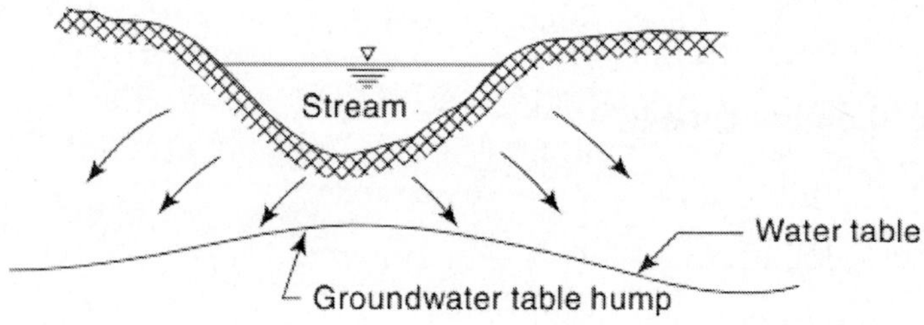

Figure 4.4: Influent stream

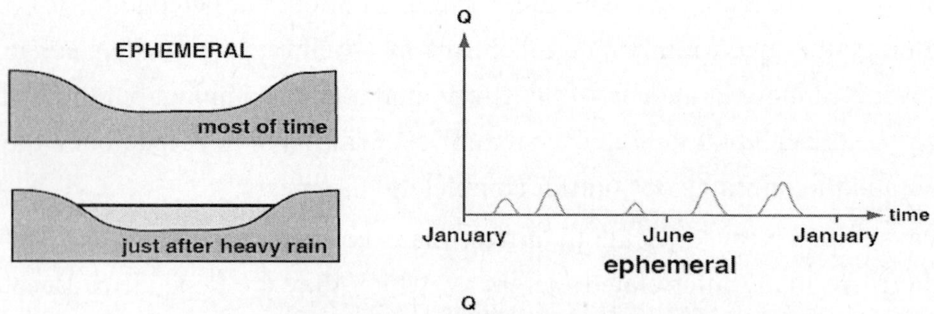

Figure 4.5: Hydrograph of an ephemeral stream

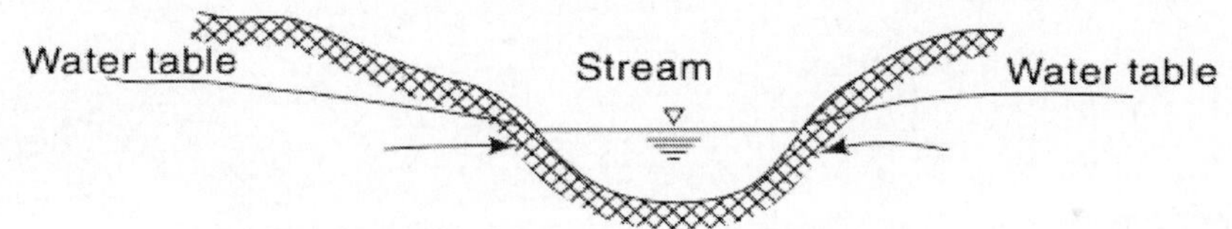

Figure 4.6: Effluent stream

4.2.4 Factors Affecting Hydrographs

Hydrographs, which represent the flow or discharge of water in a river or stream over time, are influenced by various factors. These factors can be natural or human-induced and can impact the shape, size, magnitude, and duration of hydrographs. These factors contribute to the rising limb, peak flow, recession limb, and overall form of the hydrograph. Understanding the complex interactions between these factors is crucial for predicting hydrological responses to rainfall events, assessing flood risk, effectively managing water resources, forecasting, and conserving ecosystems. Hydrological modelling techniques are often used to simulate hydrograph shapes under different scenarios and identify the dominant controls on hydrological processes. Here are some key factors affecting hydrographs:

a. **Precipitation:** The amount, intensity, duration, and spatial distribution of precipitation events directly affect the shape of hydrographs. Higher intensities or longer durations result in steeper rising limbs and higher peak flows, while shorter or less intense rainfall events produce gentler hydrograph responses. Heavy rainfall can lead to rapid increases in river flow, resulting in steep rising limbs on hydrographs.

b. **Antecedent Conditions**: The moisture status of the catchment prior to a rainfall event, known as antecedent conditions, influences how water is partitioned between infiltration, surface runoff, and groundwater recharge. Wet antecedent conditions lead to faster runoff response times and higher peak flows compared to dry antecedent conditions.

c. **Catchment Characteristics:** The physical characteristics of the catchment, including size, shape, slope, soil type, land use, and vegetation cover, influence the hydrograph shape. Catchments with steep slopes, impermeable soils, and limited vegetation tend to generate flashier hydrographs with rapid responses to rainfall. The geological composition of the catchment area and its topographical features influence how water moves through the landscape. Figure 4.7 shows the effect of catchment shape on the hydrograph.

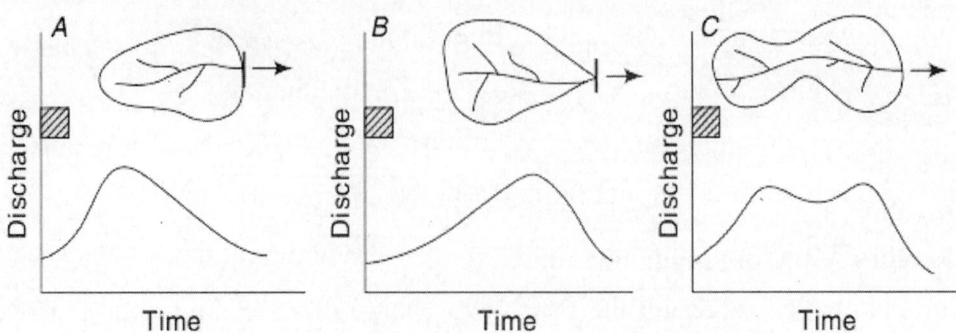

Figure 4.7: Effect of catchment shape on the hydrograph

d. **Vegetation Cover:** Vegetation plays a crucial role in regulating the hydrological cycle. Vegetative cover affects infiltration rates, evapotranspiration rates, and interception of rainfall. Changes in land use, such as deforestation or afforestation, can alter hydrographs significantly.

e. **Land Use Changes**: Human activities such as urbanization, agriculture, deforestation, and construction of impervious surfaces alter the natural hydrological processes. Urbanization, for example, increases surface runoff and reduces infiltration rates, leading to flashier hydrographs with faster response times.

f. **Drainage Density and Channel Network:** The drainage density is defined as the ratio of the total channel length to the total drainage area. The density and connectivity of the drainage network within the catchment affect the flow pathways and travel times of runoff. Higher drainage densities and well-connected channels lead to faster conveyance of water and shorter response times in hydrographs. A skewed hydrograph with a gently rising limb is the result of overland flow predominating in basins with lower drainage densities (Figure 4.8).

g. **Storage Capacities**: Natural and artificial storage features such as lakes, wetlands, reservoirs, and detention basins within the catchment can attenuate or delay peak flows, leading to modified hydrograph shapes. Storage capacities influence the recession limb of hydrographs by regulating base flow contributions.

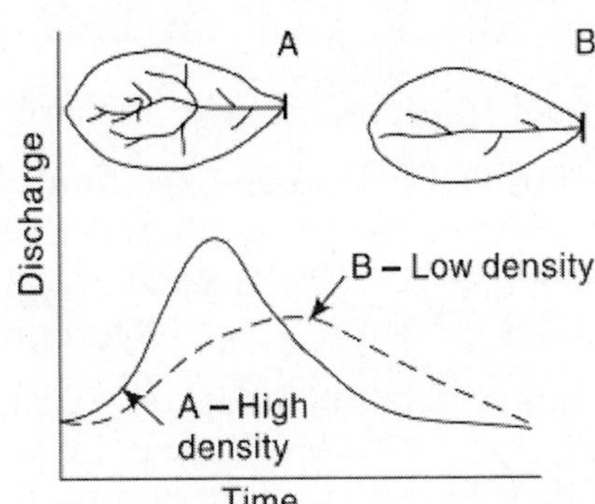

Figure 4.8: Role of drainage density on the hydrograph

h. **Human Modifications:** Engineering structures such as dams, reservoirs, levees, and channel modifications alter the natural flow regime of rivers. These structures can attenuate or amplify hydrograph peaks, change flow velocities, and modify base flow dynamics. Anthropogenic alterations to the landscape, such as urbanization, deforestation, channelization, and construction of drainage infrastructure, can significantly impact hydrograph shapes.

i. **Climate Variability**: Climate variability and long-term climatic trends influence precipitation patterns, temperature regimes, and hydrological processes. Changes in climate can lead to alterations in the frequency, magnitude, and timing of rainfall events, thereby affecting hydrograph shapes.

j. **Snowmelt Dynamics**: In regions where snowmelt contributes to streamflow, the timing, rate, and duration of snowmelt events influence hydrograph shapes. Rapid snowmelt can produce distinct peaks in hydrographs, while prolonged melt periods result in more gradual responses.

k. **Hydrological Connectivity**: The connectivity between surface water and groundwater systems within the catchment affects base flow contributions and recession limb characteristics of hydrographs. Changes in groundwater levels and flow paths influence the shape of hydrographs, particularly during the dry periods.

- **Extreme Events and Natural Disasters**: Extreme weather events such as hurricanes, tropical storms, and prolonged droughts can produce hydrograph shapes that deviate from typical patterns. Natural disasters such as landslides or dam failures can also alter hydrograph shapes abruptly.

- **Soil Moisture**: Soil moisture content influences the partitioning of rainfall into infiltration, surface runoff, and groundwater recharge. Wet or saturated soils can lead to increased surface runoff and faster response times in hydrographs.
- **Natural Events:** Natural phenomena such as wildfires, earthquakes, volcanic eruptions, and landslides can influence hydrographs by altering land surface characteristics, increasing sediment loads in rivers, or changing drainage patterns temporarily or permanently.

4.3 BASE FLOW SEPARATION

Base flow, also known as groundwater flow or interflow, refers to the portion of streamflow that is sustained by groundwater discharge. It represents the slow-moving, relatively constant component of streamflow during dry periods between precipitation events. Base flow is vital for maintaining streamflow during dry seasons, regulating stream temperature, providing habitat for aquatic organisms, and supporting ecosystem functions.

The technique of separating surface runoff from base flow or groundwater runoff on the streamflow hydrograph is known as base flow separation, sometimes referred to as hydrograph analysis. Hydrograph analysis benefits from this sort of division despite the fact that it is somewhat arbitrary and subjective. Numerous methods have been devised for executing base flow separation. Some of the techniques separate the hydrograph into direct runoff and groundwater runoff and some into surface runoff, interflow, and base flow.

4.3.1 Factors affecting Base flow generation

Here are some common factors influencing base flow generation:

a. **Infiltration and Recharge**: Infiltration of precipitation into the soil and subsequent recharge into the groundwater system contribute to base flow generation. The rate and extent of infiltration depend on soil properties, land cover, slope, and antecedent moisture conditions.

b. **Groundwater Seepage and Springs:** Groundwater discharge through seepage zones, springs, and gaining streams directly contribute to base flow. These features represent direct pathways for groundwater to enter surface water bodies, sustaining streamflow during dry periods.

c. **River-Aquifer Interaction**: The exchange of water between rivers and aquifers through hypothetic zones or the loss/gaining of stream reaches influences base flow dynamics. During high-flow periods, rivers can recharge adjacent aquifers, which subsequently discharge groundwater to sustain the base flow during low-flow periods.

d. **Watershed Characteristics**: Watershed characteristics, including geology, topography, soil type, and vegetation cover, influence base flow generation. Watersheds with permeable soils, high recharge rates, and extensive groundwater storage tend to have higher base flow contributions.

e. **Land Use and Land Cover Changes**: Alterations in land use and land cover, such as urbanization, deforestation, agriculture, and construction of impervious surfaces, can affect base flow generation. Changes in land cover alter infiltration rates, groundwater recharge patterns, and streamflow dynamics.

f. **Aquifer Properties**: Hydrogeological properties of aquifers, such as porosity, hydraulic conductivity, and storage capacity, control the movement and storage of groundwater. Aquifers with high storage capacity and slow hydraulic conductivity tend to sustain base flow for a longer duration.

g. **Climate Variability:** Climatic factors, including precipitation patterns, temperature regimes, evapotranspiration rates, and drought frequency, influence base flow generation. Changes in climate can alter groundwater recharge rates and seasonal patterns of base flow.

4.3.2 Methods for base flow generation

The methods for base flow generation are often used in combination to improve accuracy and reliability in estimating base flow contributions and understanding the hydrological processes controlling base flow generation within watersheds. Several Methods for estimating or quantifying base flow contributions to streamflow include:

a. **Hydrograph Separation:** Hydrograph separation techniques separate total streamflow into baseflow and stormflow components based on the recession limb characteristics of hydrographs. Methods include graphical methods (e.g., recession curve analysis), digital filtering (e.g., Eckhart filter, Web-based filter), and automated techniques (e.g., PART, HYSEP).

b. **Chemical and Isotopic Tracers:** Chemical and isotopic tracers such as stable isotopes of water (e.g., $\delta^{18}O$, $\delta^{2}H$), dissolved ions (e.g., chloride, sulfate), and environmental tracers (e.g., tritium, CFCs) can be used to distinguish between different water sources (e.g., groundwater vs surface water) and quantify base flow contributions.

c. **Hydrological Modeling:** Hydrological models, such as conceptual, semi-distributed, or physically-based models, can simulate groundwater-surface water interactions and estimate base flow contributions to streamflow. Models integrate various data sources and hydrological processes to simulate base flow dynamics at different spatial and temporal scales.

d. **Field Measurements:** Direct field measurements, such as groundwater level monitoring, streamflow gauging, streamflow recession analysis, and dye tracing studies, provide valuable data for understanding base flow dynamics and estimating base flow contributions.

e. **Two-component separation methods**

- the area method and
- the subjective method

I. **Area Method:** Base flow separation using the area method is based on a nonlinear relationship between time and area (Linsley et. al., 1958) as given in Equation 4.3. Smaller watersheds should not use this equation, and it should be verified for a variety of streamflow hydrographs. Typically, it provides a longer time base.

$$N = bA^{0.2} \quad \text{............(4.3)}$$

where,

A = drainage-basin area (km^2);

b = coefficient with a value of 0.83; and

N = time in days from the hydrograph peak, which, as Figure 4.9 illustrates, marks the start of groundwater flow.

Example 4.2: If the basin area of a catchment (A)= 1000 km^2, find out the time in days from the hydrograph peak.

Solution

Here, ,

A = drainage-basin area (km^2) = 1000 km^2

b = coefficient = 0.83;

time in days from the hydrograph peak

$N = bA^{0.2}$ = 0.83x $(1000)^{0.2}$

= 3.30 days

This shows that if rainfall occurs for 6 hours, its effects will be felt for more than 3 days. Therefore, it is preferable to employ an arbitrary approach by looking at the hydrograph.

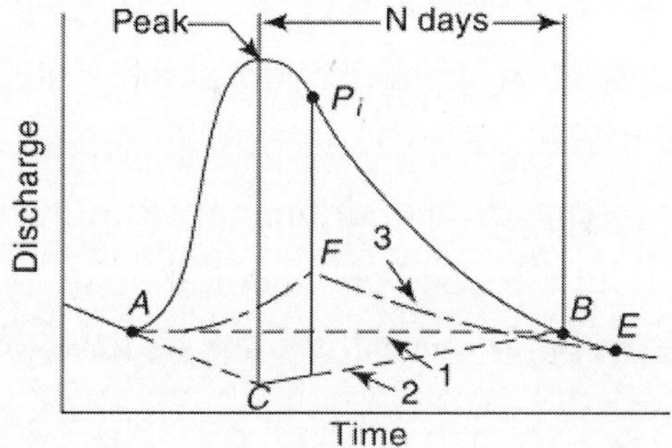

Figure 4.9: Base flow separation based upon area

II. **Subjective Method:** There are several subjective methods of base flow separation. One simple method is to visually inspect the hydrograph and choose a discharge on the recession that appears to represent the beginning of base flow based upon inspection of previously recorded base flow hydrographs. A base flow hydrograph has a slow recession because the rate of movement of groundwater is slow. The slope of the recession is low and is similar for all preceding recessions. Drawing a straight line from the chosen position on the hydrograph recession to the place on the hydrograph where the rise starts or the peak is located allows for the separation. This linear separation may be theoretically objectionable because channel banks become saturated as water rises in the channel, and this same bank storage water drains back into the channel as the stage recedes. This process is not linear, and hence, linear separation is not correct. Because the actual contribution of base flow to the hydrograph is not known, many hydrologists object to any base flow separation.

f. **Three-Component Separation Method:** Separating groundwater flow, interflow, and surface runoff is the three-component separation process. Barnes (1940) created a system that is shown in Figure 4.10. The approach is predicated on Equation (4.4).

$$Q_t = Q_0 K_r^t \qquad \text{.............(4.4)}$$

where,

Q_0 = Initial discharge at any time,

Q_t = discharge at time interval t later and

K_r = recession constant dependent upon the units of time and less than unity.

Considerable smoothing of the runoff hydrograph is necessary for the application of this method. The hydrograph recession frequently has numerous humps considering the direct runoff from different areas of the drainage basin, which makes it impossible to determine surface flow and interflow. This method may typically be used to achieve groundwater flow from these irregular recessions despite these issues.

4.4 EFFECTIVE RAINFALL

The effective rainfall is that portion of rainfall that contributes to direct runoff. In a similar vein, rainfall excess is the amount of rainfall that ends up in surface runoff. The difference between effective rainfall and rainfall excess is that the former includes the latter plus some abstractions. However, the terms effective rainfall and rainfall excess are freely interchanged in hydrologic usage. Thus, a rainfall storm is considered to be composed of two portions: one that contributes to runoff and the other that contributes to abstractions, including interception, evaporation, transpiration, depression and detention storage, and infiltration. Because the effective rainfall entirely becomes direct runoff, its volume must equal to the volume of direct runoff.

By isolating the base flow, the volume of direct runoff for a particular rainfall-runoff event can be calculated. This yields the volume of effective rainfall and, in turn, the volume of rainfall that is used up by abstractions. The distribution of this volume in time (determination of loss function) and then its subtraction from the rainfall hyetograph yields effective rainfall. This is illustrated in Figure 4.11.

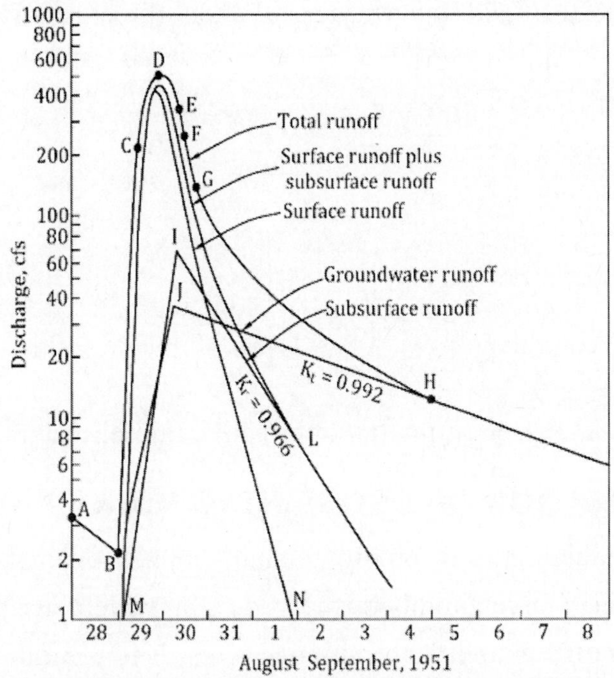

Figure 4.10: Three-component hydrograph separation

However, due to extreme space-time heterogeneity in antecedent conditions, extrapolation of this loss function for any other rainfall event is extremely difficult. Let us briefly pause for a moment. If we neglect all other abstractions than infiltration for simplicity, it is immediately seen that during and after a rainfall episode, runoff and infiltration occur simultaneously. Indeed, infiltration continues to occur as long as surface runoff occurs or there is water over the ground. It is normally true that surface runoff has a much greater duration than rainfall when comparing the lengths of rainfall and runoff. However, infiltration is allowed to occur only during the period of rainfall. The question is: What happens to the infiltration during the period of surface runoff and in excess of rainfall? One might argue that infiltration may be very small after the cessation of rainfall and hence neglected. This may not be true. Besides, a physical standpoint, in order to compensate for infiltration accounted for after the period of rainfall, infiltration has to be allowed to occur at a rate higher than that at which it actually occurs. This is inevitable to ensure the volume continuity of rainfall excess and surface runoff. Therefore, the concept of rainfall excess or effective rainfall is an artificial one can be said about direct runoff. In nature, rainfall occurs as a continuum. And so does runoff.

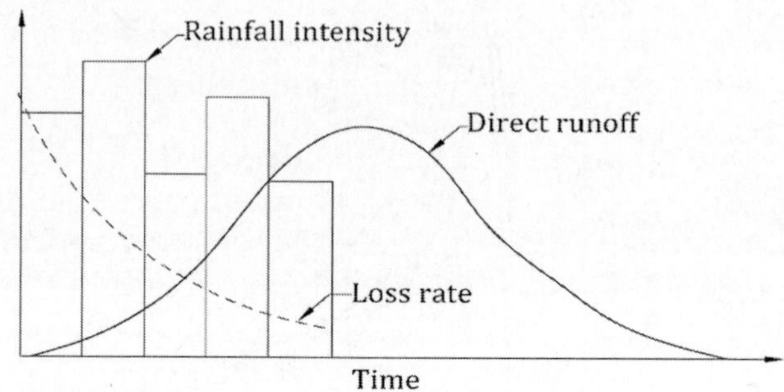

Figure 4.11: Determination of effective rainfall from rainfall-runoff

4.5 SCS-CN METHOD OF ESTIMATING RUNOFF VOLUME

In hydrologic engineering and environmental impact evaluations, the SCS-CN (Soil Conservation Service Curve Number) method is a widely used empirical approach for estimating direct runoff volume from rainfall events. It is commonly used in hydrology and watershed modelling. The method was developed by the USDA Soil Conservation Service, and it is documented in the Soil Conservation Service (SCS) National Engineering Handbook Section 4: Hydrology (NEH-4) (SCS 1985). Its simplicity, authority, ease of use, and responsiveness to four easily comprehended catchment properties: soil type, land use/treatment, surface condition, and antecedent condition have made it extremely popular. Simplexes, predictability, stability, reliance on a single parameter, and reactivity to key runoff-producing watershed factors (soil type, land use/treatment, surface condition, and antecedent condition) are among the technique's alleged benefits of SCS curve number method.

SCS-CN (Soil Conservation Service Curve Number) is a method used in hydrology to estimate direct runoff from rainfall events in small, ungauged watersheds. It is a widely used method due to its simplicity and effectiveness in estimating runoff. Here is a basic method to obtain the SCS-CN:

a. **Understand the Variables**: The SCS-CN method requires three main parameters:

- **Land Use and Land Cover (LU/LC) Classification:** The type of land cover affects runoff characteristics. Each land cover type has an associated curve number.
- **Hydrologic Soil Group (HSG):** Soils are classified based on their infiltration characteristics, ranging from A to D, where A represents the highest infiltration rate and D represents the lowest.
- **Rainfall Characteristics:** The amount and intensity of rainfall also influence runoff.

b. **Determine Curve Numbers (CN)**: The curve number is a key parameter representing the hydrological properties of a watershed. It depends on land use, soil type, and antecedent moisture conditions. You can obtain curve numbers from lookup tables provided by the Soil Conservation Service or other relevant sources. These tables categorize land uses and soil types and provide corresponding curve numbers.

c. **Determine Hydrologic Soil Group (HSG)**: Determine the hydrologic soil group (A, B, C, or D) for the area of interest based on soil type and infiltration characteristics. Soil surveys or soil data maps can provide this information.

d. **Applying Antecedent Moisture Conditions**: If the watershed has experienced previous rainfall events, you may need to adjust the curve number based on antecedent moisture conditions. SCS-CN method provides methods for this adjustment.

e. **Estimating Runoff:** Once the curve number is determined for the specific area of interest, it is used in conjunction with the rainfall characteristics to estimate the direct runoff volume. The SCS-CN method consists of a Water balance equation (Equation 4.5).

$$P = I_a + F + Q \qquad \textbf{............(4.5)}$$

where,

P = total precipitation,

I_a = initial abstraction,

F=cumulative infiltration excluding I_a and

Q= direct surface runoff.

In addition, two significant concepts are utilized in SCS-CN method as follows:

- the amount of actual infiltration (F) divided by the amount of potential maximum retention (S) is equal to the ratio of the actual amount of direct surface runoff (Q) to the maximum potential surface runoff (= P- I_a), given as equation 4.6; and

$$\frac{Q}{P-I_a}=\frac{F}{S}$$
............(4.6)

- the amount of initial abstraction (I_a) is a portion of the potential maximum retention (S), given as equation 4.7.

$$I_a=\lambda S$$
............(4.7)

Combining eqs. (4.6) and (4.7) with equation (4.5), we get equation 4.8 as below.

$$Q=\frac{(P-I_a)^2}{P-I_a+S}$$
............(4.8)

where,

S= potential maximum retention or infiltration.

λ= constant ranging from 0.1 to 0.4 (Refer to section 4.5.4)

when P >I_a, equation (4.8) is true; in other cases, Q = 0. For λ = 0.2, equation (4.8) can be rewritten as equation 4.9.

$$Q=\frac{(P-0.2S)^2}{P+0.8S}$$
............(4.9)

The catchment's soil, vegetation, and land use complex, as well as the antecedent soil moisture condition in the catchment immediately before the start of the rainfall event, determines the parameter S, which represents the potential maximum retention. The Soil Conservation Services (SCS) of the USA have expressed S (in mm) in terms of a dimensionless parameter CN (the Curve number) for simplicity of use in practical applications as given in equation 4.10.

$$S=\frac{25400}{CN}-254=254\left(\frac{100}{CN}-1\right)$$
............(4.10)

To express S in mm, use the constant 254. Now, the curve number CN with S can be written as equation 4.11.

$$CN = \frac{25400}{S+254} \qquad \text{.............(4.11)}$$

Moreover, its range is 100 > CN > 0. A CN value of 100 is represented by a condition of zero potential retention or an impermeable catchment. An infinitely abstracting catchment with $S = \infty$ represents a CN value of 0. This curve's CN value is dependent on (a) Soil type, (b) Antecedent moisture condition, (c) Land use/cover, described in the following sections.

4.5.1 Soils

The hydrological soil classification is used to determine CN. Here, the infiltration and other features are used to categorize soils into four classes: A, B, C, and D. The effective soil depth, average clay content, infiltration properties, and permeability are significant soil attributes that affect the hydrological categorization of soils. Table 4.1 represents minimum infiltration rates for all four hydrologic soil groups. A synopsis of the four hydrologic soil groups is provided below:

Table 4.1: Description of hydrologic groups

Hydrologic Soil Group	Minimum infiltration rate (inch/hr.)
A	>0.30
B	0.15-0.30
C	0.05-0.15
D	0-0.05

a. **Group-A (Low Runoff Potential):** Soils mostly consisting of deep, well-too-excessively-drained sands or gravels, having high infiltration rates even when heavily wetted. There is a high rate of movement of water within these soils. For instance, aggregated silt, deep loess, and deep sand.

b. **Group-B (Moderately Low Runoff Potential):** Soils with moderately fine to moderately coarse textures, moderately deep to deep, moderately well to well-drained soils, and moderate infiltration rates when thoroughly wetted comprise most of these soils. The rate of water conveyance in these soils is moderate. For instance, red loamy soil, red sandy soil, and red loess.

c. **Group-C (Moderately High Runoff Potential):** Soils with relatively low rates of infiltration under thorough wetting, mostly composed of moderately deep to deeply drained, moderately fine to moderately coarse-textured soils that are moderate to deeply embedded. The rate of water conveyance in these soils is moderate. Examples: Reddish-black soils with a high clay content, shallow sandy loams, clayey loams.

d. **Group-D (High Runoff Potential):** Soils with extremely low rates of infiltration upon complete wetting, consisting mainly of clay soils with a high potential for swelling, soils with a permanent high-water table, soils with a clay pan or layer at or near the surface, and shallow soils covering nearly impermeable material. For instance: Deep black soils, some salty soils, and heavy plastic clays.

4.5.2 Antecedent Moisture Condition (AMC)

The moisture content of the soil at the start of the rainfall-runoff event under investigation is referred to as the antecedent moisture condition (AMC). It is commonly known that AMC controls both initial abstraction and penetration. SCS recognizes three degrees of AMC, which are as follows for practical application:

- **AMC-I:** The soils are dry but not completely up to the wilting point. A satisfactory level of cultivation has occurred.
- **AMC-II:** Average conditions
- **AMC-III:** Over the last five days, there has been enough rainfall. Soil conditions are saturated.

Table 4.2 lists the limitations of these three AMC classes based on the total amount of rainfall during the preceding five days. It should be mentioned that the restrictions vary depending on the two seasons (i) the growth season and (ii) the dormant season. Table 4.3 shows curve numbers for all three antecedent moisture conditions.

Table 4.2: Antecedent Moisture (AMC) for Obtaining the CN

AMC Type	Total Rain in Previous 5 Days Dormant Season	Growing Season
I	Less than 13 mm	Less than 36 mm
II	13 to 28 mm	36 to 53 mm
III	More than 28 mm	More than 53 mm

Table 4.3: Curve numbers for three antecedent moisture conditions

AMC II	**AMC I**	**AMC III**	**AMC II**	**AMC I**	**AMC III**
100	100	100	60	40	78
99	97	100	59	39	77
98	94	99	58	38	76
97	91	99	57	37	75
96	89	99	56	36	75
95	87	98	55	35	74
94	85	98	54	34	73
93	83	98	53	33	72
92	81	97	52	32	71
91	80	97	51	31	70
90	78	96	50	31	70
89	76	96	49	30	69
88	75	95	48	29	68
87	73	95	47	28	67
86	72	94	46	27	66
85	70	94	45	26	65
84	68	93	44	25	64
83	67	93	43	25	63
82	66	92	42	24	62
81	64	92	41	23	61
80	63	91	40	22	60
79	62	91	39	21	59
78	60	90	38	21	58

AMC II	**AMC I**	**AMC III**	**AMC II**	**AMC I**	**AMC III**
77	59	89	37	20	57
76	58	89	36	19	56
75	57	88	35	18	55
74	55	88	34	18	54
73	54	87	33	17	53
72	53	86	32	16	52
71	52	86	31	16	51
70	51	85	30	15	50
69	50	84			
68	48	84	25	12	43
67	47	83	20	9	37
66	46	82	15	6	30
65	45	82	10	4	22
64	44	81	5	2	13
63	43	80	0	0	0
62	42	79			
61	41	78			

4.5.3 Land Use

In the SCS-CN method, the type of land use is a critical parameter that influences the runoff characteristics of a watershed. Different land uses have different abilities to absorb water and generate runoff. The Soil Conservation Service (SCS), now known as the Natural Resources Conservation Service (NRCS), has categorized land uses into various classes, each assigned a specific curve number (CN) that represents the runoff potential. These land use categories typically include:

a. **Urban or Built-up Areas**: This category includes developed areas such as cities, towns, and industrial zones with impervious surfaces like roads, buildings, and pavements. Runoff from these areas tends to be high due to limited infiltration.

b. **Agricultural Land**: Agricultural land includes crop land, pasture land, and orchards. The runoff potential varies based on factors such as crop type, tillage practices, and soil management techniques.

c. **Forest and Woodland**: Forested areas generally have lower runoff potential compared to developed areas or agricultural land due to the presence of vegetation and relatively high infiltration rates.

d. **Grassland and Open Spaces**: These areas, including meadows, grasslands, and parks, typically have moderate to low runoff potential depending on vegetation cover and soil characteristics.

e. **Water Bodies**: Lakes, ponds, rivers, and streams are considered water bodies. In the SCS-CN method, they are usually not assigned a curve number as they represent the endpoint of the runoff process rather than a source of runoff.

f. **Barren Land**: Barren land includes areas with little or no vegetation cover, such as deserts, sand dunes, or rocky terrain. Runoff potential from barren land can vary widely depending on soil characteristics and surface conditions.

g. **Mixed Land Use**: Some areas may have a mix of different land use types, resulting in varied runoff characteristics. In such cases, a weighted average of curve numbers corresponding to the predominant land uses may be used.

These are general categories, and specific land use classifications may vary depending on the region and the purpose of the analysis. It is essential to select the most appropriate land use category or combination of categories to accurately represent the runoff characteristics of the watershed under consideration. The runoff curve numbers for hydrological cover complexes for selected land use categories may be taken from Table 4.4. Additionally, adjustments may be made to account for land management practices, soil conservation measures, and other factors that influence runoff.

Table 4.4: Runoff curve numbers for Hydrological cover conditions

SN	Land use Description/ Treatment	Hydrologic Condition / Percentage impervious Area	Hydrologic soil groups			
			A	B	C	D
Urban						
1	Residential:					
	Average lot size - 1/8 acre or less	65	77	85	90	92
	¼ acre	38	61	75	83	87
	1/3 acre	30	57	72	81	86
	½ acre	25	54	70	80	85
	1 acre	20	51	68	79	84
	2 acre	12	46	65	77	82
2	Paved parking lot, roofs,		98	98	98	98
3	Streets and roads:					
	Paved with curbs, storm sewers		98	98	98	98
	Paved, open ditches		82	89	92	93
	Gravel (including right-of-way)		76	85	89	91
	Dirt (including right-of-way)		72	82	87	89
4	Western desert areas:					
	Natural desert landscaping areas		63	77	85	88
5	Urban districts:					
	Commercial and business areas	85	89	92	94	95
	Industrial districts	72	81	88	91	93
6	Developing areas:					

SN	Land use Description/ Treatment	Hydrologic Condition / Percentage impervious Area	Hydrologic soil groups			
			A	B	C	D
	Graded areas, No vegetation		77	86	91	94
7	Open spaces, lawns, parks					
	Grass cover on >75% area	Good	39	61	74	80
	Grass cover in 50% -75% area	Fair	49	69	79	84
	Agricultural					
	Cultivated lands:					
8	Fallow:					
	Bare soil Straight row	----	77	86	91	94
	Crop residue cover	Poor	76	85	90	93
		Good	74	83	88	90
9	Row crops:					
	Straight row	Poor	72	81	88	91
	Straight row	Good	67	78	85	89
	Crop residue cover straight row	Poor	71	80	87	90
	Crop residue cover straight row	Good	64	75	82	85
	Contoured	Poor	70	79	84	88
	Contoured	Good	65	75	82	86
	Crop residue cover Contoured	Poor	69	78	83	87
	Crop residue cover Contoured	Good	64	74	81	85
	Contoured & terraced	Poor	66	74	80	82
	Contoured & terraced	Good	62	71	78	81

SN	Land use Description/ Treatment	Hydrologic Condition / Percentage impervious Area	Hydrologic soil groups			
			A	B	C	D
	Crop residue cover Contoured	Poor	65	73	79	81
	Crop residue cover Contoured	Good	61	70	77	80
10	Small grain:					
	Straight row	Poor	65	76	84	88
	Straight row	Good	63	75	83	87
	Crop residue cover straight row	Poor	64	75	83	86
	Crop residue cover straight row	Good	60	72	80	84
	Contoured	Poor	63	74	82	85
	Contoured	Good	61	73	81	84
	Crop residue cover Contoured	Poor	62	73	81	84
	Crop residue cover Contoured	Good	60	72	80	83
	Contoured & terraced	Poor	61	72	79	82
	Contoured & terraced	Good	59	70	78	81
	Crop residue cover Contoured	Poor	60	71	78	81
	Crop residue cover Contoured	Good	58	69	77	80
11	**Close-seeded legumes or rotation meadow:**					
	Straight row	Poor	66	77	85	89
	Straight row	Good	58	72	81	85
	Contoured	Poor	64	75	83	85
	Contoured	Good	55	69	78	83
	Contoured & terraced	Poor	63	73	80	83

SN	Land use Description/ Treatment	Hydrologic Condition / Percentage impervious Area	Hydrologic soil groups			
			A	B	C	D
	Contoured & terraced	Good	51	67	76	80
12	**Pasture or range:**	Poor	68	79	86	89
		Fair	49	69	79	84
		Good	39	61	74	80
	Contoured	Poor	47	67	81	88
	Contoured	Fair	25	59	75	83
	Contoured	Good	6	35	70	79
13	Meadow- grass, protected from	Good	30	58	71	78
	Brush- weed grass with brush	Poor	48	67	77	83
	being the major element	Fair	35	56	70	77
14	**farmsteads-building, lanes, driveways, and surrounding lots**	----	59	74	82	86
	Woods and forests					
15	Woods and forests land	Poor	45	66	77	83
		Fair	36	60	73	79
		Good	25	55	70	77
16	Woods-grass combination (orchard)	Poor	57	73	82	86
		Fair	43	65	76	82
		Good	32	58	72	79
Arid and Semiarid rangelands:						

SN	Land use Description/ Treatment	Hydrologic Condition / Percentage impervious Area	Hydrologic soil groups			
			A	B	C	D
17	Herbaceous	Poor		80	87	93
		Fair		71	81	89
		Good		62	74	85
18	Oak-aspen	Poor		66	74	79
		Fair		48	57	63
		Good		30	41	48
19	Pinyon-juniper	Poor		75	85	89
		Fair		58	73	80
		Good		41	61	71
20	Sagebrush with grass understory	Poor		67	80	85
		Fair		51	63	70
		Good		35	47	55
21	Desert shrub	Poor	63	77	85	88
		Fair	55	72	81	86
		Good	49	68	79	84

4.5.4 SCS-CN Equation for Indian Conditions

Numerous researchers from different geographical regions, including the USA and many other countries, have documented values of λ varied in the range $0.1 < \lambda < 0.4$. after extensive research in small catchments, the value of $\lambda = 0.2$ has been adopted as a standard value. It has also been suggested that $\lambda = 0.1$ and 0.3 be used in Indian settings, subject to specific limitations related to soil type and AMC type (Refer to equation 4.12 and 4.13).

- **for black soil under AMC-II & III**

$$Q = \frac{(P-0.1S)^2}{P+0.9S} \text{ for } P > 0.1S; \qquad \text{...........(4.12)}$$

- **for black soils under AMC-I of type I and for all soils having AMC-I, II and III**

$$Q = \frac{(P-0.3S)^2}{P+0.7S} \text{ for } P > 0.3S \qquad \text{...........(4.13)}$$

4.5.5 Procedure for Estimating Runoff Volume from a Catchment

The following is the process for estimating runoff volume from a catchment:

- The catchment under study's land use/cover data is obtained through the interpretation of multi-season satellite imagery. Establishing a GIS database for the watershed and connecting it with land use/cover data is very beneficial.
- The National Bureau of Soil Survey and Land Use Planning (NBSS & LUP) (1966) soil maps are used to obtain the soil information of the catchment. The identification of pertinent soil data for the catchment, the creation of an appropriate hydrological soil classification, and the spatial storage of this data in a GIS database are all performed.
- The available rainfall data of the catchment is collected by multiple rain gauge stations, verified for quality and consistency, and linked to the GIS database. A rainfall record spanning at least 25 years is ideal for accurate estimation of catchment yield.
- Thiessen polygons are generated for every identified rain gauge station.
- Appropriate area-weighted CN values for each Thiessen cell are determined by considering the spatial variance in soil types, land use, and/or cover. Additionally, equivalent AMC-I and AMC-III values have been obtained for every cell.
- For each cell, the appropriate daily runoff series is produced sequentially with the rainfall data using the pertinent SCS-CN equations. This technique is utilized to derive the required weekly, monthly, and annual runoff time series. Furthermore, the relevant catchment runoff time series is created by merging the data from several cells that make up the catchment.
- By appropriately summing up the aforementioned time series, seasonal and annual runoff volume series are generated and the necessary, consistent catchment yield can be obtained from there. A graph illustrating the SCS-CN approach is presented in Figure 4.12.

4.5.6 Advantages of SCS-CN method

- The SCS-CN method is applied at the watershed scale, where land use, soil types, and other relevant factors vary across the area. The method provides a quick and relatively simple way to estimate runoff volumes for a variety of storm events and watershed conditions.
- Possibly the most widely used technique for estimating the amount of surface runoff that small agricultural, forest and urban watersheds will produce during a specific rainfall event.
- Considers most of the factors of watersheds that cause runoff, such as soil type, land use, hydrologic state, and antecedent moisture condition.
- A conceptually sound, stable, and straightforward one-parameter rainfall-runoff model that can be used in ungauged watersheds.

4.5.7 Limitations

- Does not contain any expression for time.
- Ignores the impact of rainfall intensity and its temporal distribution.
- A lack of clear guidance on the variation of AMC for lower curve numbers and/or rainfall amounts.
- Works best on agricultural watersheds, well on urban watersheds, fairly well on rangelands, and poorly on forest sites.
- No explicit provision for spatial scale effects. Literature warns against its use on watersheds greater than 250 km^2.
- It is important to note that while the SCS-CN method is widely used and provides reasonable estimates for many situations, it has its limitations, particularly in areas with highly variable soils or land cover types. Additionally, calibration and validation with local data are often necessary to ensure accurate results.
- Disregard the influence of the amount and timing of rainfall.
- A lack of precise instructions on how to adjust the AMC for smaller curve numbers and/or rainfall totals.
- Does not specifically account for the impacts of spatial scale; it performs best on agricultural watersheds, well on urban watersheds, mediocrely on rangelands, and poorly on forest sites. Research advises against applying it to watersheds larger than 250 km^2.

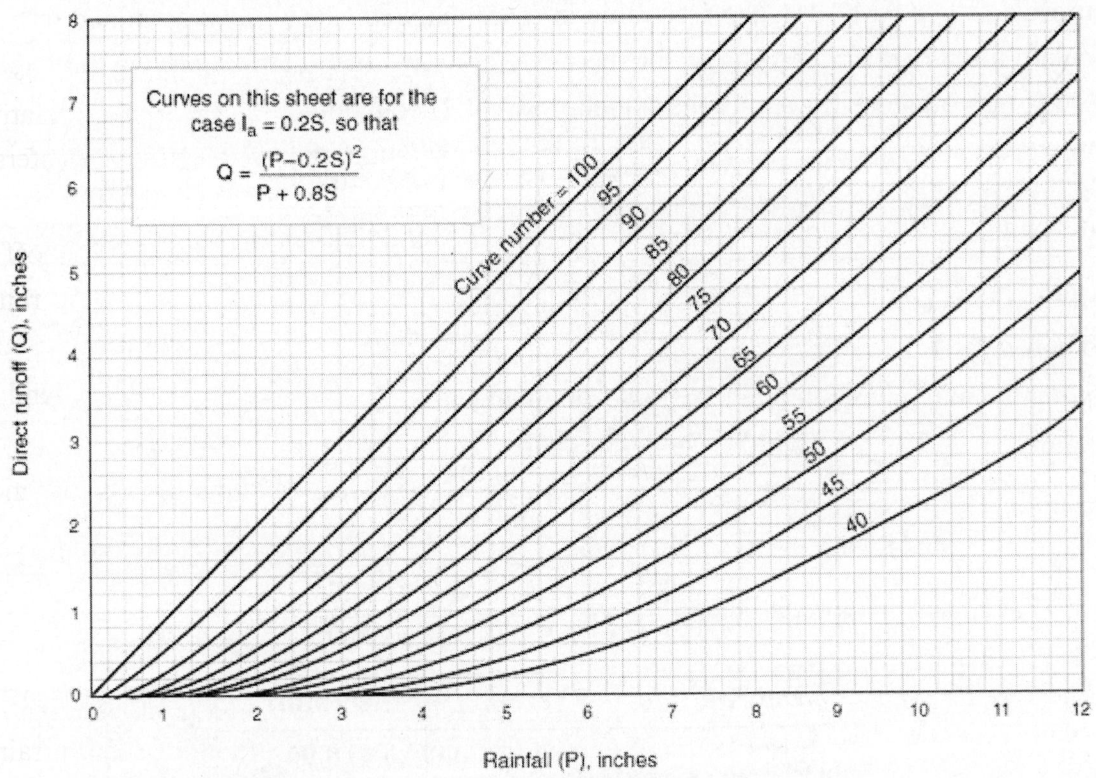

Figure 4.12: SCS-Runoff Curve Number Method

Example 4.3: Calculate the direct runoff for a storm that had an average rainfall depth of 129.22 mm on a watershed with good pasture cover, soil type B, and antecedent moisture condition (AMC II).

Solution

From Table 4.2, the curve number for soil type Cis equal to 61.

The potential maximum retention S can be obtained as:

$$S = \frac{25400}{CN} - 254 = \frac{25400}{61} - 254 = 162.39\ (mm)$$

Compute the amount of initial abstraction I_a as

I_a =0.2 x S=0.2 x 162.39 = 32.48 (mm)

The direct runoff Q is:

$$Q = \frac{(P - 0.2S)^2}{P + 0.8 * S} == \frac{(129.22 - 0.2 * 162.39)^2}{129.22 + 0.8 * 162.39} = \frac{9358.63}{259.13} = 36.11\ (mm)$$

Example 4.4: For AMC-1I1, the CN value in a 450-ha watershed was determined to be 79. (Calculate the direct runoff volume for the next four days of precipitation. On July 15, the AMC fell into category III. Apply the standard SCS-CN equation.

Date	July 15	July 16	July 17	July 18
Rainfall (mm)	60	45	12	20

Solution

Given $CN_{III} = 79$;

The potential maximum retention S can be obtained as:

$$S = \frac{25400}{CN} - 254 = \frac{25400}{79} - 254 = 67.52\ (mm)$$

The direct runoff Q is:

$$Q = \frac{(P - 0.2S)^2}{P + 0.8 * S} == \frac{(P - 0.2 * 67.52)^2}{P + 0.8 * 67.52} = \frac{(P - 13.5)^2}{P + 54.02} for\ P > 13.5\ mm$$

Date	**P (mm)**	**Q (mm)**
July 15	60	18.96
July 16	45	10.02
July 17	12	0
July 18	20	0.57
Total	137	29.55

Total runoff volume over the catchment

$V_r = 450 \times 10^4 \times 29.55/(1000)$

= 132,975 m³

4.6 FLOW-DURATION CURVES

Flow-duration curves (FDCs) are vital tools in hydrological analysis, encapsulating the variability of water flow in streams and rivers over time. Essentially, an FDC is a graphical representation that ranks streamflow data from highest to lowest, indicating the percentage of time during which specific flow rates are met or exceeded. A typical Flow Duration Curve is shown in Figure 4.13.

As a cornerstone in water resource management, FDCs offer insights into the frequency and magnitude of flows, which are critical for designing infrastructure, managing water supply, and assessing ecological impacts. Their role extends to diverse applications such as predicting the availability of water for hydropower, determining the reliability of water supply for urban and agricultural needs, and understanding the potential for dilution of pollutants.

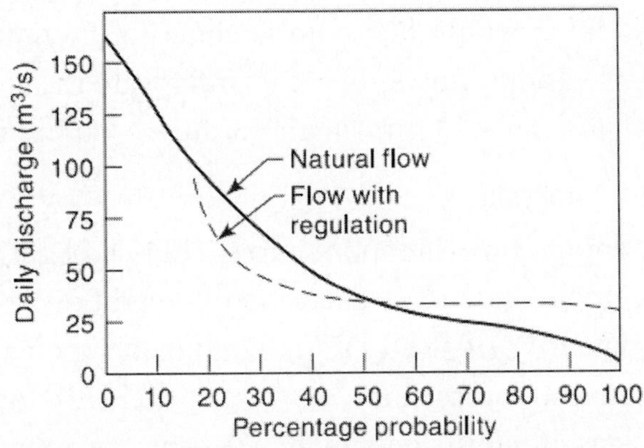

Figure 4.13: A typical Flow Duration Curve

Through FDCs, engineers and scientists can forecast streamflow patterns, aiding in the strategic planning and operation of water resource projects. They bridge the gap between the theoretical aspects of hydrology and the practical, on-the-ground decision-making necessary for sustainable water management.

In educational settings, the study of FDCs equips students with an understanding of both the natural variability of river systems and the human implications of flow patterns, preparing them to tackle complex challenges in the field of water resources engineering.

4.6.1 Theoretical Foundations

The theoretical basis of flow-duration curves is rooted in the fundamental principles of hydrology, particularly the analysis of streamflow as part of the water cycle. At the core of FDC analysis is the concept of frequency and magnitude of streamflow, which is crucial for understanding the behavior of watercourses over time.

Streamflow is inherently variable, influenced by meteorological, hydrological, and geological factors. The theoretical underpinning of FDCs provides a framework for quantifying this variability in a meaningful way. It encompasses probability theory and statistics, as these curves represent the cumulative probability distribution of streamflow.

The basis of an FDC is the order of daily flow measurements from highest to lowest, which creates a correlation between flow volume and frequency of occurrence. This frequency analysis is not concerned with the sequential order of flow events but rather with the likelihood of occurrence of different flow magnitudes.

Furthermore, the FDC encapsulates the continuity equation of hydrology, which states that the change in storage within a system is equal to the input minus the output. By observing the frequency of various flows, one can infer patterns in watershed storage and release mechanisms, which are crucial for managing water resources effectively.

4.6.2 Data Collection and Analysis

The process of constructing a flow-duration curve (FDC) begins with meticulous data collection. Hydrologists compile extensive datasets of streamflow, typically measured daily at gauging stations. The integrity of the FDC is contingent upon the quality of this data, which must be both accurate and representative over a sufficiently long period to encapsulate various hydrologic conditions. Analysis entails the statistical examination of this data, where flows are ranked and their frequencies calculated. The FDC is then derived by plotting these frequencies against the corresponding flow rates, revealing the percentage of time each flow rate is equaled or exceeded. This data not only reflects the physical characteristics of the catchment area but also embodies the climatic and weather patterns impacting the streamflow. The resultant curve is a comprehensive synthesis of the temporal distribution of flows within a river or stream, indispensable for effective water resource management.

4.6.3 Curve Construction

- **Data Compilation:** Begin by gathering a comprehensive set of streamflow data. This typically involves daily flow measurements over several years to ensure variability in hydrologic conditions is captured.
- **Data Sorting:** Rank the streamflow data from the highest to the lowest values. This ranking does not consider the time sequence but only the magnitudes of flow.
- **Frequency Analysis:** For each flow magnitude, calculate the exceedance probability. This frequency can be determined by the equation 4.14:

$$P = \frac{m}{n+1} \times 100 \quad \text{...........(4.14)}$$

where,

P = exceedance probability,

m = rank order, and

n = total number of flow events

- **Plotting the Curve:** Create a graph with the flow magnitudes on the y-axis (logarithmic scale) and the exceedance probabilities on the x-axis (linear or logarithmic scale). The plotted points should form a smooth curve that represents the flow-duration relationship.
- **Curve Smoothing:** Apply curve smoothing techniques if necessary, to ensure the curve is a fair representation of the ranked data, removing outliers that might skew the analysis.

4.6.4 Interpretation of Curve Segments

- **High Flow Segment:** The leftmost part of the curve represents high flows that occur infrequently. Analysis of this segment is crucial for flood risk management.
- **Middle Segment:** This portion reflects the flows that are exceeded between 10% to 70% of the time and can be vital for designing water supply systems to ensure reliability.
- **Low Flow Segment**: The rightmost part indicates low flows, which is crucial for drought management, environmental flow assessments, and pollution dilution analysis.

By dissecting the curve into these segments, hydrologists and engineers can draw specific inferences about water availability, ecosystem health, and the potential for resource development, providing a robust foundation for managing water resources efficiently and sustainably.

4.6.5 Geological Influences on Streamflow

The geological characteristics of a watershed play a crucial role in influencing streamflow patterns, which are reflected in the flow-duration curves. The permeability and porosity of the underlying soil and rock determine how quickly and how much rainfall will infiltrate into the groundwater system versus becoming surface runoff. For instance, watersheds underlain by highly permeable geological formations, like sandstone or fractured bedrock, tend to have higher base flows and less pronounced peak flows, as a significant portion of the water is absorbed and slowly released into stream channels. Conversely, regions with impermeable layers, such as clay or fractured bedrock, often experience rapid runoff and higher peak flows, leading to steeper curves on the high-flow end. Additionally, geological formations influence the soil's water-holding capacity and, by extension, its drought resilience. Understanding these geological impacts is essential for accurate water resource assessment and the design of flood mitigation and water conservation strategies.

4.6.6 Applications in Water Resources Engineering

In water resources engineering, flow-duration curves (FDCs) are pivotal for a multitude of applications. They are instrumental in designing and managing water supply systems, particularly in determining the reliability and capacity of reservoirs. FDCs are also crucial in flood risk management, aiding in the prediction and planning for extreme flow events. Furthermore, their role in environmental flow assessments is significant, as they help in maintaining ecological balance in riverine systems. Additionally, in the realm of hydropower, FDCs guide the optimization of water resources for energy generation, ensuring both efficiency and sustainability.

4.6.7 Advanced Topics in Flow-Duration Curves

Advanced topics in flow-duration curves delve into areas like climate change impact analysis, where FDCs are used to assess how shifting climatic patterns influence streamflow variability. There's growing interest in applying machine learning algorithms to predict and analyze FDCs, harnessing large datasets to enhance prediction accuracy. Another emerging area is the integration of remote sensing and GIS technologies, offering more comprehensive watershed analysis by incorporating spatial variability and land-use changes. These advanced methodologies are crucial for developing adaptive and resilient water resource management strategies in an era of increasing environmental uncertainties.

Flow-duration curves (FDCs) are invaluable in water resources engineering, offering insights into streamflow variability for effective management and planning. They facilitate understanding of high, average, and low flow conditions, aiding in water supply design, flood management, and ecological assessments. The application of FDCs is poised to evolve with advancements in data analysis techniques, including machine learning and GIS, and their role will become increasingly significant in addressing the challenges posed by climate change on water resources.

4.7 FLOW-MASS CURVE

The plotting of cumulative discharge (Q) against time in chronological order is represented by the flow-mass curve (Equation 4.15). It is apparent that the flow-mass curve is an integral curve, or summation curve, of the hydrograph as the hydrograph is a plot of Q vs. t. Rippl (1882) was the one who first proposed the flow-mass curve, and it is often referred to as Rippl's mass curve. A typical flow-mass curve is shown in Figure 4.14.

$$V = \int_{t_0}^{t} Q dt \qquad \text{.........(4.15)}$$

where,

Q = Rate of discharge rate;

t_0 = time at the beginning of the curve,

V = ordinate of the mass curve at any given time t.

Keeping in mind that in this picture, the abscissa represents chronological time in months. It can also be expressed in weeks, months, or days, based on the analysis of the data. The ordinate is expressed in million m³ of volume. Additional units used for ordinate are mm over a catchment area, ham, and m³/day (cumec day).

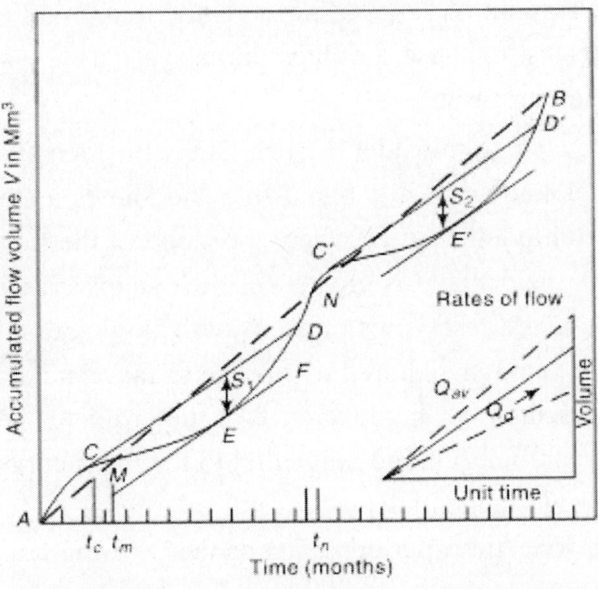

Figure 4.14: A typical Flow-Mass Curve

Any particular point slope on the mass curve indicates the current flow rate, which is equal to dV/dt = Q. In a scenario in which a reservoir of sufficient size is readily available, the average rate of flow that can be sustained between times t_1 and t_2 is represented by the slope of a straight line connecting two points, M and N. The average discharge for the whole duration of the depicted record is therefore represented by the slope of the line AB connecting a mass curve's beginning and ending points.

4.7.1 Calculation of Storage Volume

Plot the mass curve of a reservoir on the stream, as demonstrated in Figure 4.14. Assuming that the reservoir is full at the start of a dry season, that is, when the rate of inflow is lower than the rate of withdrawal, the maximum volume of water that can be extracted from storage equals the difference between the total volumes of supply and demand since the beginning of dry season. As a result, the needed storage S is represented as Equation 4.16.

$$S = \text{maximum of}\left(\sum V_D - \sum V_S\right) \qquad \text{...........(4.16)}$$

where,

V_S = supplied volume; and

V_D = demanded volume

S = The largest difference in the ordinate between the mass curves of supply and demand is applied to determine the amount of storage, which is the maximum cumulative efficiency in each dry season. The greatest of these S values during various dry seasons is the minimum storage volume that a reservoir needs.

At a high point on a ridge, consider the line CD with slope Q_d that is drawn tangential to the mass curve. This is known as a demand line and shows a steady rate of withdrawal (Q_d) from a reservoir. From point C to E, where the slope of the flow-mass curve is smaller than the demand line CD, the demand is greater than the supply rate if the reservoir is full at C (at time t_0). As a result, the reservoir's capacity will be decreasing, with E marking the lowest point. The volume of water required as storage to meet the demand starting when the reservoir was full is represented by the difference in the ordinates between the demand line CD and a line EF drawn parallel to it and tangential to the mass curve at F, S in Figure 4.14. The demand lines are drawn tangentially at several additional ridges (such as C'D' in Fig. 4.14), if flow data for a long time period is available. The highest amount of the storage values obtained is determined as the minimum storage requirement for a reservoir.

4.8 SURFACE WATER RESOURCES OF INDIA

Stream flows measured at a river's terminal site, typically near the sea, are used to evaluate a basin's water resources. It is obvious that the length of the data series and the precision of the discharge observations affect the assessment's accuracy. This duration is insufficient in many cases in India, and it is unknown how accurate the observations were made; the time spans for which data are available vary greatly throughout basins. Longer rainfall data are commonly available in India and are often used to prolong the runoff series. The water resources of the basin are typically identified by the average annual flow at a river's terminal point. Keep in mind that this is referring to a 50% chance of water availability. Projects involving water resources must be planned with reliability at other levels, such as 90% and 75%. The time series of observations is used to assess such reliability.

Worldwide, most of the basins are no longer virgin; upstream use has a major impact on the flow at the terminal site. Add to the surface flow measured at the terminal site the net export of surface water out of the basin; the net increase in surface water storage, additional evapotranspiration caused by the use or storage of surface water; direct Groundwater flow from the river basin below or along the terminal site; the net export of Groundwater out of the basin; the net increase in Groundwater storage and soil moisture storage; and additional evapotranspiration caused by the use or storage of Groundwater are some ways to calculate a basin's natural runoff. This is the general water balance method, which may be used in any basin at any time. However, the storage change would be zero or very small if averages over a lengthy period of time were used. Additionally, a simplification is achievable if no export or import occurs and the Groundwater flow beneath or along the terminal site is ignored. This simplification allows the average annual natural flow to be calculated by adding the average annual extra evaporation/evapotranspiration from the use or storage of surface water and the average annual extra evaporation/evapotranspiration from the storage or use of Groundwater to the average annual surface flow measured at the terminal site.

India receives 4,000 km^3 of precipitation annually on average, of which 700 km^3 are instantly lost to the atmosphere, 2,150 km^3 seep into the earth, and 1,150 km^3 flow as surface runoff. The projected total amount of water resources in the nation is 1,953 km^3. The Ganga-Brahmaputra-Meghna basin contains 1,202 km^3, or nearly 62%, of the world's total water resources. Of the total water resources, 751 km^3 are found in the remaining 23 basins. In terms of useable water resources, India has an annual water availability of 1,122 km^3. In addition, by 2050, there will be an additional 123 km^3 to 169 km^3 of return flow available due to increased utilization for irrigation, residential, and industrial uses. From around 3,000 m^3 in 1951 to 1,100 m^3 in 1998, the per capita availability of usable water is predicted to drop to 687 m^3 by 2050. Table 4.5 represents the surface water resources at a glance of India.

The four to five-month monsoon season accounts for the majority of the surface water flow in Indian rivers. The nation's yearly surface runoff was estimated by Rao (1973) to be 1,645 km^3. The Central Water Commission estimated that the nation has 1,869 km^3 of surface water resources. These are approximations derived from statistics using current river flows. NCIWRD (1999) has also calculated the water resources of many basins. They claim that India has 1,953 km^3 of surface water resources in total. Table 4.6 provides an overview of the average annual flow in Indian rivers across several basins.

Table 4.5: India's water resources at a glance

S. No.	**Water Resource at a Glance**	**Quantity (km^3)**	**Percentage**
1	Annual precipitation (Including snowfall)	4000	100
2	Precipitation during monsoon	3000	75
3	Evaporation + Soil water	2131	53.3
4	Average annual potential flow in rivers	1869	46.7
5	Estimated utilizable water resources	1123	28.1
	Surface water	690	17.3
	Replenishable groundwater	433	10.8
	Storage created of utilizable water	253.381	22.52
	Storage (under construction) of utilizable water	50.737	4.5
6	Estimated water needs in 2050	1450	129
7	Estimated deficit	327	29
	Interlinking can give us	200	17.8

(Source: https://indiawris.gov.in/wiki/doku.php?id=india_s_water_wealth)

Table 4.6 shows that while some rivers in the northern region, such as the Ganga and Brahmaputra, have a significant quantity of water that is not utilizable, some rivers in the southern peninsula, including the Pennar and Cauvery, are able to utilize practically all of their capacity. Furthermore, it is demonstrated that the available flow in the Pennar and east-flowing rivers between Pennar and Kanyakumari exceeds the potential.

Table 4.6 demonstrates that the Brahmaputra, which has the largest average annual runoff of 585 km^3, gives only 4% of the utilizable flow of surface water, whereas the Ganga basin has the maximum amount of surface water or around 50% of the average annual runoff of 525 km^3. Additionally, it demonstrates that almost half of the Sabarmati River's 3.812 km^3 yearly runoff is usable flow. It is almost the same story with Subarnarekha.

Table 4.6: Surface water resources potential of river basins (km^3) of India

S. N.	Name of the River Basin	Average annual potential in the river	As per NCIWRD (1999)	Estimated utilizable flow, excluding Groundwater	Cultivable area (thousands ha)
1.	Indus (Area in Indian Territory)	73.31	73.31	46.00	9,638
2.	a) Ganga	525.02	525.02	250.00	60,161
	b) Brahmaputra, Barak, and others	585.60	677.41	24.00	6,145
3.	Godavari	110.54	110.54	76.30	18,931
4.	Krishna	78.12	69.81	58.00	20,299
5.	Cauvery	21.36	21.36	19.00	5,523
6.	Pennar	6.32	6.32	6.86	3,539
7.	East flowing and rivers from Mahanadi to Godavari and Krishna to Pennar	22.52	22.52	13.11	
8.	East-flowing rivers between Pennar and Kanyakumari	16.46	16.6	16.73	
9.	Mahanadi	66.88	66.88	49.99	7,994

S. N.	Name of the River Basin	Average annual potential in the river	As per NCIWRD (1999)	Estimated utilizable flow, excluding Groundwater	Cultivable area (thousands ha)
10.	Brahmani & Baitarani	28.48	28.48	18.30	2,360
11.	Subarnarekha	12.37	12.37	6.81	1,194
12.	Sabarmati	3.81	3.81	1.93	1,548
13.	Mahi	11.02	11.02	3.10	2,210
14.	West flowing rivers of Kutch & Saurashtra, including Luni	15.10	15.10	14.98	
15.	Narmada	45.64	45.64	34.50	5,901
16.	Tapi	14.88	14.88	14.50	4,292
17.	West-flowing rivers from Tapi to Tadri	87.41	87.41	11.94	
18.	West-flowing rivers from Tadri to Kanyakumari	113.53	113.53	24.27	
19.	Area of inland drainage in Rajasthan desert	~ 0	~ 0		
20.	Minor rivers draining to Myanmar (Burma) & Bangladesh	31.00	31.00		
Total		**1869.00**	**1952.8**	**690.00**	

4.9 ENVIRONMENTAL FLOWS

The realization that rivers' survival is essential because of the ecosystem services they provide is the concept of environmental flows or EFs. The amount, timing, and quality of water flows required to maintain freshwater and estuarine ecosystems and the livelihoods that depend on them are known as ecosystem functions (EFs). The concept evolved to include broader issues like geomorphology, sediment movement, freshwater habitats, and non-fish species requirements.

The health and integrity of the entire ecosystem is fundamental to sustained human well-being. Environmental flows are necessary to maintain the health and biodiversity of water bodies, including rivers, coastal waters, wetlands (mangroves, sea grass beds, floodplains) and estuaries. According to the Brisbane Declaration (2007), environmental flows (EFs) are the quantity, timing, duration, frequency, and quality of flows required to sustain freshwater, estuarine, and nearshore ecosystems and the human livelihoods and well-being that depend on them. Note that besides the amount, one should also specify the temporal pattern of the flows.

The environmental flow requirement of a river depends on the properties of the aquatic ecosystem, the development stage of the area and the societal requirements. Exact values of EFs for a project can only be established by using the detailed hydrologic data, river cross-sections, needs of the biotic life and how it is likely to respond to the reduction in river flow after commissioning of the project and the preferences of all stakeholders.

The current attention on EFs has emerged because some people believe that the dams and diversions constructed to regulate rivers for societal needs have significantly and (mostly) adversely impacted the rivers. However, besides the dams and diversions, many other changes in the catchment affect the flow regime of a river. Since a considerable part of the flow in a river during the lean season comes from Groundwater, large-scale Groundwater pumping and consequent lowering of the water table will have a bearing on it. Extensive land use changes in the catchment also affect river flows. Further, as stated in the definition, the quality of water is an integral part of EFs. River water quality is chiefly damaged by the disposal of untreated municipal and industrial waste in the river, and the return flows from those agricultural areas where high quantities of chemical fertilizers and pesticides are applied.

4.9.1 Trade-offs in Development and Conservation

- There's a balance between using water for development (e.g., agriculture, industry) and conserving natural water bodies.
- The total benefits from natural and managed systems reach a maximum point that can be considered optimal resource development.

4.9.2 Estimation of Environmental Flows

- EFs depend on factors like the river's physical, chemical, and biological characteristics, its natural state, desired state, and water uses.

- EF assessment (EFA) involves several stages, including defining the issue, scope, and objective, deciding on estimation sites, collecting data, and analyzing it.
- Methodologies for EF estimation vary in complexity and include hydro-biology, hydrology and hydraulics, and hydrological methods. We now briefly review the methodologies to assess EFs.

4.9.3 Environmental Flow Assessment Methodologies

a. Hydro-biology Methodologies

These methods integrate hydrological, hydraulic, and biological data to assess environmental flows. Key approaches include:

- **Building Block Method (BBM):** This method determines the flow requirements for key ecosystem components, such as low flows, high flows, and floods. BBM considers the river as a series of building blocks, each with specific flow requirements for different ecological functions or species. It is a detailed approach that requires extensive ecological and hydrological data.
- **Ecological Limits of Hydrological Abstractions (ELOHA):** ELOHA is a framework that combines regional-scale ecological classification with local-scale empirical analysis. It begins with the classification of river types based on physical and ecological characteristics. Then, for each river type, the relationship between flow alteration and ecological response is assessed. This method is useful for large-scale applications and offers a balance between ecological protection and water resource development.

b. **Hydrology and Hydraulics-Based Methodologies**

These methodologies rely on the relationship between river flows and simple hydraulic characteristics. They are less complex than hydro-biological methods and include:

- **Flow Duration Curve (FDC) Method:** This approach uses flow duration curves, which depict the percentage of time a certain flow is equaled or exceeded. The FDC method identifies specific flow percentiles that are crucial for maintaining ecological health. This method is already discussed above.
- **Tennant Method (Montana Method):** The Tennent Method, also known as the Tennent-Eade Method, is primarily used in hydraulic engineering, particularly for estimating the flow discharge required to maintain a specific environmental flow

regime in rivers or streams. Environmental flows, also known as ecological flows or eco-hydrology, refer to the water flows necessary to sustain freshwater and estuarine ecosystems and the services they provide. Here is an overview of how the Tennent Method is applied to estimate environmental flows:

- **Characterization of the River System**: Before applying the Tennent Method, it is essential to characterize the river or stream system under consideration. This involves gathering data on the hydrological, hydraulic, geomorphological, and ecological characteristics of the river reach.
- **Determination of Environmental Flow Requirements**: Based on ecological considerations, such as the needs of aquatic habitats, riparian vegetation, and other ecosystem components, environmental flow requirements are established. These requirements often vary seasonally and may depend on factors such as species diversity, life stages of aquatic organisms, and habitat types.
- **Calculation of Flow Components:** The Tennent Method involves dividing the total flow discharge into various flow components, including base flow, low-flow components, and high-flow components. Base flow represents the flow sustained by groundwater discharge and is crucial for maintaining streamflow during dry periods. Low-flow components may include drought flows, while high-flow components may include flood flows necessary for ecosystem health and habitat maintenance.
- **Iterative Process:** Estimating environmental flows using the Tennent Method often involves an iterative process of refining flow requirements, adjusting management strategies, and reassessing the predicted ecological outcomes. This process may involve stakeholder engagement, ecological monitoring, and adaptive management approaches to ensure that environmental flow objectives are met effectively.

Overall, the Tennent Method provides a structured approach to estimating environmental flows, helping water resource managers and environmental practitioners balance the needs of ecosystems with human water use requirements in river systems.

c. **Hydrological Methods:** Hydrological methods play a crucial role in determining environmental flows, ensuring that water resources are managed sustainably to support freshwater ecosystems. By employing these hydrological methods, water resource managers and environmental practitioners can assess the ecological implications of water management decisions and design environmental flow regimes that sustainably meet the needs of freshwater ecosystems. These models can help predict river flows under different scenarios, including changes in land use, climate, and water management practices. Hydrological models provide valuable information for estimating baseline flow conditions and assessing the impacts of water withdrawals and infrastructure development on river flows.

d. **Rainfall-Runoff Models:** Rainfall-runoff models simulate the transformation of rainfall into streamflow through various hydrological processes such as infiltration, runoff, and groundwater recharge. These models estimate the contribution of different hydrological pathways to streamflow, including surface runoff, interflow, and baseflow. Rainfall-runoff models help assess the sensitivity of streamflow to changes in precipitation patterns, land cover, and land use practices, which is essential for understanding the potential impacts of climate change and land management on environmental flows.

e. **Flow Routing Models:** Flow routing models simulate the movement of water through river channels and reservoirs, considering channel characteristics, hydraulic properties, and flow regulation structures. These models estimate flow attenuation, dispersion, and timing along river reaches and can assess the effects of flow regulation on downstream flow conditions. Flow routing models are used to design flow releases from dams and reservoirs to maintain environmental flow requirements downstream while meeting water supply and flood control objectives.

f. **Hydrological Indices:** Hydrological indices, such as baseflow index, recession constant, and runoff coefficient, provide quantitative measures of hydrological processes and streamflow characteristics. These indices are used to assess the hydrological regime of river systems and identify trends and anomalies in flow patterns over time. Hydrological indices help characterize the natural variability of streamflow and inform the development of environmental flow targets and management strategies.

4.9.4 Implementation of EF

Implementing environmental flow involves several steps and considerations to ensure the sustainable management of water resources while protecting freshwater ecosystems. Implementing EFs is challenging due to factors like coordination among different regulatory agencies and existing water allocations. Adaptive management, involving estimating and implementing EFs, monitoring ecosystem health, and revising decisions, is recommended. Here is a generalized outline of how environmental flow implementation can be approached:

- **Stakeholder Engagement and Collaboration**
  - Identify and engage stakeholders, including government agencies, water users, indigenous communities, environmental organizations, and local communities.
  - Foster collaboration and dialogue among stakeholders to understand diverse perspectives, interests, and concerns related to water management and environmental conservation.
- **Legislative and Policy Framework**
  - Develop or revise relevant laws, policies, and regulations to integrate environmental flow considerations into water management frameworks.
  - Establish legal mechanisms for allocating and managing water resources to ensure the provision of environmental flows.
- **Scientific Assessment and Monitoring**
  - Conduct scientific assessments of hydrological, ecological, and geomorphological conditions of river systems to determine environmental flow requirements.
  - Monitor key indicators of ecosystem health, such as water quality, aquatic biodiversity, habitat availability, and species populations, to assess the effectiveness of environmental flow regimes.
- **Flow Regime Design**
  - Design flow regimes that mimic natural flow patterns and meet the ecological needs of freshwater ecosystems.
  - Determine flow requirements for different seasons, flow regimes (e.g., base flows, flood flows), and ecological zones within river systems.

- **Water Allocation and Management**
  - Allocate water rights or permits that include provisions for maintaining environmental flows.
  - Implement water management strategies, such as water conservation measures, water use efficiency improvements, and water allocation prioritization, to ensure the provision of environmental flows.
- **Infrastructure and Engineering Solutions**
  - Design and implement infrastructure projects, such as dam operations, flow releases, and water diversions, to regulate flow regimes and provide environmental flows.
  - Incorporate environmental considerations into engineering solutions to minimize adverse impacts on freshwater ecosystems.
- **Adaptive Management and Monitoring**
  - Adopt adaptive management approaches to iteratively adjust environmental flow regimes based on new information, changing conditions, and stakeholder feedback.
  - Implement robust monitoring programs to assess the ecological outcomes of environmental flow implementation and inform adaptive management decisions.
- **Capacity Building and Education**
  - Build institutional capacity and expertise among water managers, policymakers, and stakeholders to effectively implement environmental flow programs.
  - Raise awareness and provide education and training on the importance of environmental flows and freshwater conservation to foster community support and engagement.
- **Evaluation and Reporting**
  - Evaluate the effectiveness of environmental flow implementation through periodic assessments and reporting.
  - Communicate findings and lessons learned to stakeholders, policymakers, and the public to promote transparency and accountability in water management decision-making.

By following these steps and considering local context and stakeholder input, environmental flow implementation can contribute to the sustainable management of freshwater resources and the protection of aquatic ecosystems.

4.9.5 Future Challenges

- Many rivers are already heavily utilized, and returning to historical conditions is not feasible. The goal should be to develop feasible flow regimes.
- Environmental flows are based on equitable water sharing, but there is often a disconnect between water management and policy-making.

4.10 UNIT HYDROGRAPH APPROACH

A unit hydrograph is a hydrological tool used to predict the runoff response of a watershed to a given amount of rainfall over a specific duration. It is a fundamental concept in hydrology and particularly useful for estimating the timing and magnitude of streamflow resulting from rainfall events. A drainage basin's unit hydrograph is a hydrograph showing direct runoff produced by one centimeter (or one inch) of effective rainfall over a given period of time that is distributed evenly over the basin's area at a constant rate. The portion of the total storm rainfall that enters the stream as direct runoff is known as the effective rainfall or rainfall surplus. The time frame that is given is the amount of time that is thought to have a uniform distribution of effective rainfall. It is believed that the effective rainfall is evenly distributed over the drainage region, causing direct runoff to start at the commencement of the excess rainfall. According to the unit hydrograph principle, it is stated that:

- The time base of the hydrographs showing direct runoff from effective rainfalls of similar length is the same.
- The total amount of direct runoff shown by each hydrograph is proportionate to the ordinates of the direct runoffs coming from effective rainfalls of equal duration but varying intensity.
- The combined physical features of a basin are reflected in the hydrographs of direct runoff from the basin during a specific time of effective rainfall.

The fundamental idea behind the unit hydrograph principle is that a direct runoff hydrograph with the same time basis as the unit hydrograph but R times the unit hydrograph's ordinates will be produced by R inches of effective rainfall over the basin in the same unit duration. If the idea is expanded, a sequence of overlapping hydrographs with ordinates proportionate to the unit hydrograph, resulting from distinct periods of uniform effective rain, can be used to construct the hydrograph of direct runoff that would arise from a series of bursts of variable intensity from continuous effective rain of variable intensity. The ordinates of each individual hydrograph are added to produce the hydrograph of total direct runoff. The ordinates of direct runoff must be raised by the predicted base flow in order to obtain the total runoff hydrograph. Here is an overview of the unit hydrograph, its method, and uses:

4.10.1 Method of Constructing a Unit Hydrograph

- **Selection of a Design Storm:** The first step in constructing a unit hydrograph is to select a design storm that represents the rainfall event for which the hydrograph is to be developed. The design storm should be representative of the rainfall characteristics of the region or watershed under study.
- **Derivation of the Hydrograph:** The unit hydrograph is derived by dividing the hydrograph of the design storm into a series of evenly spaced increments, typically representing time intervals of one hour or less. The total area under the hydrograph curve is normalized to one unit of runoff per unit of rainfall depth.
- **Application of the S-Hydrograph Theory (Clark's Method)**: One of the commonly used methods for deriving a unit hydrograph is the S-hydrograph theory, also known as Clark's method. This method assumes that the watershed response to rainfall can be represented by a linear reservoir system. The unit hydrograph is derived by convolving the excess rainfall hyetograph with the unit hydrograph of the watershed.
- **Adjustments and Calibration**: Depending on the characteristics of the watershed and the available data, adjustments may be made to the unit hydrograph to account for factors such as basin size, shape, land use, soil type, and antecedent moisture conditions. Calibration of the unit hydrograph may be necessary to match observed hydrological responses.

4.10.2 Uses of Unit Hydrographs

Unit hydrographs are versatile tools in hydrology and water resources management, providing insights into the watershed response to rainfall and supporting various applications in flood forecasting, water resources planning, and environmental management. The major uses of unit hydrograph are described below:

- **Hydrological Modeling:** Unit hydrographs are commonly used in hydrological modeling to simulate the response of watersheds to different rainfall events. They provide a simplified representation of the runoff process, allowing hydrologists to predict streamflow volumes and timing for various scenarios.
- **Flood Forecasting:** Unit hydrographs are used in flood forecasting systems to estimate the magnitude and timing of peak flows in rivers and streams. By combining rainfall forecasts with unit hydrographs, flood forecasters can issue timely warnings and implement flood mitigation measures.
- **Water Resources Management:** Unit hydrographs are valuable tools for water resources management, helping planners and engineers assess the impacts of land development, land use changes, and infrastructure projects on watershed hydrology. They are used to design stormwater management systems, reservoirs, and drainage infrastructure.
- **Environmental Impact Assessment:** Unit hydrographs are used in environmental impact assessments to evaluate the potential impacts of land development and land use changes on streamflow patterns, aquatic habitats, and water quality. They help assess the sustainability of development projects and inform decision-making processes.
- **Hydraulic Design**: Unit hydrographs are used in hydraulic design to estimate peak flows for the design of bridges, culverts, and other hydraulic structures. They provide engineers with essential information for designing infrastructure that can safely convey runoff during storm events.

4.10.3 Synthetic Unit Hydrograph Parameters

Deriving a unit hydrograph for the basin of interest is the initial step in the synthetic unit hydrograph process. The hydrograph is then estimated using this unit hydrograph and the design rainfall. Synder put out two fundamental equations to define the synthetic unit hydrograph. The first equation (Equation 4.17) represents the basin's lag time in terms of the time of peak (t_p), which is the duration from the unit storm's centre to the unit hydrograph's peak in the synthetic unit hydrograph procedure:

$$t_p = C_t \left(LL_{ca}\right)^{0.3} \qquad \text{.........(4.17)}$$

where,

t_p= duration of peak hydrograph from the midpoint of hourly rainfall

L= length in kilometers along the stream from the study point to the basin's upstream limits,

L_{ca}= distance in kilometers from the study point to the basin centroid,

C_t = a regional constant representing watershed slope and storage (varies between 1.35 to 1.65 as per Synder's study). However, variations ranging from 0.3 to 6.0 have been reported.

Linsey et al. discovered a stronger correlation between the basin lag t_p and the catchment parameter $\frac{LL_{ca}}{\sqrt{S}}$; where S represents the basin slope. As a result, the following amended form for t_p was proposed (Equation 4.18).

$$t_p = C_t L\left(\frac{LL_{ca}}{\sqrt{S}}\right)^n \qquad \text{...........(4.18)}$$

Where C_t and n are basin constants. The values of C_t L for basins in the United States were 1.03 for foothill areas, 0.5 for valley areas, and 1.715 for mountainous drainage areas. The value of n was found to be 0.38. Synder used the typical length of t_r hours of precipitation that was provided by equation 4.19.

$$t_r = \frac{t_p}{5.5} \qquad \text{...........(4.19)}$$

The second equation (Equation 4.20) represents the unit hydrograph peak as:

$$Q_{ps} = \frac{2.78 C_p A}{t_p} \qquad \text{...........(4.20)}$$

Where:

Q_{ps} = peak discharge in m^3/sec of a unit hydrograph of standard duration t_r h

C_p = a regional constant

A = catchment area in km^2.

This equation assumes that the peak discharge is proportional to the mean discharge of $\frac{1cm * catchment\ area}{Duration\ of\ rainfall\ excess}$

For Synder's research regions, the coefficient C_p values vary from 0.56 to 0.69, and this is regarded as a measure of the watershed's capacity for retention and storage. Similar to C_t, values of C_p can differ significantly based on the features of the area; values ranging from 0.31 to 0.93 have been documented.

The magnitude of the basin lag is impacted if a non-standard rainfall duration, t_r h, is used to create a unit hydrograph in place of the standard value, t_r. The adjusted basin latency is provided by equation 4.21 as follows:

$$t_p' = t_p + \frac{t_R + t_r}{4} = \frac{21}{22} t_p + \frac{t_R}{4} \quad \text{...........(4.21)}$$

Hence, the peak discharge for a non-standard ER of time duration t_R can be given by equation 4. 22 as follows:

$$Q_{ps} = \frac{2.78 C_p A}{t_p} \quad \text{...........(4.22)}$$

Synder gives the time base (t_b) of a unit hydrograph as equation 4.23.

$$t_b = 3 + \frac{t_p}{8} \; days \quad \text{........... (4.23)}$$

For tiny catchments, this could result in excessively high time base values. Taylor and Schwartz advise avoiding using tiny catchments and suggested the revised equation 4.24.

$$t_b = 5\left(t_p + \frac{t_R}{2}\right) hours \quad \text{...........(4.24)}$$

The US Army Corps of Engineers determined the width of unit hydrographs at 50% and 75% of the peak for US catchments when they were sketching the unit hydrograph. These widths, which are determined by, have a relationship to the peak discharge intensity and are given by equations 4.25 and 4.25a.

$$W_{50} = \frac{5.87}{q^{1.08}}, \quad \text{and} \quad \text{...........(4.25)}$$

$$W_{75} = \frac{W_{50}}{1.75} \quad \text{...........(4.25a)}$$

Since the values of the coefficients C_t and C_p differ between regions, it is best to use established unit hydrographs of a meteorologically homogeneous catchment to identify the values of these coefficients before applying them to the basin under investigation in practical applications. In this manner, scaling hydrograph data from one similar catchment to another is made possible using Synder's equations.

SUMMARY

Runoff is defined in engineering hydrology as the quantity of precipitation that emerges a catchment area through surface drains. The intensity and length of rainfall, soil type, land slope, vegetation cover, and other anthropogenic activities including urbanization and changes in land use and land cover are several of the storm and catchment factors that affect it. The current lesson covers the runoff process, the variables that influence it, and techniques for estimating runoff volume, such as the SCS-CN strategy. Additionally, the module explains the runoff hydrograph's components, elements that affect it, flow mass curve, flow duration curve, etc. The curriculum also discusses India's varied surface water resources and effective rainfall. Finally, environmental flow techniques are also described.

EXERCISE

Revision Questions

1. Give a brief overview of runoff process.
2. Describe the factors affecting the seasonal and annual runoff of a catchment.
3. Why the study of runoff process is essential in hydrology?
4. Write a short note on (i) infiltration (ii) surface runoff.
5. With the help of annual hydrographs, describe the salient features of perennial, intermittent, and ephemeral streams.
6. Write a short note on (i) direct runoff (ii) base flow (iii) natural flow
7. What do you understand by yield?
8. Define a water balance equation.

9. What do you understand by 75% dependable yield of a catchment? Describe the method to estimate the same by using annual runoff volume time series.
10. Define various components of a natural hydrograph. How can we generate a hydrograph?
11. Describe various factors affecting a hydrograph.
12. Define base flow. Explain the method for generating base flow.
13. Describe the factors affecting base flow generation.
14. Write a short note on effective rainfall.
15. Explain briefly the *SCS-CN* method of estimation of the yield of a catchment using daily rainfall records.
16. Explain the flow-duration curve with a neat diagram. What information can be gathered from a study of the flow duration curve of a river at a site?
17. Explain applications of flow-duration curves (FDCs) in water resources engineering.
18. What do you understand by flow-mass curve? Explain the limitations of the flow mass curve.
19. Describe briefly India's surface water resources.
20. Describe briefly the concept of environmental flows. Explain various methods of obtaining environmental flow.
21. Define Unit Hydrograph. Explain the method to construct a unit hydrograph.
22. Describe the various uses of Unit Hydrograph.
23. Describe various methods to obtain synthetic unit hydrograph. Also, explain various parameters of the synthetic unit hydrograph.

Numerical Problems

1. The observed discharge values for the year 2001 at a river gauging station are displayed in the table given below. A barrage constructed across the river, upstream of the gauging site, diverts 4.5 Mm^3 and 2.00 Mm^3 of water every month for irrigation and industrial usage, respectively. Estimated return flows to the river upstream of the gauging location are 0.95 Mm^3 from irrigation and 0.80 Mm^3 from industry. Calculate the natural flow. Find the runoff-rainfall ratio if the catchment area is 240 km^2 and the average annual rainfall is 201 cm.

Month	1	2	3	4	5	6	7	8	9	10	11	12
Gauged Flow (Mm^3)	2.5	1.8	1.2	1.1	2.1	9.0	18.4	23.2	13.1	10.1	7.2	3.9

2. If the basin area of a watershed (A)= 950 km^2, find out the time in days from the hydrograph peak. Assume the value of coefficient (b) = 0.83.
3. Estimate the direct runoff for a storm with an average rainfall of 132.35 mm on a watershed with good grass cover, soil type C, AMC II condition.
4. For a 550-ha watershed in Andhra Pradesh with predominantly non-black cotton soil, the CNII has been estimated as 68. (a) If the total rainfall in the past five days is 28 cm and the season is dormant, estimate the runoff volume due to 90 mm of rainfall in a day? (b) What would the runoff volume be if the rainfall in the past five days was 45 mm?
5. For AMC-III, the CN value in a 522-ha watershed was determined to be 81. (Calculate the direct runoff volume for the next four days of precipitation. On July 20, the AMC fell into category III. Apply the standard SCS-CN equation.

Date	July 20	July 21	July 22	July 23
Rainfall (mm)	58	48	11	22

6. A watershed has the following land use: (i) 450 ha of row crop with poor hydrologic condition and (ii) 150 ha of good pasture land. The soil is of hydrologic soil group B. Estimate the runoff volume for the watershed under antecedent moisture category III when 2 days of consecutive rainfall of 110 mm and 95 mm occur. Use standard *SCS-CN* equations.

Multiple Choice Questions

1. What is runoff?

A) Water that infiltrates the soil
B) Water that flows over the ground surface
C) Water that evaporates from plants
D) Water stored in aquifers

2. Which of the following factors influences runoff?

A) Soil type
B) Vegetation cover
C) Rainfall intensity
D) All of the above

3. What type of land surface typically produces the most runoff?

A) Sandy soil B) Grassy fields

C) Urban areas with concrete D) Forested land

4. Which of the following is a consequence of increased runoff?

A) Increased groundwater recharge B) Soil erosion

C) Improved water quality D) Decreased flood risk

5. How can vegetation help reduce runoff?

A) By increasing soil compaction B) By promoting evaporation

C) By enhancing soil structure and absorption D) By blocking precipitation

6. What is the term for the area where runoff collects before it flows into a body of water?

A) Watershed B) Drainage basin

C) Floodplain D) Riparian zone

7. Which of the following best describes "stormwater runoff"?

A) Water that evaporates during storms

B) Water from rainfall that flows into sewers and drains

C) Water that is absorbed by the soil during storms

D) Water that is stored in ponds after a storm

8. What practice can help manage and reduce runoff in urban areas?

A) Increasing paved surfaces B) Implementing green roofs

C) Deforestation D) Building more parking lots

9. Which of the following best describes "surface runoff"?

A) Water that infiltrates into the ground B) Water that flows over the land surface

C) Water that evaporates into the atmosphere D) Water that is stored in lakes and reservoirs

10. The volume of runoff can be influenced by which of the following factors?

A) Land use and vegetation cover B) Soil type and saturation

C) Precipitation intensity and duration D) All of the above

11. What does the rising limb of a hydrograph represent?

A) The period of base flow

B) The increase in streamflow due to runoff

C) The decrease in streamflow after a rain event

D) The total volume of runoff

12. Which type of runoff is primarily associated with precipitation falling on impervious surfaces?

A) Infiltration runoff
B) Saturation excess runoff
C) Hortonian runoff
D) Groundwater runoff

13. If a watershed has an area of 500 acres and receives 2 inches of rainfall, what is the approximate volume of runoff (in acre-feet) if the runoff coefficient is 0.5?

A) 20 acre-feet
B) 25 acre-feet
C) 30 acre-feet
D) 35 acre-feet

14. In a hydrograph, what does the "peak flow" indicate?

A) The minimum flow recorded
B) The highest flow rate after a rainfall event
C) The total volume of water in the watershed
D) The base flow level before rainfall

15. Which of the following is the primary purpose of a hydrograph?

A) To measure water temperature
B) To show changes in water flow over time
C) To assess soil moisture levels
D) To predict weather patterns

16. What does the base flow of a hydrograph represent?

A) Flow resulting from surface runoff
B) The normal, non-storm flow of a stream
C) Flow during heavy rainfall events
D) Evaporation losses from the stream

17. A steep rising limb in a hydrograph indicates:

A) Low intensity of rainfall
B) Rapid runoff response to rainfall
C) High infiltration rates
D) Long duration of rainfall

18. Which of the following is NOT a type of streamflow classification?

A) Perennial streams
B) Intermittent streams
C) Ephemeral streams
D) Oscillating streams

19. What is the significance of the lag time on a hydrograph?

A) It measures the duration of rainfall

B) It indicates the time taken for runoff to reach the stream

C) It shows the rate of evaporation

D) It determines the volume of groundwater

20. Which classification describes streams that flow continuously throughout the year?

A) Intermittent streams B) Perennial streams

C) Ephemeral streams D) Seasonal streams

21. When scaling a hydrograph, what does the area under the curve typically represent?

A) Maximum flow rate

B) Total volume of flow over a specified period

C) Average rainfall intensity

D) Depth of water in the stream

22. Base flow is primarily derived from which of the following sources?

A) Direct precipitation B) Surface runoff

C) Groundwater discharge D) Snowmelt

23. Which of the following factors can significantly influence base flow in a river?

A) Soil permeability B) Vegetation cover

C) Precipitation patterns D) All of the above

24. The duration of base flow can be affected by:

A) Rainfall intensity B) Land use changes

C) Evapotranspiration rates D) All of the above

25. Base flow primarily contributes to streamflow during which of the following conditions?

A) During heavy rainfall events B) In dry seasons

C) Immediately after snowmelt D) During flooding

26. How does urbanization typically affect base flow?

A) Increases base flow due to reduced evaporation

B) Decreases base flow by increasing impervious surfaces

C) Has no effect on base flow

D) Increases base flow due to increased vegetation

27. What does the Base Flow Index (BFI) indicate?

A) The ratio of base flow to total streamflow B) The amount of surface runoff

C) The level of groundwater recharge D) The average rainfall in a watershed

28. Base flow tends to be higher during which season?

A) Winter
B) Spring
C) Summer
D) Fall

29. Which of the following methods is commonly used to separate baseflow from total streamflow?

A) Total runoff analysis
B) Filter methods
C) Regression analysis
D) Storm event analysis

30. What is the primary purpose of the SCS method in hydrology?

A) To calculate sediment transport
B) To estimate surface runoff from rainfall
C) To assess groundwater recharge
D) To measure water quality

31. The SCS method uses the Runoff Curve Number (CN) to estimate runoff. What does a higher CN value indicate?

A) Higher infiltration rates
B) Greater potential for runoff
C) More vegetation cover
D) Lower soil moisture

32. Which of the following factors does NOT affect the SCS Runoff Curve Number?

A) Land use
B) Soil type
C) Temperature
D) Hydrologic condition

33. Which technique is often used to model baseflow generation in watersheds?

A) Rainfall-runoff modelling
B) Soil moisture accounting
C) Hydrologic response modelling
D) All of the above

34. The SCS method is particularly useful in which type of watershed?

A) Urbanized watersheds
B) Rural agricultural watersheds
C) Forested watersheds
D) Glacial watersheds

35. One common method for separating baseflow from total streamflow is:

A) The logarithmic method
B) The digital filter method
C) The time series analysis
D) The evaporation method

36. What does a flow duration curve (FDC) represent?

A) The average flow rate over time
B) The frequency of flow rates over a specified period

C) The total volume of flow in a watershed

D) The seasonal variations in flow

37. On a flow duration curve, the x-axis typically represents:

A) Flow rates B) Time

C) Percent of time exceeded D) Cumulative flow volume

38. What is the primary purpose of a flow mass curve?

A) To show the total discharge over time B) To illustrate seasonal flow variations

C) To calculate sediment transport D) To assess groundwater levels

39. Flow duration curves are commonly used in which of the following applications?

A) Water resource management B) Flood forecasting

C) Environmental impact assessments D) All of the above

40. In a flow mass curve, what does a steep slope indicate?

A) Low flow conditions B) High flow conditions

C) Increased consistency in flow D) Decreased variability in precipitation

41. If a flow duration curve shows that a certain flow rate is exceeded 70% of the time, this means:

A) The flow rate is very low B) The flow rate is frequently met or exceeded

C) The flow rate is rare D) The flow rate is constant

42. Flow mass curves are particularly useful for:

A) Analysing historical flood events B) Determining water supply for reservoirs

C) Assessing evaporation rates D) Predicting land use changes

43. What is the primary goal of maintaining environmental flows in rivers?

A) To maximize agricultural irrigation

B) To support ecosystem health and biodiversity

C) To increase industrial water supply

D) To enhance recreational opportunities

44. Which river is considered one of the largest sources of surface water in India?

A) Ganges B) Yamuna

C) Brahmaputra D) Godavari

45. Which of the following is a benefit of maintaining environmental flows in rivers?

A) Improved fish migration
B) Increased sedimentation
C) Enhanced urban development
D) Reduced flood risk

46. What does a unit hydrograph represent?

A) The total runoff from a watershed
B) The response of a watershed to 1 inch of effective rainfall
C) The maximum discharge of a river
D) The average annual rainfall in a region

47. Unit hydrographs are primarily used for:

A) Estimating sediment transport
B) Predicting river flow rates during storms
C) Designing flood control structures
D) Analysing groundwater levels

48. Which of the following factors is crucial for assessing environmental flows in Indian rivers?

A) Climatic variations
B) Socio-economic factors
C) Water quality
D) All of the above

49. What does the shape of a unit hydrograph indicate?

A) The average annual precipitation
B) The rate of runoff response over time
C) The seasonal variation in temperature
D) The sediment load in the river

50. What is a synthetic unit hydrograph?

A) A hydrograph derived from actual observed data
B) A model used to predict streamflow based on rainfall
C) A theoretical representation of runoff response for a given watershed
D) A hydrograph that only considers base flow

51. Synthetic unit hydrographs are primarily used for:

A) Estimating groundwater levels
B) Designing drainage systems and flood control measures
C) Measuring evaporation rates
D) Calculating sediment transport

52. Which of the following methods is commonly used to create synthetic unit hydrographs?

A) Rational method
B) SCS (Soil Conservation Service) method
C) Modified Blaney-Criddle method
D) Manning's equation

Answer: 1-B; 2-D; 3-C; 4-B; 5-C; 6-B; 7-B; 8-B; 9-B; 10-D; 11-B; 12-C; 13-B; 14-B; 15-B; 16-B; 17-B; 18-B; 19-B;20-B; 21-B; 22-C; 23-D; 24-D; 25-B; 26-D; 27-A; 28-B; 29-B; 30-B; 31-B; 32-C; 33-D; 34-B; 35-B; 36-B; 37-C; 38-A; 39-D; 40-B; 41-B; 42-B; 43-B; 44-A; 45-A; 46-B; 47-B; 48-D; 49-B; 50-C; 51-B; 52-B.

KNOW MORE

For more information related to this topic scan the QR code.

OR

Type this link in your browser

https://drive.google.com/file/d/15Vm03qNH_TtYGffXC8MzCe
mPj8wsmu2W/view?usp=sharing

BIBLIOGRAPHY

1. Asawa, G.L. (1993), "Irrigation Engineering", Published by Willey Eastern Limited, ISBN 81-224-0457-X.
2. Asawa, G.L. (2005), "Irrigation and Water Resources Engineering, New Age International (P) Ltd., ISBN-13: 978-8122416732
3. Chow. V.T, (Ed), (1964), "Handbook of Applied Hydrology", McGraw-Hill, New York, NY, 1964.
4. Jain S. K., Agarwal, P. K. And Singh, V. P. (2007), Hydrology and Water Resources of India, Water Science and Technology Library, Vol 57, Springer, The Netherland, ISBN-10 1-4020-5179-4(HB).
5. Punmia, B.C. and Lal, Pande B.B., (1992), "Irrigation and Water Power Engineering", Twelfth Edition. Laxmi Publications (P) Ltd.
6. Singh, V. P. (1992). Elementary hydrology. Prentice Hall, New Jersy, 973.
7. Subramanya, K. (2015), "Engineering Hydrology", McGraw-Hill Education (India) Private Limited, New Delhi, 2015, ISBN(13): 978-1-25902997-4.

5 Groundwater and Well Hydrology

UNIT SPECIFICS

Groundwater has been a vital water resource throughout the ages. It is a significant water source for many municipalities, industries, irrigation, sub-urban homes, and farms. Like other natural resources, groundwater supplies are not unlimited. The present unit deals with the occurrence and form of groundwater, aquifer properties, and geological formation of aquifers. It also deals with well hydraulics, including steady-state flow in wells, methodologies for estimating yield for wells, equilibrium equations for confined and unconfined aquifers and aquifer tests.

RATIONALE

To learn about the groundwater and well hydrology, aquifer properties, equilibrium equations for confined and unconfined aquifers, and various aquifer tests.

PRE-REQUISITE

Nil

UNIT OUTCOMES

The list of outcomes of this unit is as follows:

U5-O1: Basics of groundwater and well hydrology

U5-O2: Aquifer properties, equilibrium equations for confined and unconfined aquifers

U5-O3: Various aquifer tests

Unit Outcomes	Expected Mapping with Course Outcomes *(1- Weak Correlation; 2- Medium correlation; 3- Strong Correlation)*							
	CO-1	**CO-2**	**CO-3**	**CO-4**	**CO-5**	**CO-6**	**CO-7**	**CO-8**
U5-O1	3	3	3	1	3	3	3	3
U5-O2	3	3	3	1	3	3	1	3
U5-O3	1	1	3	3	3	3	3	3

5.1 INTRODUCTION

Groundwater is that portion of the water beneath the surface of the earth that can be collected with wells, tunnels, or drainage galleries or that flows naturally to the earth's surface via seeps or springs. Groundwater has been an important water resource throughout the ages. Old dug wells can be found along the wadis of the Middle East, the cradle of our civilization. Some of the ancient tunnels or 'ghanats" in Iran are still in use. Today, groundwater is a major source of water for many municipalities and industries, as well as irrigation, suburban homes, and farms. Like other natural resources, groundwater supplies are not unlimited. They must be wisely managed and protected against undue exploitation and contamination by pollutants or salt water.

5.2 DEFINITION OF GROUNDWATER

Not all underground water is groundwater. If a hole is dug, moist or even saturated soil may be encountered. As long as this water does not seep freely into the hole, it is not groundwater. True groundwater is reached only when water begins to flow into the hole. Since the air in the hole is at atmospheric pressure, the pressure in the groundwater must be above atmospheric pressure if it is to flow freely into the hole. By the same token, the underground water that did not flow into the hole must be at less than atmospheric pressure. Thus, what distinguishes groundwater from the rest of the underground water is that its pressure is greater than atmospheric pressure. Since such water moves freely under the force of gravity into wells, it is also called free water or gravitational water. Groundwater depths can vary from less than 1 m to more than 1000 m. Additionally, there are locations where groundwater has not been at all accessed.

5.3 FORMS OF SUBSURFACE WATER

To understand the occurrence of groundwater and its vertical distribution, let us first consider the hydrological zones present below the ground. The zone between the ground surface and the top of the groundwater is called the vadose zone or unsaturated zone (Figure 5.1). This zone still contains water (very little in dry climates), but the water is held to the soil particles or other underground material by capillary forces. As a result, although the water can still travel inside the vadose zone, it is unable to do so in penetrating wells or other locations that experience atmospheric pressure.

The term vadose zone should be preferred over unsaturated zone. This is because portions of the vadose zone may actually be saturated, even though the pressure of the water is below atmospheric pressure. Examples of regions in vadose zones are the capillary fringe above groundwater, topsoil, and saturated layers of clay or other fine materials that hold water more tightly than underlying coarser material. Thus, the zone above the groundwater should not be called the unsaturated zone. By the same token, it is not correct to refer blankly to the groundwater region as the saturated zone because air bubbles may remain entrapped in this zone and prevent complete saturation. For these reasons, *the terms vadose zone and groundwater zone rather than unsaturated and saturated zone are used.* Atmospheric pressure is the dividing line between the two, with the pressure of vadose water being below atmospheric pressure and that of groundwater above atmospheric pressure.

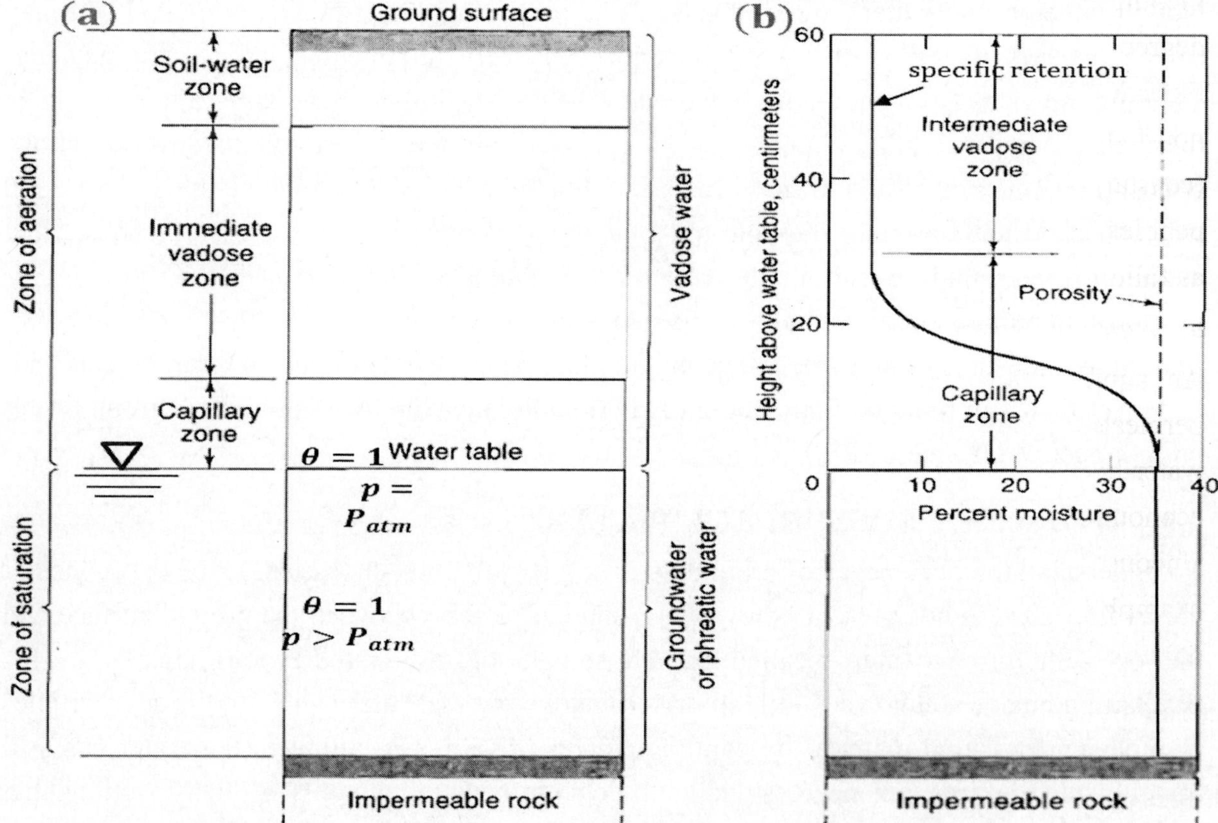

Figure 5.1: Classification and distribution of subsurface water.

5.4 TYPES OF SUBSURFACE FORMATION

The substance found beneath the surface is typically referred to as a subsurface deposit or subsurface formation in subsurface hydrology. Soil and geologic formations are the two main categories into which subsurface formations are classified. As it is widely understood, the weathering of rocks forms soil. Geologic formations, on the other hand, are classified as unconsolidated, semi-consolidated, or consolidated. Meanwhile, rocks with closely spaced grains that are created by cementation, compaction, and re-crystallization are known as consolidated geologic formations. 'Hard rocks' are a substitute for them. Igneous and metamorphic rocks like granite, basalt, and schist, as well as indurated sedimentary rocks like sandstone, shale, and limestone, are examples of consolidated geologic formations. Semi-consolidated gcologic formations are sedimentary rocks wherein the induration process is incomplete, and the primary porosity (inter-granular porosity) is preserved to a varying degree. Sandstone is the most productive of semi-consolidated sedimentary rocks because, like sand, it has a very high primary porosity in the early stages of cementation. Conversely, non-indurated colluvium, alluvial, Aeolian (wind-borne sediments), lacustrine, marine (coastal), and glacial deposits make up unconsolidated geologic formations. There are pebbles, gravel, clay, sand, and silt in these deposits. The subsurface formations are classified as follows:

a. Aquifer

An aquifer is a formation or a geological structure which contains sufficient saturated permeable materials to yield a significant quantity of water to a well and springs. Or the water-bearing geologic formations or strata which yield a significant quantity of water for economic extraction from wells. This indicates the capacity to transfer and store water. Unconsolidated sedimentary formations like gravel and sand are typical and excellent examples of aquifers. Fractured igneous and metamorphic rocks and carbonate rocks with solution cavities also form good aquifers.

b. Aquiclude

It is a geological feature that essentially prohibits water from flowing through it. This formation can only store water but cannot transmit a significant amount. It may be considered as closed to water movement even though it may contain large amounts of water due to its high porosity. Clay is an example of an aquiclude.

c. Aquitard

It is a saturated formation but poorly permeable stratum that impedes groundwater movement and does not yield water free to the well that may transmit appreciable water to or from the adjacent aquifer. This formation through which only seepage is possible, and thus, the water yield is insignificant compared to an aquifer. It is partly permeable. An aquitard is an example of sandy clay. An aquitard with an appreciable quantity of water may leak into an aquifer below it.

d. Aquifuge

A relative impermeable geological formation neither contains nor transmits water. It is a formation which is neither porous nor permeable,, e.g. massive igneous, solid granite or metamorphic rocks.

5.5 AQUIFER TYPES AND THEIR HYDRAULIC PROPERTIES

The lateral continuity and vertical boundaries are often not well defined. The aquifers may be either localized in nature or may extend over a distance of several hundred kilometers. Aquifers in the Ganga Basin in India; the Great Australian Artesian Basin, and the Sahara (Nubian sandstone) have been traced over distances of several hundred kilometers. Aquifers are generally aerially extensive and may be underlain or overlain by a confined bed which may be defined as a relatively impermeable material strati-graphically adjacent to one or more aquifers. Based on the hydraulic characteristics, the aquifers can be classified into the following four types:

i. Confined Aquifers

These are also termed as artesian aquifers. A confined aquifer is overlain and underlain by a confining layer (aquiclude or aquifuge). Water in the confined aquifers occurs under pressure that is greater than atmospheric pressure. The piezometric surface, which is an imaginary surface to which the water will rise in wells penetrating the confined aquifer, should lie above the top of the aquifer, i.e. above the base of the overlying confining layer. A particular aquifer at one place may be a confined aquifer, while at another place, it may behave as an unconfined aquifer where the water level falls below the base of the overlying confining layer. Similar to the above, over time, an aquifer at a specific location may transform from confined to unconfined.

ii. Semi-confined Aquifers

In nature, truly confined aquifers are rare because the confining layers are not exactly impervious. In semi-confined or leaky confined aquifers, the aquifer is overlain or underlain by an aquitard or semi-pervious layer through which vertical leakage takes place due to head difference. Generally, the permeability of the semi-confining layer is small so that any horizontal component of flow in it can be neglected.

iii. Unconfined Aquifers

This is the topmost water-bearing stratum, having no confined impermeable cover burden laying over it, but it has a confining layer at its bottom. It is normally partly saturated with water, and the upper surface of saturation is termed the water table, which is under atmospheric pressure. Water in unconfined aquifer is called unconfined or phreatic water. In unconfined aquifer, the gravity drainage is often not instantaneous, and therefore, there is some time lag in the lowering of the water table and the drainage of the aquifer. The delay effect is greater in the fine-grained aquifers than in the coarse-grained aquifers.

iv. Semi Unconfined Aquifers

These aquifers exhibit characteristics between semi-confined and unconfined aquifers as the permeability of the fine-grained overlying layers is greater than that of a semi-confined aquifer, and the horizontal flow component in it cannot be neglected.

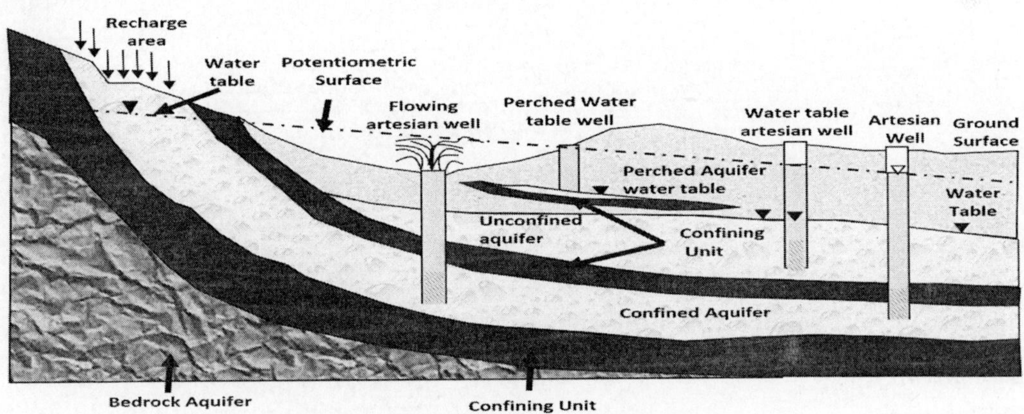

Figure 5.2: Type of Aquifers.

The distinction between different types of aquifers is sometimes difficult. The subsurface lithology, water levels and other hydrological parameters of both the aquifers and confining layers should be studied carefully to ascertain the nature of the aquifers. The distinction between different types of aquifers is shown in Figure 5.2. The important hydraulic properties of aquifers and confining layers are:

a. Porosity (n)

It is an important hydrological characteristic of a formation. It is the major criteria for the occurrence of groundwater. The portion of rock or soil not occupied by a solid mineral matter can be occupied by Groundwater. The porosity as shown in Figure 5.3 is the measure of the interstices present in a formation. Porosity is defined as the ratio of the volume of voids (V_v) to its total volume (V_0) including volume of void and volume of solid and can be expressed either as a percentage or as decimal fraction (Equation 5.1 to 5.3).

In percentage form, it is the percentage of voids present in each volume of aggregate. It depends on the shape, packing, and degree of sorting of the component grains in each material. Generally, ranges from 0 to 50% for most of the rock materials.

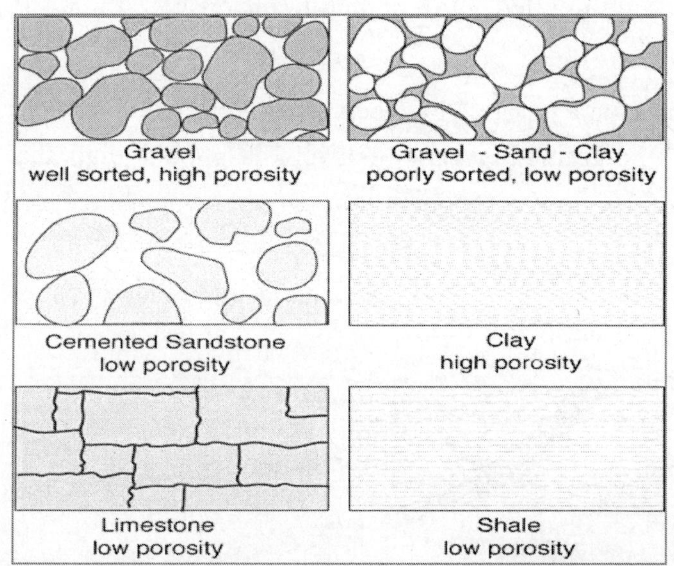

Figure 5.3: Pore space in various geologic formations.

Mathematically,

$$Porosity\ (n) = \frac{Total\ volume\ of\ voids\ in\ soil\ (Vv)}{Total\ volume\ of\ soil\ (V)}$$

$$n = \frac{Vv}{v} * 100\%$$

$$n = \frac{V-Vs}{V} * 100\% \qquad \text{.............(5.1)}$$

Also, porosity can be expressed as

$$n = \frac{\rho_m - \rho_d}{\rho_m} = 1 - \frac{\rho_d}{\rho_m} \qquad \text{............. (5.2)}$$

where,

ρ_m = density of mineral particles (grain particles),

ρ_d = bulk density.

The porosity of an aquifer is the sum of specific retention (S_r) and specific yield (S_y). Mathematically:

Porosity = Specific Yield + Specific Retention.

$$n = S_y + S_r \qquad \text{.............(5.3)}$$

- **Specific retention**: a measure of the volume of water which is retained by the aquifer material against gravity on account of cohesive and intergranular forces.
- **Specific yield**: The water-yielding capacity is also termed as effective porosity. Specific yield is expressed quantitatively as the percentage of the total volume of rock occupied by the water, which can be drained out by gravity. Specific yield increases with an increase in grain size and sorting, while specific retention increases with a decrease in grain size and assortment.

***b.* Hydraulic conductivity (k)**

Hydraulic conductivity, also known as permeability, is a measure of how easily a fluid can traverse from a formation. It is defined as the quantity of flow per unit cross-sectional area under the influence of a unit hydraulic gradient. It has the dimensions of velocity (LT^{-1}) and is usually expressed in m/day.

The hydraulic conductivity depends both upon the properties of the fluid and on the properties of the aquifer. The specific permeability, k', which is the permeability of the porous medium, depends only on the property of the medium and it is independent of the fluid properties. Specific permeability is related with hydraulic conductivity by the relation given as in Equation 5.4:

$$k = k'\frac{\gamma}{\mu} \qquad \text{.............. (5.4)}$$

where,

μ= co-efficient of viscosity; and

γ= specific weight of the fluid.

Specific permeability has the dimensions of $\mathbf{L^2}$ and is usually expressed in Darcy units.

The **Darcy** is defined as follows. A porous medium is said to have a permeability of one Darcy if a single-phase fluid of one centipoise viscosity that completely fills the pore space of the medium will flow through it at a rate of 1 cm^3/sec per cm^2 of cross-sectional area under a pressure gradient of 1 atm. per cm.

1 Darcy = 0.987 x 10^{-8} cm^2.

For water at 20°C, a medium of permeability of 1 Darcy would have a hydraulic conductivity (k) of 9.613 x 10^{-4} cm/sec.

c. Transmissivity (T)

Transmissivity or coefficient of transmissibility is a hydraulic characteristic of the aquifer which was first introduced in Groundwater literature by C.V. Theis in 1935. It is defined as the rate of flow of water at the prevailing field temperature under a unit hydraulic gradient through a vertical strip of the aquifer of unit width and extends through the complete saturated thickness of the aquifer. It is, therefore, a product of the average permeability and the saturated thickness of the aquifer (Equation 5.5), i.e.

$$T = Kb \qquad \text{.............. (5.5)}$$

where, b is the thickness of the aquifer. Transmissivity has the dimensions of L^2T^{-1} and is usually expressed in m^2/day. The concept of transmissivity holds good in confined aquifers but in unconfined aquifers, as the saturated thickness of the aquifer changes with time, the T will also change accordingly.

d. Coefficient of Storage or Storability (S)

The storage coefficient of an aquifer is defined as the volume of water that a vertical column of the aquifer of unit cross-sectional area releases from storage or takes into storage as the average head within this column declines or rises a unit distance. or simply can be defined as the volume of water an aquifer would release from or take into storage per unit surface area of the aquifer for a unit change in head. It is dimensionless. In artesian aquifers, where water is released from or taken into storage entirely due to the compressibility of aquifer and water, the storage co-efficient S is given in Equation 5.6 below.

$$\mathbf{S = bS_s} \qquad \text{.............. (5.6)}$$

where,

b = thickness of the aquifer and

S_s = specific storage, which has the dimensions of L^{-1}.

The quantity of water that an aquifer's unit volume releases from storage as an outcome of aquifer compression and expansion under a unit decline in the average head can be defined as specific storage.

The storage coefficient in confined aquifers has the order of magnitude of 10^{-3} to 10^{-6}. The storage coefficient S_w for a water table aquifer is given by Equation 5.7 as follows.

$$S_w = S_y + bS_s \qquad \text{..............(5.7)}$$

where, b= height of water table above the base of the free aquifer, and

S_y= specific yield of the aquifer.

Usually $S_y \geq b\ Ss$, thus S_w for all practical purposes be regarded as the specific yield. The volume of water that a rock or soil will yield by gravity divided by its own volume is known as the specific yield. Alternatively, it is a very close approximation of effective porosity. The storage coefficient in unconfined aquifers, S_w, ranges from 0.05 to 0.30. In a confined aquifer, the storage coefficient depends on the compressibility of the aquifer and the expansion of water. Since an unconfined aquifer is not bounded by confining layers, the specific yield or storage coefficient does not depend upon the compressibility of either the aquifer or the fluid. The specific yield for all practical purposes is the same as effective porosity or drainable porosity. Both S and S_y are important hydrological properties, and their accurate determination is important for Groundwater balance studies.

e. Hydraulic Diffusivity (D)

Hydraulic diffusivity is defined as the ratio of transmissivity (T) and storability (S). Diffusivity has the dimensions of L^2T^{-1} and is generally expressed in m^2/day. For unconfined conditions, the hydraulic diffusivity term is directly proportional to the transmissivity of the aquifer, obtained as the product of the hydraulic conductivity of the water bearing material and the average saturated thickness b_{av} of the aquifer and is inversely proportional to the storage coefficient. In an unconfined aquifer, transmissivity can be expressed as Equation 5.8 as follows:

$$T = kb_{av} \qquad \text{.............(5.8)}$$

where, b_{av}, = average saturated thickness of the unconfined aquifer.

f. Leakage Coefficient or Leakance (1/C)

It is the property of semi confining layer. It is the ratio of the vertical permeability of semi-confining layer to its thickness i.e. k/b. It has dimensions of T^{-1}.

g. Hydraulic Resistance (C)

It is a characteristic of confining layers of leaky aquifers and is also known as reciprocal leakage coefficient or resistance against vertical flow. It is equal to b/k. It characterizes tile resistance of the semi-pervious layer to upward or downward leakage. It has the dimensions of time. If hydraulic resistance C = ∞, the aquifer is confined.

h. Leakage Factor (L)

The leakage factor $L = \sqrt{T.C}$ determines the distribution of the leakage into the leaky (semi confined) aquifer. High value of L indicates a great resistance of semi-pervious strata to flow. Factor L has length dimensions and is usually expressed in meters.

i. Delay Index (-1/α)

Is a measure of the delayed drainage of an unconfined aquifer and has the dimension of time (T).

j. Drainage Factor (B)

The drainage factor $B = \sqrt{kb/\alpha S_y}$ is a property of unconfined aquifer. Large values of B indicate a fast drainage. The drainage factor has length dimensions (L) and is expressed in meters.

k. Specific Capacity

It is a measure of both the effectiveness of a well and of the aquifer characteristics (T and S). It is defined as the ratio of the pumping rate and the drawdown, and it is often reported in liters per minute per meter of drawdown for a certain pumping interval.

I. Specific Capacity Index

It is a measure of the formation characteristics. It is obtained by dividing the specific capacity by the saturated thickness of the aquifer. The specific capacity index values are of use in determining the relative productivity of different units in a multi-unit aquifer and also in predicting well yield from a given thickness of the aquifer. Unit area-specific capacity values are obtained by dividing the specific capacity by the cross-sectional area of the well. The specific capacity can also be divided by $2\pi r_w b$ (where r_w, is the radius of the well) to account for variation in the well radius and depth.

5.6 DARCY'S LAW FOR GROUNDWATER VELOCITY

Darcy based on experiments through sand columns, found that discharge is directly proportional to the head loss (ΔH) and area of cross-section (A) of the soil and inversely proportional to the length of the soil sample (L) (See Equation 5.9).

$$Q \propto \frac{\Delta H}{L} A \qquad \text{.......... (5.9)}$$

$$Q \propto iA$$

$$Q = KiA$$

$$\frac{Q}{A} = V = Ki \qquad \text{............ (5.9a)}$$

where,

K = proportionality constant, (changing with the type of soil and hence represent a property of soil, called permeability or coefficient of permeability.

A = area of the soil medium, through which the flow is taking place.

Thus, Darcy's law states that; velocity is directly proportional to the hydraulic gradient. It may be noted that Darcy's velocity is a fictitious velocity and not the actual Groundwater flow velocity through soil medium. Since the actual flow velocity $\mathbf{V_a}$ occurs through the voids of cross-sectional area V_v and not through the actual area A. Therefore, equation 5.9 can be rewritten as (Equation 5.9b) below:

$$A_V * V_a = A * V$$

$$V = \frac{A_V}{A} V_a \qquad \text{............ (5.9b)}$$

When A is large in comparison A_v, we can satisfy that the ratio of the area of the voids to the total area (A) is the same as the ratio of the voids (V_v) to the total volume V (Equation 5.10) as;

$$\frac{A_V}{A} \approx \frac{V_V}{V} = \eta \ (Porosity)$$

$$V = \frac{V_V}{V} = \eta * V_a \qquad \text{............. (5.10)}$$

Knowing the value of V from Darcy's equation, and dividing it by porosity η, the actual velocity of flow V_a through the soil can be worked out.

Example 5.1: In an aquifer whose area is 100 ha, the water table has dropped by 3.0 m. Assuming porosity and specific retention of the aquifer material to be 30% and 10%, respectively. Determine the specific yield of the aquifer and the change in the groundwater storage.

Solution:

Porosity = ***Specific Yield* + *Specific retention*.**

$n = S_y + S_r$

$S_y = n - S_r = 30\text{-}10 = 20\%$

Change in groundwater storage = ***Area* × *Depth fluctuation of GWT* × *Specific Yield***

$$\Delta GWS = A \times \Delta H \times S_y = 100 \times 10^4 \times 3 \times 0.2$$

$$= 60 \times 10^4 m^3$$

Example 5.2: In a certain place in Andhra Pradesh, the average thickness of the confined aquifer is 30m and extends over an area of 800km^2. The piezometric surface fluctuates annually from 19m to 9m above the top of aquifer. Assuming a storage coefficient of 0.0008, what groundwater storage can be expected annually? Assuming the average well yield of 30m^3/hr. and about 200 days of pumping in a year. How many wells can be drilled in the area?

Solution

$$\Delta GWS = A \times \Delta H \times S_y = 800 \times 10^6 \times (19 - 9) \times 0.0008$$

$$= 6.4 \times 10^6 \ \mathbf{m^3} \text{ or } \mathbf{6.4\ Mm^3}$$

Annual draft (V) = $30m^3/hr \times 200 \times 24hr = 144 \times 10^5 m^3 = 0.144 Mm^3$

$$Number\ of\ well\ that\ can\ be\ drilled\ in\ area = \frac{\Delta GWS}{Annual\ Draft\ volume}$$

$$= \frac{6.4}{0.144} = 44.5 \text{ say } \mathbf{44\ wells.}$$

5.7 EMPIRICAL FORMULAS FOR ESTIMATING GROUNDWATER VELOCITY

Before Darcy came into the picture, certain empirical formulae, based on the experimental results, was being used to find out the velocity of the Groundwater flow. The formulae which were in common use are given below:

a. **Slichter's formula (Equation 5.11):** According to which

$$V_a = K'i\frac{D_{10}^2}{\mu} \qquad \text{.............. (5.11)}$$

where,

V_a= Velocity of Groundwater flow in m/day

K' = A constant whose value is about 400

i = The slope of the hydraulic gradient line.

D_{10} = The effective size of the particles in mm (i.e., the hypothetical size which is larger than 10% of the particles in the sample, i.e., only 10% of the particles will pass through this size).

μ = Viscosity of water depending on temperature

b. **Hazen's formula.** Formula (Equation 5.12) in M.K.S. system is

$$V_a = K''i\frac{D_{10}^2}{60} \times (1.8T + 42) \qquad \textbf{.............. (5.12)}$$

where,

V_a=velocity of Groundwater flow in $\frac{m}{day}$

T = Temperature in ^0C

K''=A constant whose value is about 1000

Example 5.3: Find out the velocities of the Groundwater flow with the following data using Slichter's and Hazen's constants as 400 and 800, respectively; viscosity coefficient of water at Groundwater temperature of 10^0C=1; Effective size of the particles in the aquifer = 0.1 mm; Hydraulic gradient= 1 in 80.

Solution

a. Using Slichter's formula

$$V_a = K'i\frac{D_{10}^2}{\mu}$$

$$= 400 \times (\frac{1}{80})\frac{0.1^2}{1}$$

$$= \boldsymbol{0.05}\frac{m}{day}$$

b. Using Hazen's formula

$$V_a = K''i\frac{D_{10}^2}{60} \times (1.8T + 42)$$

$$= 800 \times (\frac{1}{80})\frac{0.1^2}{60} \times (1.8 \times 10 + 42)$$

$$= \frac{0.1\,m}{day}$$

5.7.1 Field Measurement of the Velocity of Groundwater

The actual flow velocity of Groundwater (v_a) can be computed by using above methods, or it can be measured in the field by using chemical tracers or electrical resistivity methods. The known time (t) taken by a chemical tracer, such as dye, to a given known distance (S) between the two observation wells will directly indicate the ground flow velocity as equal to $V_a = \frac{S}{t}$ Such a test, in which a chemical tracer is introduced into one observation well, and its entry into the second observation well is observed, will also help to compute K, since $V = K\frac{H_L}{S}$

where,

H_L= Difference of water levels in the two observation wells and

S = Distance between two observation wells.

$V = nV_a \frac{H_L}{S}$ and

$K = nV_a \frac{S}{H_L}$ .... **(5.13)**

Example 5.4: In a field test, a time of 6 hours was required for a tracer to travel through an aquifer from one well to another. The observation wells were 42 m apart and the difference in their water levels was found to be 0.42 m. Compute (i) the discharge velocity (ii) the coefficient of permeability (K). Given the porosity of the soil medium as 20% (iii) Also compute the value of the coefficient of intrinsic permeability for the aquifer in Darcy's if the viscosity of water v = cm^2/sec.

Solution:

Velocity of Groundwater $V_a = \frac{S}{t} = \frac{42\ m}{6\ h} \frac{cm}{s}$

$$= \frac{4200}{6*60*60} \frac{cm}{s} = 0.196 \frac{cm}{s}$$

Discharge velocity $V = nV_a = 0.2 * 0.194 \frac{cm}{s} = \frac{0.0388\ cm}{s}$

Hydraulic gradient between the wells $i = \frac{H_L}{S} = = \frac{0.42\ m}{42\ m} = \frac{1}{100}$

Using $V = Ki = K\frac{1}{100}$

$$0.0388 = \frac{K}{100}$$

$$K = 3.88 \frac{cm}{s}$$

Intrinsic permeability coefficient

$$K_O = \frac{K_V}{g} = \frac{3.88 \frac{cm}{sec} \times 0.01 \frac{cm^2}{sec}}{\frac{981\ cm}{sec^2}}$$

$$= \frac{3.96 \times 10^{-5}\ cm}{sec^2}$$

Since $9.87 \times 10^{-9} cm^2 = 1\ Darcy$, we have

K_O In Darcy $= \frac{3.96*10^{-5}}{9.87*10^{-10}}$ Darcy's = **4007 Darcy's**

5.8 GROUNDWATER HYDRAULICS

5.8.1 Analytical methods for estimation of Yield from wells and tube wells

a. Thiem's Equilibrium formulas for unconfined aquifer under steady-state conditions

Various assumptions made in its derivations are:

i. The aquifer is homogeneous, isotropic, and of infinite and areal extent so its coefficient of transmissivity or permeability is constant everywhere.

ii. The well has been sunk through the full depth of the aquifer, and it receives water from the entire thickness of the aquifer.

iii. Pumping has continued for a sufficient time at a uniform rate so that the equilibrium stage or steady flow conditions have been reached.

iv. Flow lines are radial and horizontal, and flow is laminar.

v. The inclination of the water surface is small, so its tangent can be used in place of the sine for the hydraulic gradient in Darcy's equation.

The following figure (Figure 5.4) describes the unconfined aquifers. Also, assuming that the inclination of the water surface is small, so that the tangent can be used in place of sine for the hydraulic gradient in Darcy's law, one has

$$I = \frac{dh}{dl} \approx \frac{dh}{dr}$$

$$I = \frac{dh}{dr}$$

Small element of drawdown curve.

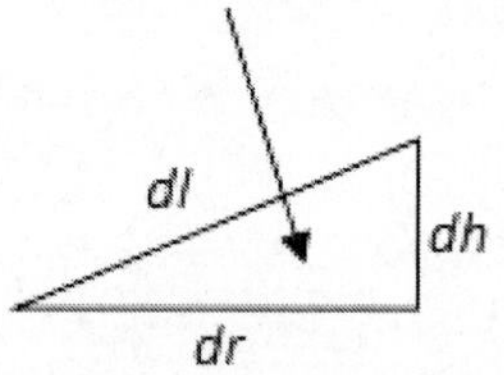

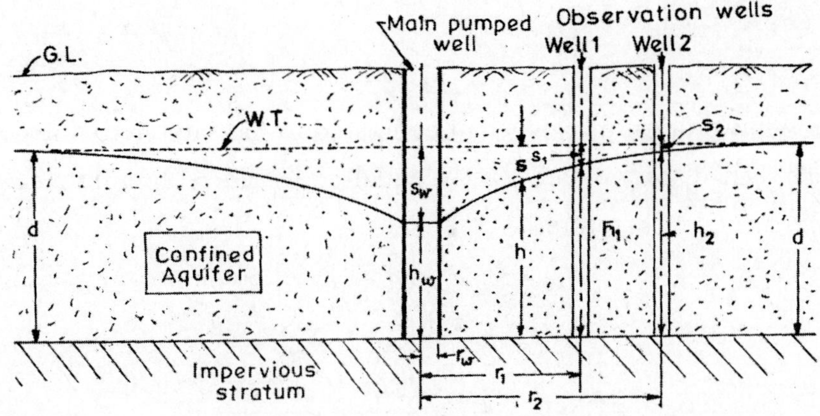

Figure 5.4: Unconfined aquifer case for derivation of Thiem's formula.

The area of flow (A) is equal to 2πrh, assuming that water flows through the full height of the aquifer, and that the flow is radial and horizontal.

$$\boldsymbol{Q = KiA = K\frac{dh}{dr}2\pi rh}$$

$$\boldsymbol{\frac{dr}{r} = \frac{2\pi Khdh}{Q}}$$

where,

Q = constant when steady conditions have reached.

K = soil permeability

Integrating between the limits r_1 to r_2, and h_1 to h_2, we get:

$$\int_{r_1}^{r_2}\frac{dr}{r} = \int_{h_1}^{h_1}\frac{2\pi K}{Q}hdh$$

Therefore: $log_e\left(\frac{r_2}{r_1}\right) = \frac{2\pi K}{Q}\left[\frac{{h_2}^2-{h_1}^2}{2}\right]$

$$Qlog_e\left(\frac{r_2}{r_1}\right) = \pi K({h_2}^2 - {h_1}^2)$$

$$K = \frac{Qlog_e\left(\frac{r_2}{r_1}\right)}{\pi({h_2}^2-{h_1}^2)} \quad or \quad Q = \frac{\pi K({h_2}^2-{h_1}^2)}{log_e\left(\frac{r_2}{r_1}\right)} = \frac{\pi K({h_2}^2-{h_1}^2)}{2.3log_{10}\left(\frac{r_2}{r_1}\right)}$$

This is important Thiem's formula which can be further simplified, if required, as follows

$$(h_2{}^2 - h_1{}^2) = (h_2 + h_1)(h_2 - h_1)$$

$$(h_2 - h_1) = (S_1 - S_2)$$

If the amount of drawdown is small compared to the saturated thickness of the water bearing material, then h_1 and h_2 are nearly equal and each is approximately equal to this saturated thickness say d. Therefore,

$$(h_1 + h_2) \approx (d + d) = 2d$$

$$h_2{}^2 - h_1{}^2 \approx (S_2 - S_1)2d = 2d(S_2 - S_1)$$

$$Then, \quad Q = \frac{2\pi Kd(S_2 - S_1)}{2.3log_{10}\left(\frac{r_2}{r_1}\right)}$$

Introducing the coefficient of transmissivity T in place of k, i.e. $T = Kd$, then

$$Q = \frac{2\pi T(S_2 - S_1)}{2.3log_{10}\left(\frac{r_2}{r_1}\right)} \qquad \text{..............(5.14)}$$

Equation 5.14 is the form of Thiem's formula for unconfined aquifers.

b. **Thiem's Equation for Confined Aquifers:**

In a confined aquifer, the flow is actually radial and horizontal and, therefore, it has not to be assumed as such, as was done in the unconfined case. Rest of the assumptions remain the same and hold good in this case also. The following figure (Figure 5.5) represents the drawdown curve in a confined aquifer. In this case, the discharge equation is

$$Q = KiA = K\frac{dh}{dr}2\pi rH$$

As water is drawn from the cylinder of radius r and height H, where H = height of the confined aquifer. Therefore,

$$Q = 2\pi KHr\frac{dh}{dr}$$

$$Or$$

$$\frac{dr}{r} = \frac{2\pi KH}{Q}dh$$

Integrating between the limits of r_1 and r_2 for r, and h_1 to h_2 for h.

$$\int_{r_1}^{r_2} \frac{dr}{r} = \frac{2\pi K}{Q} \int_{h_1}^{h_2} dh$$

It gives; $log_e\left(\frac{r_2}{r_1}\right) = \frac{2\pi KH}{Q}(h_2 - h_1)$

Or $\quad Q = \frac{2\pi KH(h_2 - h_1)}{log_e\left(\frac{r_2}{r_1}\right)} = \frac{2\pi KH(h_2 - h_1)}{2.3 log_{10}\left(\frac{r_2}{r_1}\right)}$

But $\quad (h_2 - h_1) = (S_1 - S_2)$

$Therefore$: $\quad Q = \frac{2\pi KH(S_1 - S_2)}{2.3 log_{10}\left(\frac{r_2}{r_1}\right)} \; Or \; Q = \frac{2\pi T(S_1 - S_2)}{2.3 log_{10}\left(\frac{r_2}{r_1}\right)}$ .... **(5.15)**

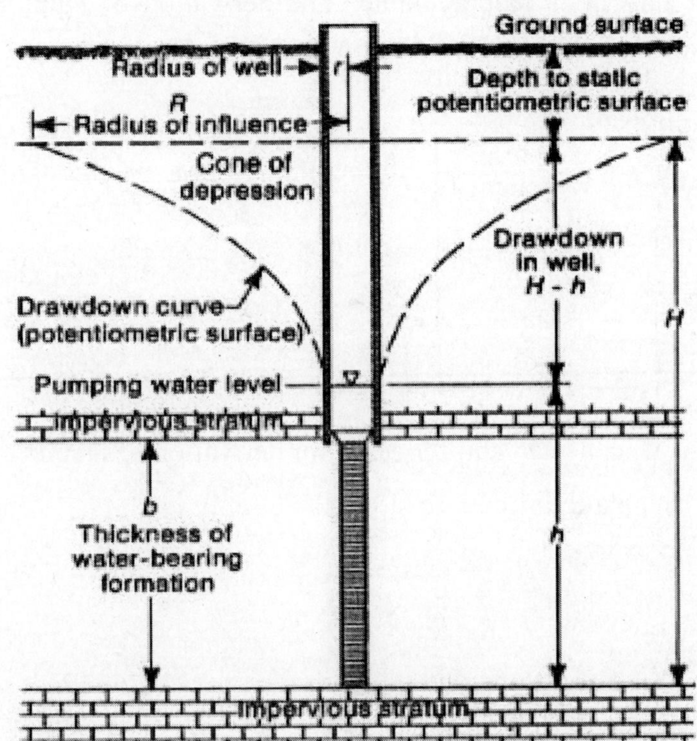

Figure 5.5: Drawdown curve in a confined aquifer.

Limitation of Thiem's formulae: Various assumptions have been made in the derivation of Thiem's formulae above. But, in actual practice, none of these conditions may not be satisfied. Further, it is very difficult to assess the effect of these assumptions. Despite the various limiting assumptions, Thiem's formulae are widely used in Groundwater problems, and many of its assumptions are taken care by appropriate adjustments.

5.8.2 Dupuit's formulae for estimating well yield under steady state condition

In Dupuit's formulae, no observation wells are required. The main well is pumped out so as to get sufficient drawdown, and then the rate of pumping is so adjusted as to establish equilibrium conditions.

All the assumptions that have been made in Thiem's formulae also hold good in Dupuit's formulae. The only difference is that the integration was carried out between the limits of r_1 and r_2 is now carried out between the limits r_w and R, where r_w is the radius of the pumped well and R is the radius of influence. The radius of influence is the distance up to the point where drawdown is zero or is inappreciable. The derivation of Dupuit's formulae for an unconfined and confined aquifer is shown below in Figure 5.6.

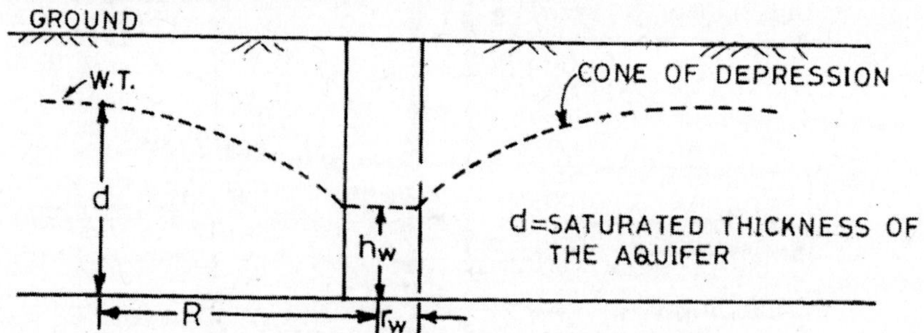

Figure 5.6: Unconfined aquifer cases for derivation of Dupuit's formula.

a. Dupuit's formula for unconfined aquifers:

As per Darcy's law; $Q = KiA$

$$Q = K\frac{dh}{dr}2\pi rh$$

$$Or$$

$$\frac{dr}{r} = \frac{2\pi K}{Q}hdh$$

Integrating between the limits r_w and R for r and $\boldsymbol{h_w}$ and $\boldsymbol{d}$ for h, we get

$$\int_{r_w}^{R}\frac{dr}{r} = \frac{2\pi K}{Q}\int_{h_w}^{d}hdh$$

$$log_e\left(\frac{R}{r_w}\right) = \frac{\pi K}{Q}[(d^2 - h_w{}^2)]$$

$$\left(\frac{R}{r_w}\right) = \frac{\pi K}{Q}[(d^2 - h_w{}^2)]$$

or $$Q = \frac{\pi K(d^2 - h_w{}^2)}{2.3031\ log_{10}\left(\frac{R}{r_w}\right)} \quad \text{.............. (5.16)}$$

Since the value of R is not easily assessable, various arbitrary values have been assigned to R by various investigators. Slitcher assumed it as 150 m, and Tolman assumed it as 300 m. But a more realistic picture is obtained, when R is taken as

$R \propto Q$ or $R = CQ$; where C is a constant.

Putting $R = CQ$ in above equation (5.15), we get

$$Q = \frac{\pi K(d^2 - h_w{}^2)}{2.303\ log_{10}\left(\frac{CQ}{r_w}\right)} \quad \text{.............. (5.16a)}$$

Q can be determined by Hit and Trial method.

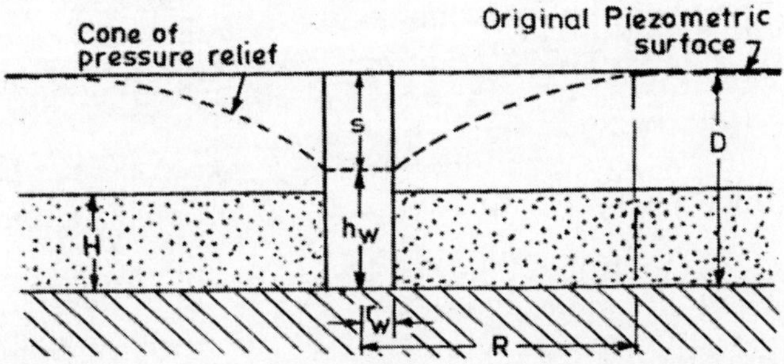

Figure 5.7: Confined aquifer case for derivation of Dupuit's formula.

b. Dupuit's formula for confined aquifers: (refer figure (5.7))

Integrating between the limits r_w and R, $\boldsymbol{h_w}$ and $\boldsymbol{D}$ for h we get

$$\int_{r_w}^{R} \frac{dr}{r} = \frac{2\pi KH}{Q}\int_{h_w}^{D} dh$$

$$log_e\left(\frac{R}{r_w}\right) = \frac{2\pi KH}{Q}[(D - h_w)]$$

$$Q = \frac{2\pi KH(D - h_w)}{log_e\left(\frac{R}{r_w}\right)} = \frac{2\pi KH(D - h_w)}{2.303 log_{10}\left(\frac{R}{r_w}\right)}$$

$$or \quad Q = \frac{2\pi KHs}{log_e\left(\frac{R}{r_w}\right)} \qquad \text{............ (5.17)}$$

where,

H = Total height of confined aquifer.

h_w = Artesian pressure in the well

r_w = Radius of the well

D= Initial artesian pressure at the bottom of the aquifer or the initial height of the piezometric surface from the bottom of the well

S= Drawdown = (D- h_w)

5.9 PARTIAL PENETRATION OF AN AQUIFER BY A WELL: FROM EQUATION

Equation (5.18); for unconfined or gravity wells

$$Q = \frac{\pi K(d^2 - h_w{}^2)}{2.303\ log_{10}\left(\frac{R}{r_w}\right)} \qquad \text{............ (5.18)}$$

Equation (5.18a); for confined wells

$$Q = \frac{2\pi KH(D - h_w)}{2.303 log_{10}\left(\frac{R}{r_w}\right)} \qquad \text{............ (5.18a)}$$

Equations 5.18 and 5.18a have been derived for gravity and artesian wells, respectively. But, if a well does not penetrate up to the bottom of the aquifer, these formulas will not be applicable, as the nature of the flow will become three-dimensional. It will not only be radial but will also have an upward component, as shown in Figure 5.8.

The measured drawdown in the zone influenced by the upward vertical flow component will be more in the case of a partially penetrating well than in a fully penetrating well for the same discharge. The discharge obtained by the fully penetrating well will of course be more than a partially penetrating well.

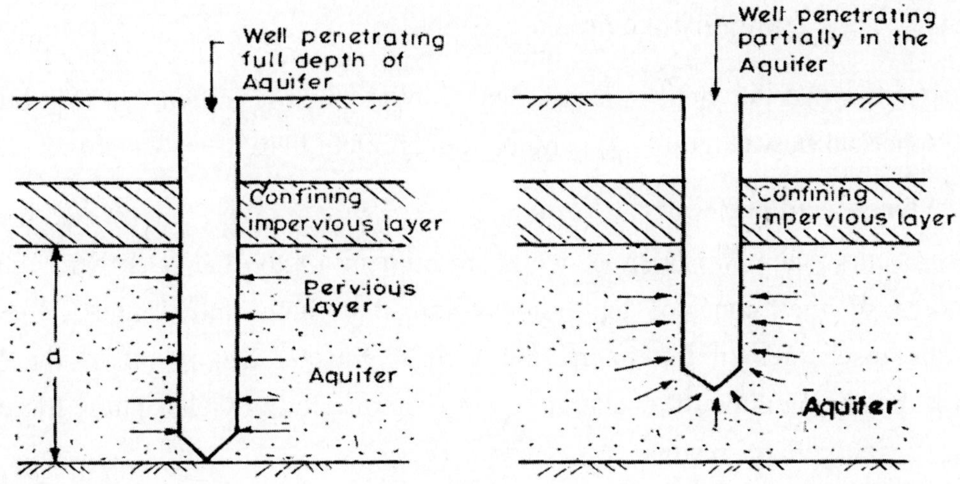

Figure 5.8: Flow lines in a well fully and partially penetrating the aquifer.

Kozeny has given a correction factor. According to him, the discharge Q_p through such a well is given as follows:

Whereas in the case of a simple radial flow in a fully penetrating artesian well, the discharge is given by the equation 5.19 as given below:

$$Q = \frac{2\pi KH(d - h_w)}{2.303 log_{10}\left(\frac{R}{r_w}\right)}$$

$$\frac{Q_s}{Q} = \frac{2\pi K r_w (D - h_w)}{\left|\frac{2\pi KH(d - h_w)}{2.303 log_{10}\left(\frac{R}{r_w}\right)}\right|} = \frac{2.303 r_w log_{10}\left(\frac{R}{r_w}\right)}{H}$$

$\frac{Q_s}{Q} = 2.303\left(\frac{r_w}{H}\right) log_{10}\left(\frac{R}{r_w}\right)$ .... **(5.19)**

For example, if

r_w = 10cm = 0.1m and R = 3000 m, then

$$\frac{R}{r_w} = 30000. \quad \textit{and let,}$$

H = Thickness of the confined aquifer = 20m (say), Then

$$\frac{Q_s}{Q} = 2.303\left(\frac{0.1}{20}\right) log_{10}(30000) = 0.05 \approx \frac{1}{20}$$

This shows that the yield in a spherical flow is much less than that in a radial flow. Hence, the spherical flow (Figure 5.9) is much less efficient than the radial flow.

5.9.1 Interference among wells

Interference occurs when two or more wells are built in a way that puts them close to one another and causes their cones of depression to intersect. Such mutual interference of wells decreases the discharges of the interfering wells. Muskat has proposed the following formulas for the computation of discharges from such interfering wells. These formulas have been found to yield reliable results.

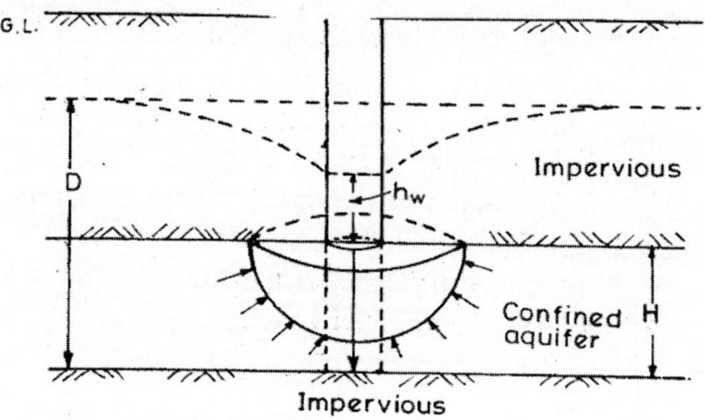

Figure 5.9: Spherical flow in a well.

a. For confined aquifers (i.e. Artesian wells)

For two artesian identical wells at a distance B apart,

$$Q_1 = Q_2 = \frac{2\pi KH(d-h_w)}{2.303log_{10}\left(\frac{R^2}{r_{w\,B}}\right)} \quad \text{........(5.20)}$$

b. Using three identical artesian wells spaced B apart in an equilateral triangle configuration,

$$Q_1 = Q_2 = Q_3 = \frac{2\pi KH(d-h_w)}{2.303log_{10}\left(\frac{R^3}{r_{w\,B^2}}\right)} \quad \text{..... (5.21)}$$

c. Using three identical artesian wells, spaced B apart in a straight line,

$$Q_1 = Q_3 = \frac{\frac{2\pi KH}{2.303}(d-h_w)log_{10}\left(\frac{B}{r_w}\right)}{log_{10}\left(\frac{R}{B}\right)log_{10}\left(\frac{B}{r_w}\right)+ log_{10}\left(\frac{B}{2r_w}\right)log_{10}\left(\frac{R}{r_w}\right)} \quad \text{.........} \ \textbf{(5.21a)}$$

and.

$$Q_2 = \frac{\frac{2\pi KH}{2.303}(d-h_w)log_{10}\left(\frac{B}{2r_w}\right)}{2log_{10}\left(\frac{R}{B}\right)log_{10}\left(\frac{B}{r_w}\right)+ log_{10}\left(\frac{B}{2r_w}\right)log_{10}\left(\frac{R}{r_w}\right)} \quad \text{..........} \ \textbf{(5.21b)}$$

where Q_1 and Q_3 are the discharges of the outer wells, and Q_2 is the discharge of the middle well.

Total discharge $= Q_1 + Q_2 + Q_3$ .... **(5.21c)**

d. For unconfined aquifers (i.e. Gravity wells)

All the three above formulas can be applied to unconfined aquifers by replacing H D by

$$\frac{d^2}{2}, and\ H, h_w\ by \frac{{h_w}^2}{2}$$

- For two identical gravity wells at a distance B apart, formula would therefore, become

$$Q_1 = Q_2 = \frac{\frac{2\pi KH}{2.303}\left(\frac{d^2}{2}-\frac{{h_w}^2}{2}\right)}{2.303log_{10}\left(\frac{R^2}{r_w\,B}\right)} \quad \text{............} \ \textbf{(5.22)}$$

$$or\ Q_1 = Q_2 = \frac{\pi K(d^2-{h_w}^2)}{2.303\ log_{10}\left(\frac{R^2}{r_w\,B}\right)} \quad \text{............} \ \textbf{(5.22a)}$$

- For three identical gravity wells at distances B apart, in a pattern of equilateral triangle, formula would be;

$$Q_1 = Q_2 = Q_3 = \frac{2\pi KH\left(\frac{d^2}{2}-\frac{{h_w}^2}{2}\right)}{2.303\ log_{10}\left(\frac{R^3}{r_w\,B^2}\right)} \quad \text{............} \ \textbf{(5.23)}$$

$$or\ Q_1 = Q_2 = Q_3 = \frac{\pi K(d^2-{h_w}^2)}{2.303\ log_{10}\left(\frac{R^3}{r_w\,B^2}\right)} \quad \text{............} \ \textbf{(5.23a)}$$

- For three identical gravity wells at distances B apart in a straight line, the formulas would become;

$$Q_1 = Q_3 = \frac{\frac{\pi K}{2.303}(d^2 - h_w{}^2) log_{10}\left(\frac{B}{r_w}\right)}{log_{10}\left(\frac{R}{B}\right) log_{10}\left(\frac{B}{r_w}\right) + log_{10}\left(\frac{B}{2r_w}\right) log_{10}\left(\frac{R}{r_w}\right)} \quad \text{..............} \ (5.24)$$

and

$$Q_2 = \frac{\frac{\pi K}{2.303}(d^2 - h_w{}^2) log_{10}\left(\frac{B}{2r_w}\right)}{2 log_{10}\left(\frac{R}{B}\right) log_{10}\left(\frac{B}{r_w}\right) + log_{10}\left(\frac{B}{2r_w}\right) log_{10}\left(\frac{R}{r_w}\right)} \quad \text{.......} \ (5.24a)$$

Example 5.5: In a constrained aquifer, three wells, each with a diameter of 10 cm, are positioned at the points of an equilateral triangle 12 meters apart. Each well has a 400 m radius of influence and a 20 m/day coefficient of permeability K. There is a 2 m drawdown in every well. The confined aquifer is 15 meters thick. Determine each well's discharge as well as the portion of the discharge that has decreased due to well interference.

Solution

r_w=10/2cm = 5cm = 0.05 m

B = 12 m, R = 400 m

H = 15 m, $K = 20 \frac{m}{day}$

Drawdown in well: $s = D - h_w = 2\,m$

$$Using\ Eqution\ Q_1 = Q_2 = Q_3 = \frac{2\pi K(d - h_w)}{2.3\ log_{10}\left(\frac{R^3}{r_w B^2}\right)}$$

$$Using\ Eqution\ Q_1 = Q_2 = Q_3 = \frac{2\pi * 20 * 15 * 2}{2.3\ log_{10}\left(\frac{400^3}{0.05 * 12^2}\right)} = 235.76 \frac{m^3}{day}$$

The discharge of each such well without interference is given by equation

$$Q = \frac{2\pi KH(D - h_w)}{2.303\ log_{10}\left(\frac{R}{r_w}\right)} = \frac{2\pi * 20 * 15 * 2}{2.303\ log_{10}\left(\frac{400}{0.05}\right)} = 419.73 \frac{m^3}{day}$$

Hence, reduction in discharge due to interference

$$= \left[\frac{419.73 - 235.76}{419.73}\right] * 100\% = 43.83\%$$

5.10 WELL LOSS AND SPECIFIC CAPACITY OF WELLS

a. **Well loss:** When water is being pumped out of an artesian well, the drawdown caused includes not only the drawdown which is given by logarithmic drawdown curve equation;

$$Q = \frac{2\pi KHs}{2.303\ log_{10}\left(\frac{R}{r_w}\right)} \quad \text{..............(5.25)}$$

But also includes a certain drawdown caused by the flow of water through the well screen and its axial movement within the well up to the pump intake.

The drawdown caused by the flow through the screen and axial movement within the well brings the water level in the well from AB to CD Figure 5.10. This vertical drawdown AC or BD is known as well loss. The magnitude of this well loss may be taken as equal to $C_2\, Q^2$, since the flow in the vicinity of the well face is turbulent. The value of C can be obtained from Figure 5.11.

The total drawdown can then be obtained by adding the two drawdowns

The first drawdown is obtained from equation

$$Q = \frac{2\pi KHs}{2.303\ log_{10}\left(\frac{R}{r_w}\right)}$$

According to which

$$s = \frac{2.303\ log_{10}\left(\frac{R}{r_w}\right)}{2\pi KH} = C_2\, Q^2 \quad \text{............. (5.26)}$$

where,

$$C_1 = \frac{2.303\ log_{10}\left(\frac{R}{r_w}\right)}{2\pi KH}$$

The total drawdown is then given by $C_1Q + C_2Q^2$

Where,

C_2Q^2 = well loss and

C_1Q = aquifer loss.

C_1 and C_2 = coefficients determined by pump test data of drawdown at various discharges. The magnitude of the well loss has an important bearing on the pump efficiency.

Abnormally high value of the well loss indicates clogging and encrustation. The well screens, needing immediate remedial action.

b. Specific capacity:

The well yield per unit of drawdown can be defined as well's specific capacity. Hence, the

$$Specific\ Capacity = \frac{Discharge\ of\ well}{Draw\ down}$$

$$= \frac{Q}{C_1 Q + C_2 Q^2} = \frac{1}{C_1 + C_2 Q}$$

$$\therefore specific\ Capacity = \frac{1}{(C_1 + C_2 Q)} \quad \text{.............. (5.27)}$$

The above equation (5.27) clearly shows that the specific capacity of the well is not constant but decreases as the discharge increases.

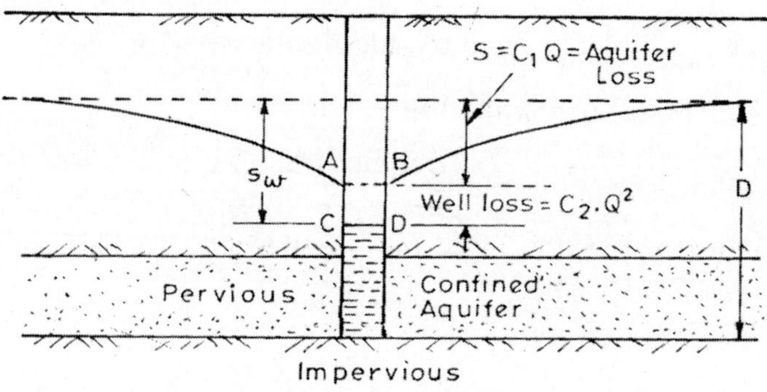

Figure 5.10: Well loss

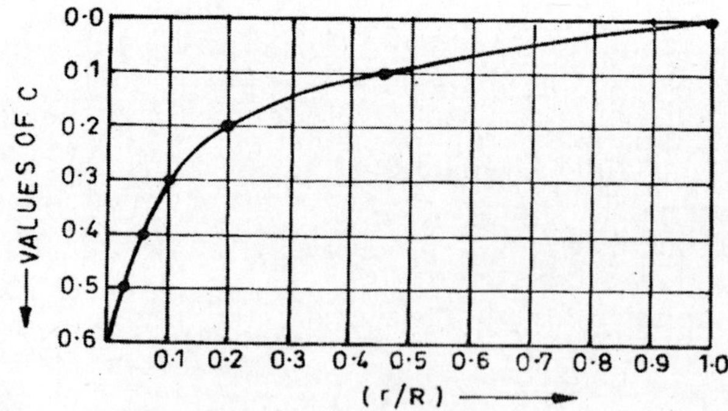

Figure 5.11: Giving values of C.

c. Efficiency of a well:

The discharge from a well is approximately proportional to the drawdown (s); whereas $s = C_1 Q$, neglecting well loss. The discharge per unit of drawdown was called specific capacity of the well. This specific capacity will be different for different well designs. A well can be operated under various drawdown settings to find the optimal drawdown discharge conditions, and a graph between drawdown and discharge may be generated, as in Figure 5.12. The curve obtained is a straight line up to a certain stage of drawdown, beyond which the drawdown increases disproportionately to yield. This place an optimum and efficient limit to the drawdown which may be allowed to be created in a well. This is generally found to be 70% of the maximum drawdown which can be created in a well.

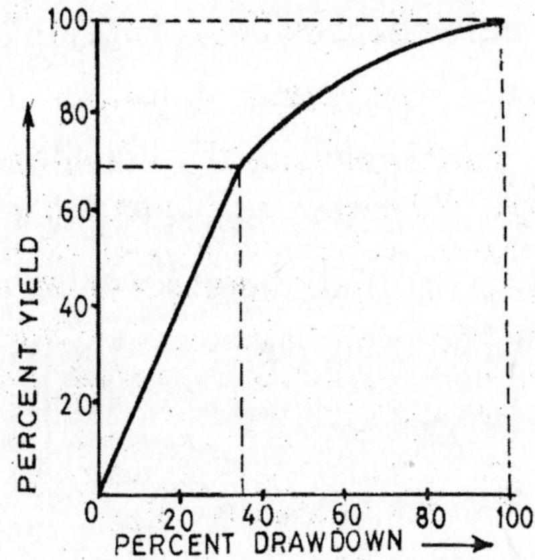

Figure 5.12: Yield-drawdown curve.

Example 5.6: A pumping test was conducted in a medium-sized sand and gravel area down to a clay bed's depth of 15 meters. The normal Groundwater level was at surface. Observation holes were located at distances of 3m and 7.5 m from the pumping well. At a discharge of 3.6 liters/sec. from the pumping well, a steady state was attained in about 24 hrs. The drawdown at 3 m was 1.65 m and at 7.5 m was 0.36 m. compute the coefficient of permeability of the soil.

Solution:

$$Q = 3.6\ \frac{litres}{sec}$$

$$r_1 = 3m;\ S_1 = 1.65m;\ h_1 = 15 - 1.65 = 13.35\ m$$

$r_2 = 7.5m;\ S_2 = 0.36\,m;\ h_2 = 15 - 0.36 = 14.64\,m$

Using equation

$$Q = \frac{\pi K(h_2{}^2 - h_1{}^2)}{2.303\ log_{10}\left(\frac{r_2}{r_1}\right)}$$

$$K = \frac{2.303\ log_{10}\left(\frac{r_2}{r_1}\right) * Q}{\pi(h_2{}^2 - h_1{}^2)} = \frac{2.303\ log_{10}\left(\frac{7.5}{3}\right) * \frac{3.6}{1000}}{\pi(14.64^2 - 13.35^2)}$$

$$K = 2.91 * 10^{-5} * 100\frac{cm}{sec}$$

$$K = 0.00219\frac{cm}{sec}$$

Example 5.7: The following measurements were taken simultaneously in a nearby test well with a 60 cm diameter well that is being pumped at a rate of 1360 liters per second. At a distance of 6 m from the well being pumped, the drawdown was 6 m, and at 15 m, the drawdown was 1.5 m. The well's bottom is 90 meters below the ground's water table. (a) Find out the coefficient of permeability. (b) If all the observation points were on Dupuit's curve, what was the drawdown in the well during pumping? (c) What is the specific capacity of a well? (d) What is the maximum rate at which water can be drawn from this well?

Solution

$$Q = \frac{\pi K(h_2{}^2 - h_1{}^2)}{2.303\ log_{10}\left(\frac{r_2}{r_1}\right)}$$

$r_1 = 6m;\ S_1 = 6.0m;\ h_1 = 90 - 6 = 84\,m$

$r_2 = 15m;\ S_2 = 15\,m;\ h_2 = 90 - 15 = 88.5\,m$

$$d = 90m, Q = 1360\ \frac{litres}{min}$$

a. $$K = \frac{2.303\ log_{10}\left(\frac{15}{6}\right) \times 1.36}{\pi(88.5^2 - 84^2)}\frac{m}{min}$$

$$K = 0.511 \times 10^{-3}\frac{m}{min}$$

$b. \quad r_w = 0.3m;\ r_2 = 15\ m; h_2 = 88.5\ m;\ h_w = ?$

$$Using\ the\ equation:\ Q = \frac{\pi K({h_2}^2 - {h_1}^2)}{2.303\ log_{10}\left(\frac{R}{r_w}\right)}$$

$$Q = \frac{\pi K({h_2}^2 - {h_w}^2)}{2.303\ log_{10}\left(\frac{R}{r_w}\right)}$$

$${h_w}^2 = {h_2}^2 - \frac{2.303Q\ log_{10}\left(\frac{R}{r_w}\right)}{\pi K}$$

$$= 88.5^2 - \frac{2.303 \times 1.36 \times\ log_{10}\left(\frac{15}{0.3}\right)}{\pi \times 0.511 \times 10^{-3}}$$

$${h_w}^2 = 88.5^2 - 3321.23$$

$$h_w = 67.2\ m$$

Drawdown in the pumped well

$$s_w = 90 - 67.2\ m = 22.8\ m$$

c. ***Specific capacity of the well. It is the discharge for a unit drawdown in the pumped well.***

Let us first find out the value of R

$$Q = \frac{\pi K(d^2 - {h_w}^2)}{2.303\ log_{10}\left(\frac{R}{r_w}\right)}$$

$$1.36 = \frac{\pi \times 0.51 \times 10^{-3}(90^2 - 67.2^2)}{2.303\ log_{10}\left(\frac{R}{0.3}\right)}$$

$log_{10}\left(\frac{R}{0.3}\right)$ =1.8335

R = 20 m

$$Specific\ Capacity = \frac{Discharge\ of\ well}{Draw\ down}$$

$$Q = 1.36\frac{m^3}{min},\ S = 22.8m$$

$$Specific\ Capacity = \frac{1.36}{22.8} = 59.65\ \frac{l}{min},$$

d. Maximum discharge will occur when $h_w = 0$

$$Q_{max} = \frac{\pi K(d^2 - {h_w}^2)}{2.303\ log_{10}\left(\frac{R}{r_w}\right)} = \frac{\pi * 0.511 * 10^{-3}(90^2 - 0^2)}{2.303\ log_{10}\left(\frac{20}{0.3}\right)}$$

$$Q_{max} = 3.09\ \frac{m^3}{min}$$

Hence the maximum discharge = $3090\ \frac{litres}{min}$

5.11 AQUIFER TEST

Determining the yield of ground-water systems and evaluating the movement and fate of ground-water pollutants require, among other information, knowledge of:

a. The position and thickness of aquifers and confining beds.
b. The transmissivity and storage coefficient of the aquifers.
c. The hydraulic characteristics of the confining beds.
d. The position and nature of the aquifer boundaries.
e. The location and amounts of ground-water withdrawals.
f. The locations, kinds, and amounts of pollutants and pollutant practices.

Investigations into hydrology and geology are necessary for acquiring knowledge about these aspects. One of the most important hydrologic studies involves analyzing the change, with time, in water levels (or total heads) in an aquifer caused by withdrawals through wells. This kind of research is called an aquifer test and typically involves pumping a well continuously for a few hours to several days while monitoring the water level changes in observation wells spaced at various intervals from the pumped well (Figure 5.13).

Successful aquifer tests require, among other things:

a. Determination of the pre-pumping water-level trend (that is, the regional trend).
b. A carefully controlled constant pumping rate.
c. Accurate water-level measurements were made at precisely known times during both the drawdown and the recovery periods.

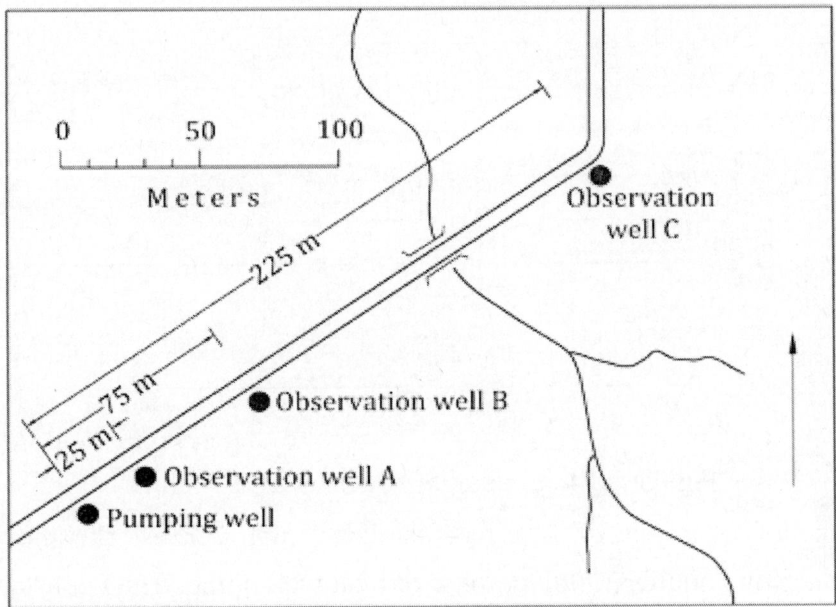

Figure 5.13: Map of the aquifer test site.

Drawdown is the difference between the water level at any time during the test and the position at which the water level would have been if withdrawals had not started. Drawdown is very rapid at first. As pumping continues and the cone of depression expands, the rate of drawdown decreases (Figure 5.14).

The recovery of the water level under ideal conditions is a mirror image of the drawdown. When a recharge well starts replenishing water at the same point and at the same rate as the pump cut-off, the change in water level throughout the recovery period is the same as if withdrawals from the pumped well had continued at the same rate. Therefore, the recovery of the water level is the difference between the actual measured level and the projected pumping level (Figure 5.14).

In addition to the constant-rate test mentioned above, analytical methods have also been developed for several other types of aquifer tests. These techniques include experiments where the rate of withdrawal is varied and tests where water is leaked into restricted aquifers by use of confining beds. The analytical methods available also permit analysis of tests conducted on both vertical wells and horizontal wells or drains.

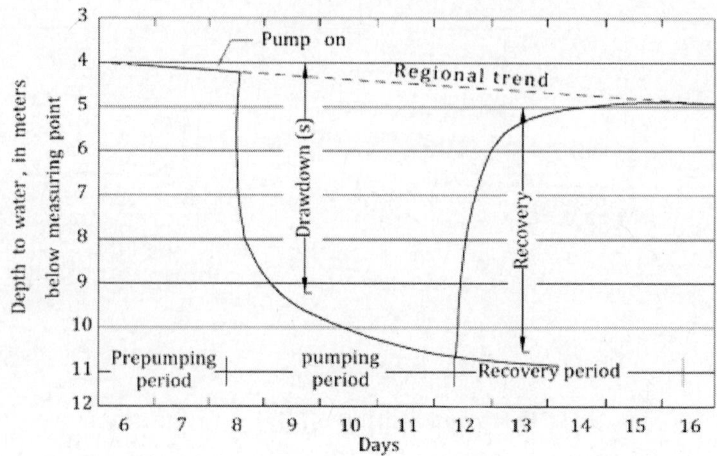

Figure 5.14: Change of water level in Well B.

The topic of "Analysis of Aquifer-Test Data" will address the most widely used technique for analyzing aquifer-test data for a vertical well pumped at a constant rate from an aquifer unaffected by lateral borders or vertical leaks. A type curve based on the values of W(u) and 1/u given in Table 5.1 is necessary for the analysis procedure. The discussion that follows covers how to prepare and use the type curve.

Table 5.1: Selected values of W(u) for the given value of 1/u

1/u	10	7.69	5.88	5.0	4.0	3.33	2.86	2.5	2.22	2.0	1.67	1.43	1.25	1.11
10^{-1}	0.219	0.135	0.075	0.049	0.025	0.013	0.007	0.004	0.002	0.001	0	0	0	0
1	1.82	1.59	1.36	1.22	1.04	0.91	0.79	0.70	0.63	0.56	0.45	0.37	0.31	0.26
10	4.04	3.78	3.51	3.35	3.14	2.96	2.81	2.68	2.57	2.47	2.30	2.15	2.03	1.92
10^2	6.33	6.07	5.80	5.64	5.42	5.23	5.08	4.95	4.83	4.73	4.54	4.39	4.26	4.14
10^3	8.63	8.37	8.10	7.94	7.72	7.53	7.38	7.25	7.13	7.02	6.84	6.69	6.55	6.44
10^4	10.94	10.67	10.41	10.24	10.02	9.84	9.68	9.55	9.43	9.33	9.14	8.99	8.86	8.74
10^5	13.24	12.98	12.71	12.55	12.32	12.14	11.99	11.85	11.73	11.63	11.45	11.29	11.16	11.04
10^6	15.54	15.28	15.01	1485	14.62	14.44	14.29	14.15	14.04	13.93	13.75	13.60	13.46	13.34
10^7	17.84	17.58	17.31	17.15	16.93	16.74	16.59	16.46	16.34	16.23	16.05	15.90	15.76	15.65
10^8	20.15	19.88	19.62	19.45	19.23	19.05	18.89	18.76	18.64	18.54	18.35	18.20	18.07	17.95
10^9	22.45	22.19	21.92	21.76	21.53	21.35	21.20	21.06	20.94	20.84	20.66	20.50	20.37	20.25
10^{10}	24.75	24.49	24.22	24.06	23.83	23.65	23.50	23.36	23.25	23.14	22.96	22.81	22.67	22.55
10^{11}	27.05	26.79	26.52	26.36	26.14	25.96	25.80	25.67	25.55	25.44	25.26	25.11	24.97	24.86
10^{12}	29.36	20.09	28.83	28.66	28.44	28.26	28.10	27.97	27.85	27.75	27.56	27.41	27.28	27.16

10^{13}	31.66	31.40	31.13	30.97	30.74	30.56	30.41	30.27	30.15	30.05	29.87	29.71	29.58	29.46
10^{14}	33.96	33.70	33.43	33.27	33.05	32.86	32.71	32.58	32.46	32.35	32.17	32.02	31.88	31.76

5.11.1 Analysis of Aquifer Test Data

C. V. Theis of the U.S. Geological Survey's New Mexico Water Resources District created the first equation in 1935 that included pumping time as a variable that could be used to assess the impact of well withdrawals. Thus, the Theis equation permitted, for the first time, the determination of the hydraulic characteristics of an aquifer before the development of new steady-state conditions resulting from pumping. The importance of this capability may be realized from the fact that, under most conditions, a new steady state cannot be developed or that, if it can, many months or years may be required. Theis assumed in the development of the equation that:

- The transmissivity of the aquifer tapped by the pumping well is constant during the test to the limits of the cone of depression.
- The water withdrawn from the aquifer is derived entirely from storage and is discharged instantaneously with the decline in head.
- The discharging well penetrates the entire thickness of the aquifer, and its diameter is small in comparison with the pumping rate, so that storage in the well is negligible.

These assumptions are most nearly met by confined aquifers at sites remote from their boundaries. Nevertheless, the equation can also be applied to the analysis of testing of unconfined aquifers, provided that appropriate safety measures are taken. The forms of the Theis equation (Equation 5.28 and 5.29) used to determine the transmissivity and storage coefficient are:

$$\mathrm{T} = \frac{QW(u)}{4\pi S} \qquad \text{.............. (5.28)}$$

$$\mathrm{S} = \frac{4Ttu}{r^2} \qquad \text{.............. (5.29)}$$

where,

T = transmissivity,

S = storage coefficient,

Q = pumping rate,

s = drawdown,

t = time,

r = distance from the pumping well to the observation well,

W(u) = well function of u =

$$-0.577216 - \log_e u + u - \frac{u^2}{2\times 2!} + \frac{u^3}{3\times 3!} - \frac{u^4}{4\times 4!} + \dots,$$

and

$$u = \frac{r^2 S}{4Tt}$$

The form of the Theis equation is such that it cannot be solved directly. To overcome this problem, Theis devised a convenient graphic method of solution that involves the use of a type curve (Figure 5.15). To apply this method, a data plot of drawdown versus time (or drawdown versus t/r^2) is matched to the type curve of W(u) versus 1/u. Values for s, t (or t/r^2), W(u), and 1/u are written at a convenient location on the overlapping portion of the sheets containing the data plot and type curve (Figure 5.16). These values are then substituted in equations 5.22 and 5.23, which are solved for T and S, respectively.

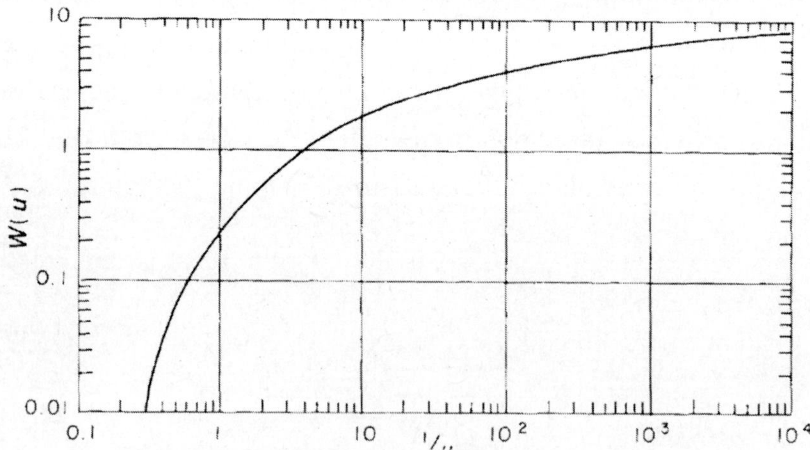

Figure 5.15: Theis type curve.

A Theis-type curve of W(u) versus 1/u can be prepared from the values given in the table contained in the preceding section, "Aquifer Tests." Plotting the data points is done on graph paper with logarithmic divisions in both the x and y directions, or logarithmic graph paper.

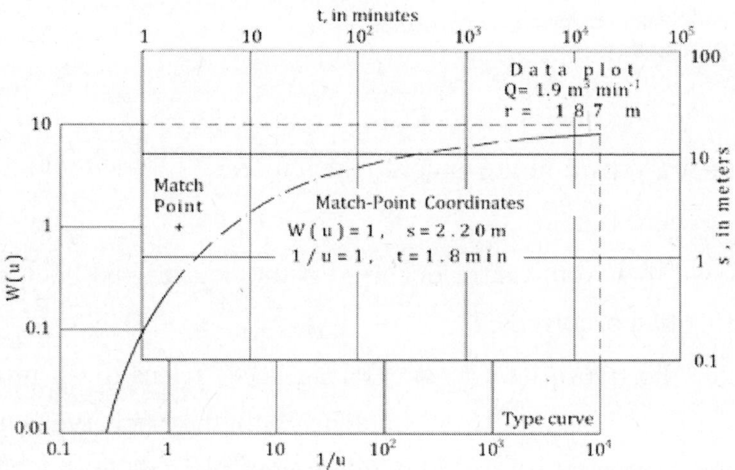

Figure 5.16: Matching of type curve.

The dimensional units of transmissivity (T) are L^2T^{-1}, where L is length and T is time in days. Thus, if Q in equation 16 is in cubic meters per day and s is in meters, T will be in square meters per day. Similarly, if, in equation 17, T is in square meters per day, t is in days, and r is in meters, S will be dimensionless.

Plotting the test data and the type curve on logarithmic graph paper is a necessary step in applying the Theis equation to analyze aquifer test results. If the aquifer and the conditions of the test satisfy Theis's assumptions, the type curve has the same shape as the cone of depression along any line radiating away from the pumping well and the drawdown graph at any point in the cone of depression.

Use of the Theis equation for unconfined aquifers involves two considerations. First, if the aquifer is relatively fine-grained, water is released slowly over a period of hours or days, not instantaneously with the decline in head. As a result, the value of S that is obtained from a short-period test can be too low.

Second, if the pumping rate is large and the observation well is near the pumping well, dewatering of the aquifer may be significant, and the assumption that the transmis of the aquifer is constant is not satisfied. The effect of dewatering of the aquifer can be eliminated with the following equation:

$$s' = s - \left(\frac{s^2}{2b}\right) \quad \text{.............. (5.30)}$$

where,

s = observed drawdown in the unconfined aquifer,

b = aquifer thickness, and

s' = drawdown that would have occurred if the aquifer had been confined (that is, if no dewatering had occurred).

To determine the transmissivity and storage coefficient of an unconfined aquifer, a data plot consisting of s' versus t (or t/r^2) is matched with the Theis type curve of W(u) versus 1/u. Both s and b in equation (5.30) must be in the same units, either feet or meters.

As noted above, Theis assumed in the development of his equation that the discharging well penetrates the entire thickness of the aquifer. Nevertheless, most discharge wells are only exposed to a portion of the aquifer they draw from, since it is not always possible or desirable to design a well that thoroughly penetrates the aquifer under development.

Such partial penetration creates vertical flow in the vicinity of the discharging well that may affect drawdowns in observation wells located relatively close to the discharging well. When observation wells approach the same zone as the discharging well but are at the same distance from it, the drawdowns in those wells will be higher compared to those in the other wells. The possible effect of partial penetration on drawdowns must be considered in the analysis of aquifer-test data. If aquifer-boundary and other conditions permit, the problem can be avoided by locating observation wells beyond the zone in which vertical flow exists.

5.12 TIME DRAWDOWN ANALYSIS

A variety of techniques has been generated for the analysis of aquifer-test results, the Theis equation being just one of them (See "Analysis of Aquifer-Test Data"). Another method, and one that is somewhat more convenient to use, was developed by C. E. Jacob from the Theis equation. The Jacob technique is more convenient because it uses semi-logarithmic graph paper rather than the logarithmic paper used in the Theis method. It is also more convenient since, in ideal circumstances, the data plot along a straight line rather than a curve.

However, it is essential to note that, whereas the Theis equation always applies and places (if the assumptions are met), Jacob's method applies only under certain additional conditions. To get trustworthy responses, these requirements must also be met.

To understand the limitations of Jacob's method, we must consider the changes that occur in the cone of depression during an aquifer test. The changes that are of concern involve both the shape of the cone and the rate of drawdown. As the cone of depression migrates outward from a pumping well, its shape (and, therefore, the hydraulic gradient at different points in the cone) changes. We can refer to this condition as an unsteady shape. At the start of withdrawals, the entire cone of depression has an unsteady shape (Figure 5.17). After a test has been underway for some time, the cone of depression begins to assume a relatively steady shape, first at the pumping well and then gradually to greater and greater distances (Figure 5.18). Drawdowns stop, and the cone of depression is considered to be in a constant state if withdrawals continue long enough for increases in recharge and/or decreases in discharge to balance the rate of withdrawal (Figure 5.19).

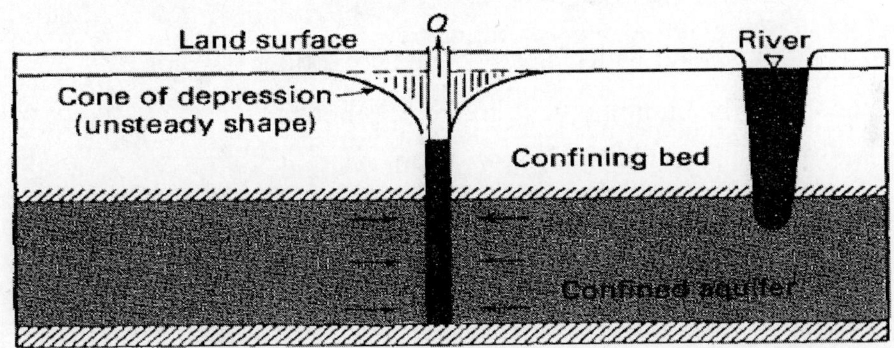

Figure 5.17: Depression cone during pumping from a confined aquifer.

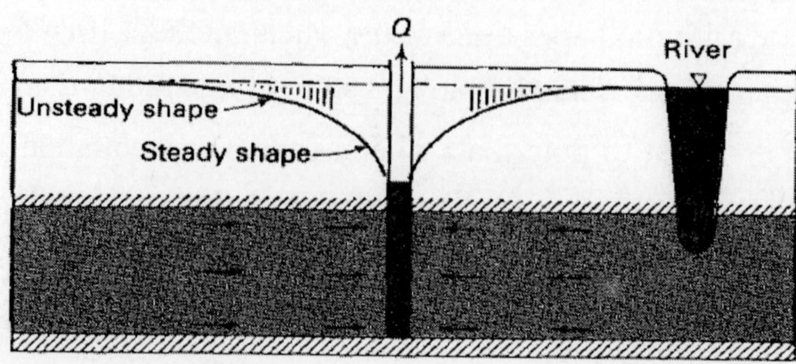

Figure 5.18: Steady and unsteady state of depression cone during pumping from a confined aquifer.

The Jacob approach can be applied to the entire cone only when steady-state conditions occur or strictly to the zone where steady-shape conditions are predominant. For practical purposes, this condition is met when $u = (r^2S)/(4Tt)$ is equal to or less than about 0.05. Substituting this value in the equation for u and solving for t, we can determine the time at which steady-shape conditions develop at the outermost observation well. Thus,

$$t_c = \frac{7200r^2S}{T} \qquad \text{.............. (5.31)}$$

where,

t_c= time, (in minutes), at which steady-shape conditions develop,

r = distance from the pumping well, in feet (or meters),

S= estimated storage coefficient (dimensionless), and

T = estimated transmissivity in square feet per day (or square meters per day).

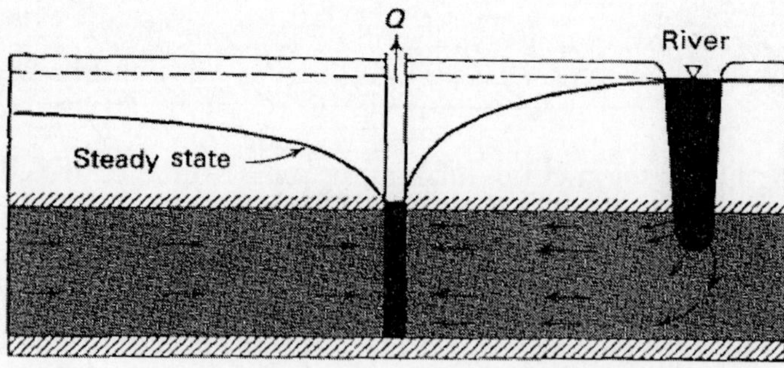

Figure 5.19: Steady state condition during pumping from a confined aquifer.

On semilogarithmic graph paper, the drawdowns at an observation well begin to decrease in a straight line after steady-shape conditions are established. Before that time, the drawdowns plot below the extension of the straight line. Plotting drawdowns on the vertical (arithmetic) axis against time on the horizontal (logarithmic) axis resulting in a time-drawdown graph.

The transmission and pumping rate have a direct relationship with the slope of the straight line. Jacob derived the following equations (Equation 5.32 and 5.33) for determination of transmissivity and storage coefficient from the time-drawdown graphs (Figure 5.20):

$$T = \frac{2.3Q}{4\pi\Delta s} \quad \text{..............} (5.32)$$

$$S = \frac{2.25Tt_0}{r^2} \quad \text{.............} (5.33)$$

where,

Q = pumping rate,

Δs=drawdown across one log cycle,

t_0= time at the point where the straight line intersects the zero-drawdown line, and

r = distance from the pumping well to the observation well.

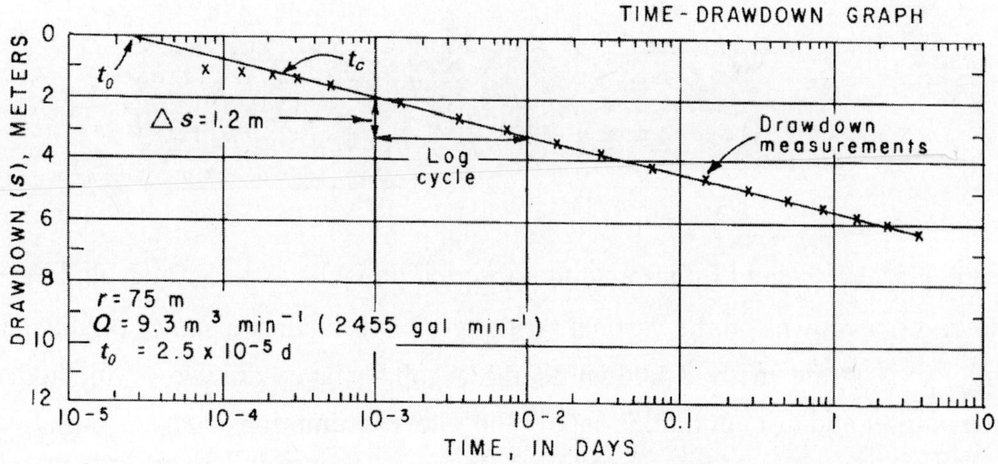

Figure 5.20: Time drawdown graph.

Equations 5.32 and 5.33 are in consistent units. Thus, if Q is in cubic meters per day and s is in meters, T is in square meters per day. S is dimensionless, so that, in equation 5.33, if T is in square meters per day, then r must be in meters and t_0 must be in days.

5.13 DISTANCE DRAWDOWN ANALYSIS

It is desirable in aquifer tests to have at least three observation wells located at different distances from the pumping well (Figure 5.21). The aquifer transmissivity and storage coefficient can be ascertained by utilizing the Theis equation and type curve to examine simultaneous drawdowns in these wells.

After the test has been underway long enough, drawdowns in the wells can also be analyzed by the Jacob method, either through the use of a time-drawdown graph using data from individual wells or through the use of a distance-drawdown graph using "simultaneous" measurements in all of the wells. To determine when sufficient time has elapsed, see "Time-Drawdown Analysis."

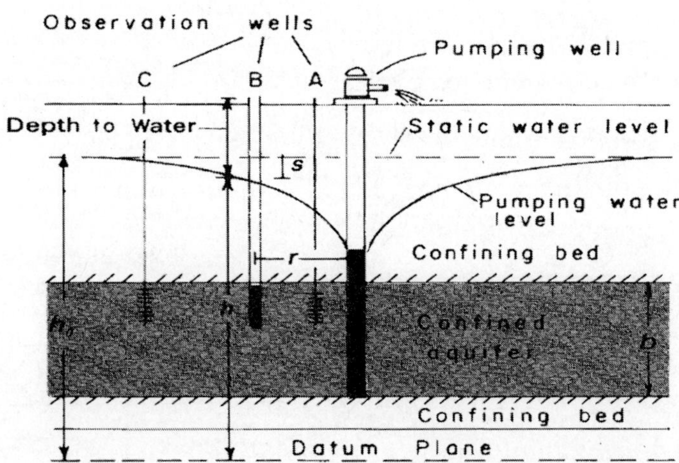

Figure 5.21: Cluster of pumping and observation wells in a confined aquifer.

Plotting drawdowns on the vertical (arithmetic) axis against distance on the horizontal (logarithmic) axis is the method known as the Jacob distance-drawdown method (Figure 5.22). If the aquifer and test conditions satisfy the Theis assumptions and the limitation of the Jacob method, the drawdowns measured at the same time in different wells should plot along a straight line (Figure 5.22).

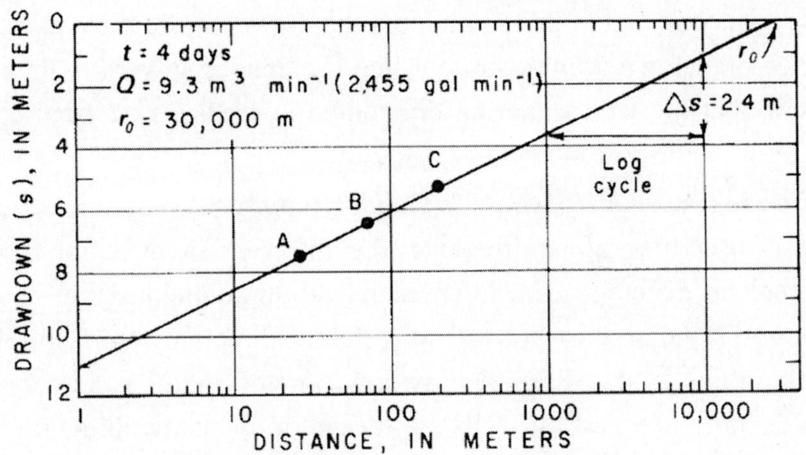

Figure 5.22: Distance-drawdown graph.

The slope of the straight line is proportional to the pumping rate and to the transmissivity. Jacob derived the following equations (Equation 5.34 and 5.35) for the determination of the transmissivity and storage coefficient from distance drawdown graphs:

$$T = \frac{2.3Q}{4\pi\Delta s} \qquad \text{............. (5.34)}$$

$$T = \frac{2.25\,\mathrm{Tt}}{r_0^2} \qquad \text{............. (5.35)}$$

where,

Q = pumping rate, ΔS is the drawdown across one log cycle,

t = time at which the drawdowns were measured and

r_0 = distance from the pumping well to the point where the straight line intersects the zero-drawdown line.

The distance r_0 does not indicate the outer limit of the cone of depression. Because non-steady shape conditions exist in the outer part of the cone, before the development of steady-state conditions, the Jacob method does not apply to that part. If the Theis equation were used to calculate drawdowns in the outer part of the cone, it would be found that they would plot below the straight line. In other words, the measurable limit of the cone of depression is beyond the distance r_0.

The drawdown observed at that moment is the drawdown in the aquifer outside of the well, if the distance-drawdown graph's straight line is extended inward to the pumping well's radius. In the event that the well's internal drawdown exceeds its external drawdown, well loss is responsible for the discrepancy.

As noted in the section on "Hydraulic Conductivity," the hydraulic conductivities and, therefore, the transmissivities of aquifers may be different in different directions. These variations may result in different drawdowns detected simultaneously in observation wells located at the same distances apart but facing different directions from the discharge well. Where this condition exists, the distance-drawdown method may yield satisfactory results only where three or more observation wells are located in the same direction but at different distances from the discharging well.

5.14 SINGLE WELL TEST

The most useful aquifer tests are those that include water-level measurements in observation wells. These tests are frequently referred to as multiple-well tests. It is also possible to obtain useful data from production wells, even where observation wells are not available. Such tests are referred to as single-well tests and may consist of pumping a well at a single constant rate, or at two or more different but constant rates (see "Well-Acceptance Tests and Well Efficiency") or, if the well is not equipped with a pump, by "instantaneously" introducing a known volume of water into the well. This discussion will be limited to tests involving a single constant rate.

To analyze the data, it is necessary to understand the nature of the drawdown in a pumping well. The total drawdown (s_t) in most, if not all, pumping wells consists of two components (Figure 5.23). There are two types of drawdowns in an aquifer: the drawdown in the aquifer (s_a) and the drawdown in the well bore (s_w) that happens when water travels from the aquifer to the pump intake. Thus, the drawdown in most pumping wells is greater than the drawdown in the aquifer at the radius of the pumping well.

The total drawdown (s_t) in a pumping well can be expressed in the form of the following equations:

$$s_t = s_a + s_w \qquad \text{.............. (5.36)}$$

$$s_t = BQ + CQ^2 \qquad \text{............. (5.36a)}$$

where,

s_a = drawdown in the aquifer at the effective radius of the pumping well,

s_w = well loss,

Q = pumping rate,

B = factor related to the hydraulic characteristics of the aquifer and the length of the pumping period, and

C = factor related to the characteristics of the well.

The factor C in equation 5.36a is normally considered to be constant so that, in a constant rate test, CQ^2 is also constant. Consequently, the well loss (s_w) raises the pumping well's overall drawdown but has no effect on how quickly the drawdown changes over time. It is, therefore, possible to analyze drawdowns in the pumping well with the Jacob time-drawdown method using semi-logarithmic graph paper. (See "Time-Drawdown Analysis."). Drawdowns are plotted on the arithmetic scale versus time on the logarithmic scale (Figure 5.24), and transmissivity is determined from the slope of the straight line through the use of the following equation (5.37):

$$T = \frac{2.3Q}{4\pi\Delta s} \qquad \text{............ (5.37)}$$

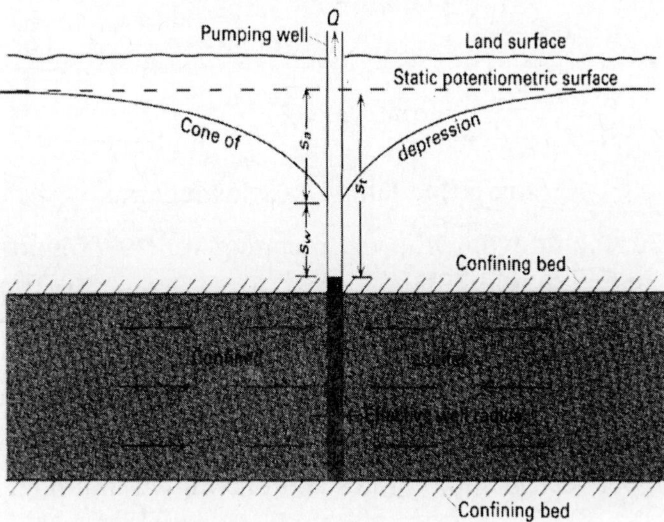

Figure 5.23: Schematic diagram of pumping test of a confined aquifer.

It is not possible to calculate the storage coefficient by extending the straight line to the line of zero drawdown when well loss exists in the pumping well. Even where well loss is not present, the determination of the storage coefficient from drawdowns in a pumping well likely will be subject to large error because the effective radius of the well may differ significantly from the "nominal" radius.

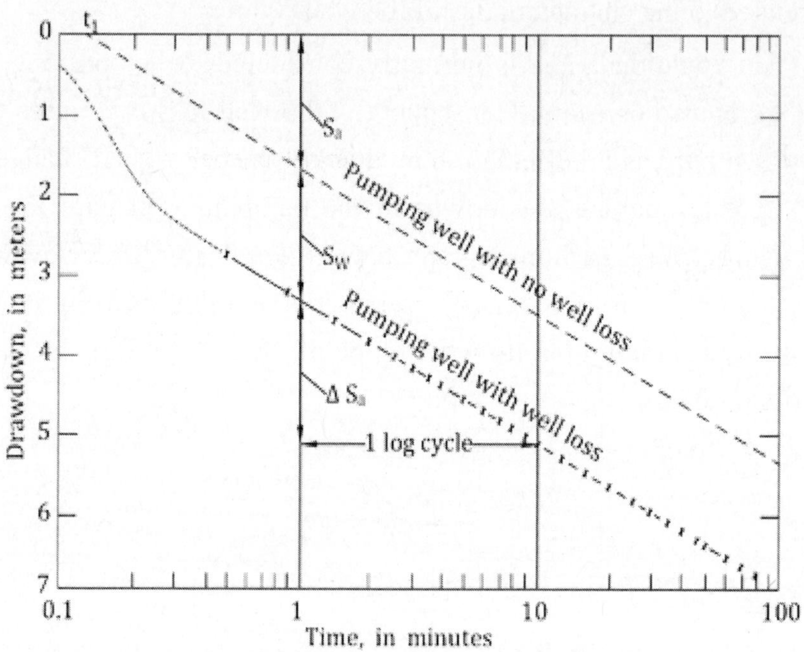

Figure 5.24: Time Drawdown curve.

In equation 5.36, the drawdown in the pumping well is proportional to the pumping rate. As long as water is drawn from the aquifer's storage, factor B in the aquifer-loss term (BQ) increases with pumping time. Factor C in the well-loss term (CQ^2) is a constant if the characteristics of the well remain unchanged, but because the pumping rate in the well-loss term is squared, drawdown due to well loss increases rapidly as the pumping rate is increased. The relation between pumping rates and drawdown in a pumping well, if the well was pumped for the same length of time at each rate, is shown in Figure 5.25. The effect of well loss on drawdown in the pumping well is important both in the analysis of data from pumping wells and in the design of supply wells.

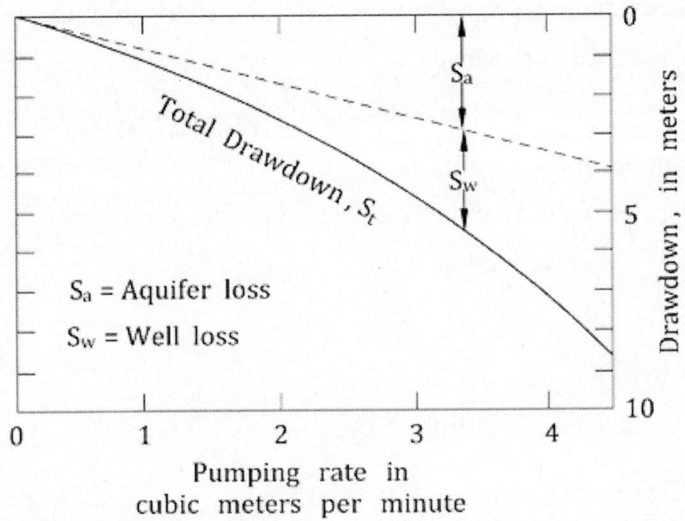

Figure 5.25: Pumping rate and drawdown relationship.

SUMMARY

Throughout history, groundwater has been a significant source of water. Numerous towns, enterprises, irrigation systems, suburban residences, and farms depends significantly on it for their water needs. Groundwater supplies are limited, just as other natural resources. The current module covers the characteristics of aquifers, their geological formation, and the occurrence and form of groundwater. Aquifer testing, equilibrium equations for confined and unconfined aquifers, methods for estimating well yield, and steady state flow in wells are among the topics covered under well hydraulics.

EXERCISE

Revision Questions

1. Define the term Groundwater. Describe the various forms of Sub-surface water.
2. Classify the various sub-surface formations.

3. What do you understand by an aquifer? Describe the types of aquifers with a neat sketch and describe their hydraulic properties.
4. Write a short note on (i) Aquifer, (ii) Aquiclude, (iii) Aquitard and (iv)Aquifuge.
5. Write a short note on (i) a Confined aquifer, (ii) a Semi-Confined aquifer, (iii) an unconfined aquifer, and (iv) a Semi-unconfined aquifer.
6. Define the following terms in brief.
   a. Porosity,
   b. Hydraulic conductivity
   c. Coefficient of storage or storability (S)
   d. Transmissivity,
   e. Hydraulic Diffusivity (D)
7. Define (i) Slichter's and (ii) Hazen's empirical formulas to obtain groundwater velocity.
8. Explain methods used in the field to obtain groundwater velocity.
9. Give a brief definition of each term used in groundwater flow studies.
   a. Porosity.
   b. Specific yield,
   c. Specific retention,
   d. Specific capacity,
   e. storage coefficient
   f. Barometric efficiency
10. What do you understand by groundwater velocity? Define Darcy's Law to obtain groundwater velocity?
11. Distinguish between
   a. Aquifer and aquitard,
   b. Unconfined aquifer and a leaky aquifer,
   c. Influent and effluent streams,
   d. Water table and piezometric surface

12. Write a short note of Thiem's Equilibrium formula for unconfined aquifers under steady-state conditions.
13. Write a short note of Thiem's Equilibrium formula for confined aquifer. Describe the assumptions and limitations of Thiem's Equilibrium formula.
14. Explain Dupuit's equilibrium formulae for estimating the yield of wells under steady-state conditions for (i) confined aquifers, (ii) unconfined aquifers.
15. With the help of a neat sketch, explain fully and partially penetration of an aquifer by a well.
16. Explain the following:
    a. Well loss
    b. Recharge
17. Why it is essential to make an aquifer test? How we can analyze the aquifer test data?
18. Explain various techniques for the analysis of aquifer-test results in brief.

Numerical Problems

1. Draw a schematic vertical cross-section from the mountains to the ocean, illustrating the hydrologic cycle. Include the following features:
    - Mountain, coastal plain, ocean, and underlying aquifers
    - Exchanges between atmospheric water, surface water, and groundwater

    Illustrate the hydrologic cycle starting with mountain rainfall, following the flow of rivers and groundwater to the ocean, and ending with ocean evaporation and precipitation returning to land.
2. In an area of 120.0 ha, the water table dropped by 5.0 m. If the porosity is 28.0% and specific retention is 9.0%, determine the specific yield of the aquifer and change in groundwater storage.
3. A phreatic aquifer extends over an area of 12.0 km^2. The water table was initially at 20.0 m below the ground level. After irrigation with a depth of 30.0 cm of water, the water table rose to a depth of 19.2 m below ground level. When 8×10^6 m^3 of water was pumped out, the groundwater table (g.w.t.) dropped to 21.5 m below ground level. Determine the specific yield of the aquifer and the deficit in soil moisture before irrigation.

4. A phreatic aquifer, extending over an area of 220.0 km², has a storability of 0.15. Estimate the amount of water lost from storage if the water level falls 0.16 m during a drought.
5. Calculate the discharge and the seepage velocities for water flowing through a pipe filled with sand with a hydraulic conductivity of 1.5×10^{-6} m/s and an effective porosity of 0.2. The hydraulic gradient is 0.01 and the cross-sectional area of the pipe is 150.0 cm².
6. A well is penetrating an aquifer with a hydraulic conductivity of 12.0 m/day and storability of 0.00735. The aquifer is 28.0 m thick and is pumped at a rate of 2800.0 m³/day. Estimate the drawdown after 1 week of pumping at a distance 16.0 m from the well.
7. A fully penetrating well is pumped at a rate of 1500 m³/day from an aquifer whose S_c and T values are 4×10^{-4} and 208.8 m²/day, respectively. Find the drawdown at a distance 3.0 m from the production well after 1 hour of pumping and at a distance 495.0 m after 2 days of pumping.
8. Find the approximate values of the radius of influence in the above problem (Problem 7) after 1 hour and 2 days of continuous pumping.
9. A well penetrating a confined aquifer is pumped at a uniform rate of 2,718.42 m³/day. An observation well is located 60.96 m away. Drawdowns measured in the observation well during the period of pumping (*t*) are given in Table 5.2 below. Compute the values of S_c and T using the Theis method.

Table 5.2: Drawdown of the water level in an observation well (r = 60.96 m)

Time from start of pumping, t (min)	Observed drawdown, s (m)	r²/t (m²/day)
1	0.2	5.35×10^6
1.5	0.27	3.57×10^6
2	0.3	2.67×10^6
2.5	0.34	2.14×10^6
3	0.37	1.78×10^6

Time from start of pumping, t (min)	Observed drawdown, s (m)	r^2/t (m^2/day)
4	0.41	1.34×10^6
5	0.45	1.07×10^6
6	0.48	8.97×10^5
8	0.53	6.73×10^5
10	0.57	5.38×10^5
12	0.6	4.48×10^5
14	0.63	3.83×10^5
18	0.67	2.99×10^5
24	0.72	2.24×10^5
30	0.76	1.79×10^5
40	0.81	1.34×10^5
50	0.85	1.07×10^5
60	0.88	8.97×10^4
80	0.93	6.73×10^4
100	0.96	5.38×10^4
120	1	4.48×10^4
150	1.04	3.57×10^4
180	1.07	2.99×10^4
210	1.1	2.55×10^4
240	1.12	2.24×10^4

10. A farm has two wells A and B located 1000 meters apart. Both wells fully penetrate a saturated thickness of 25 meters. The hydraulic conductivity is 0.001 m/s and the storage coefficient is 0.25. If well A is pumped at the rate of 0.20 m^3/s for a period of 120 days, how much decline will it cause to the water table of well B?

Multiple Choice Questions

1. What is groundwater?

A) Water found in oceans

B) Water that fills the spaces between soil particles and rocks underground

C) Water vapor in the atmosphere

D) Water that flows in rivers and lakes

2. Which of the following is a primary source of groundwater recharge?

A) Evaporation B) Precipitation

C) Surface water runoff D) All of the above

3. What is an aquifer?

A) A body of surface water

B) A geological formation that can store and transmit groundwater

C) A type of groundwater contamination

D) An area where groundwater evaporates

4. What term describes the upper surface of the saturated zone in an aquifer?

A) Water table B) Groundwater level

C) Saturation point D) Aquiclude

5. Which of the following human activities can lead to groundwater contamination?

A) Agricultural runoff B) Industrial discharges

C) Landfills D) All of the above

6. How is groundwater typically accessed for human use?

A) Through wells B) By evaporation

C) By capturing rainfall D) By creating artificial lakes

7. What is the process of groundwater moving from high to low pressure called?

A) Percolation B) Recharge

C) Flow D) Transpiration

8. Which of the following factors can affect the rate of groundwater recharge?

A) Soil type B) Vegetation cover

C) Land use practices D) All of the above

9. Which of the following describes an alluvial formation?

A) Formed from volcanic activity
B) Deposited by rivers and streams
C) Created by glacial action
D) Comprised mainly of limestone

10. What type of subsurface formation is typically characterized by layered sedimentary rocks?

A) Igneous formation
B) Metamorphic formation
C) Stratified formation
D) Unconsolidated formation

11. Which formation is primarily made up of compacted clay and is known for its low permeability?

A) Gravel formation
B) Silt formation
C) Shale formation
D) Sandstone formation

12. What type of subsurface formation is known for its high porosity and is often a reservoir for groundwater?

A) Basalt formation
B) Limestone formation
C) Granite formation
D) Sand formation

13. Which type of subsurface formation is typically associated with volcanic regions?

A) Sedimentary formation
B) Igneous formation
C) Metamorphic formation
D) Alluvial formation

14. What is the primary characteristic of a confined aquifer?

A) It is not saturated with water.
B) It is overlain by a layer of impermeable rock or clay.
C) It is always under atmospheric pressure.
D) It is easily contaminated.

15. Which type of aquifer is typically more productive?

A) Unconfined aquifer
B) Confined aquifer
C) Semi-confined aquifer
D) Perched aquifer

16. What hydraulic property measures the ability of an aquifer to transmit water?

A) Specific yield
B) Hydraulic conductivity
C) Storage coefficient
D) Porosity

17. What is the term for the volume of water that an aquifer can release per unit area of aquifer per unit decline in hydraulic head?

A) Specific yield
B) Storage coefficient
C) Porosity
D) Transmissivity

18. Which type of aquifer is characterized by having its water table at the ground surface?

A) Confined aquifer
B) Unconfined aquifer
C) Semi-confined aquifer
D) Perched aquifer

19. What is a confined aquifer?

A) An aquifer with a layer of impermeable rock above it
B) An aquifer that is open to the atmosphere
C) An aquifer located near the surface
D) An aquifer that only receives surface water

20. Which of the following describes an unconfined aquifer?

A) It is capped by an impermeable layer
B) It is directly recharged by surface water
C) It is always under pressure
D) It cannot be accessed by wells

21. How does water pressure differ between confined and unconfined aquifers?

A) Confined aquifers have no pressure
B) Unconfined aquifers have higher pressure
C) Confined aquifers are under greater pressure
D) There is no difference in pressure

22. Which type of aquifer is typically more susceptible to contamination from surface activities?

A) Confined aquifer
B) Unconfined aquifer
C) Both types equally
D) Neither type

23. What is the recharge area for a confined aquifer?

A) The area above the aquifer
B) A designated zone far from the aquifer
C) The area where surface water infiltrates
D) The area beneath an impermeable layer

24. Which of the following statements about well water in confined aquifers is true?

A) Water in the well flows freely without pumping

B) Water must always be pumped to access it

C) The water level in the well is always at the water table

D) Wells can only be drilled into unconfined aquifers

25. Which of the following best describes the pressure situation in an unconfined aquifer?

A) It is under constant pressure
B) It is subject to atmospheric pressure

C) It is isolated from atmospheric pressure
D) It has negative pressure

26. In which type of aquifer is artesian flow most found?

A) Unconfined aquifer
B) Confined aquifer

C) Both types equally
D) Neither type

27. What is the primary purpose of conducting an aquifer test?

A) To determine the hydraulic conductivity of the aquifer

B) To measure the water quality of the aquifer

C) To assess the total water storage capacity of the aquifer

D) To evaluate the age of the groundwater

Answer: 1-B; 2-D; 3-B; 4-A; 5-D; 6-A; 7-C; 8-D; 9-B; 10-C; 11-C; 12-B; 13-B; 14-B; 15-A; 16-B; 17-B; 18-B; 19-A;20-B; 21-C; 22-B; 23-C; 24-A; 25-B; 26-B; 27-A.

KNOW MORE

For more information related to this topic scan the QR code.

OR

Type this link in your browser

https://drive.google.com/file/d/1aS2zasoCSbI7k2bytRIjq9JHD13eYFSs/view?usp=sharing

BIBLIOGRAPHY

1. Bouwer, H. (1978) Groundwater Hydrology, McGraw-Hill Book, New York, 480.
2. Chow. V.T, (Ed), (1964), "Handbook of Applied Hydrology", McGraw-Hill, New York, NY, 1964.
3. Heath, R. C. (1987). Basic ground-water hydrology (Vol. 2220). US Department of the Interior, US Geological Survey.
4. Jain S. K., Agarwal, P. K. And Singh, V. P. (2007), Hydrology and Water Resources of India, Water Science and Technology Library, Vol 57, Springer, The Netherland, ISBN-10 1-4020-5179-4(HB).
5. Singh, V. P. (1992). Elementary hydrology. Prentice Hall, New Jersy, 973.
6. Subramanya, K. (2015), "Engineering Hydrology", McGraw-Hill Education (India) Private Limited, New Delhi, 2015, ISBN(13): 978-1-25902997-4.
7. Todd, D.K. and L. Mays (2004), Groundwater Hydrology, 3rd Edition, ISBN: 978-0-471-05937-0, 656 Pages, Wiley.

6 Water Withdrawals and Uses

UNIT SPECIFICS

Water withdrawals and uses refer to an essential process of extracting water from its various sources, such as rivers, lakes, and underground aquifers, to fulfil the demands of various sectors such as agriculture, domestic uses, hydropower generation, and irrigation. These withdrawals are essential to water management and play a crucial role in sustaining human activities and ecosystems. This unit describes water demand for domestic use, agriculture, water requirements for India's crops and crop seasons, energy production, and other uses. Further, surface and Groundwater supply analysis, including duty and delta, irrigation water supply, soil-water relationship, root zone soil water, infiltration, and consumptive use, has also been discussed. In addition, a description of irrigation requirements, their frequency, and various irrigation methods, such as surface, sub-surface, sprinkler, and trickle/drip irrigation, have been discussed.

RATIONALE

To learn about water usage methods, such as water for energy production, agriculture, flood control, and water supply. Also, to know about the water requirements for crops, quality of irrigation water, soil-water relationship, and the methods of irrigation.

PRE-REQUISITE

Nil

UNIT OUTCOMES

The list of outcomes of this unit is as follows:

U6-O1: Water usage such as water supply, agriculture, energy production, and flood control

U6-O2: Water requirements for crops

U6-O3: Irrigation water quality and soil-water relationship

U6-O4: Irrigation methods

Unit Outcomes	Expected Mapping with Course Outcomes (1- Weak Correlation; 2- Medium correlation; 3- Strong Correlation)							
	CO-1	CO-2	CO-3	CO-4	CO-5	CO-6	CO-7	CO-8
U6-O1	3	3	3	1	3	3	3	3
U6-O2	3	3	3	1	3	3	1	3
U6-O3	1	1	3	3	3	3	3	3
U6-O4	3	3	3	3	3	3	3	3

6.1 WATER DEMAND

The demand for water is the volume of water required to be supplied for a specific use. As an illustration, the demand in agriculture includes crop consumptive requirements, canal seepage, evaporation, and percolation losses. Fulfilling various demands, such as demand for agriculture, domestic uses, hydropower generation, irrigation, etc., as well as water withdrawals and uses, is an important process in hydrology. Water withdrawals and uses refer to the process of extracting water from its source (such as rivers, lakes, or underground aquifers) for various purposes. These withdrawals are an essential aspect of water management and play a crucial role in sustaining human activities and ecosystems. It is necessary to consider prevailing practices as well as management policies, technology availability, and other resources. Emerging technologies and techniques that are anticipated to be used in the future socioeconomic environment must be considered when projecting future demand. Water withdrawals can be categorized into various sectors, such as agriculture/irrigation, domestic water supply, hydropower generation, industrial use, recreational uses, environmental flow, aquaculture, mining etc, based on their usage. Some important sectors of water withdrawals and their uses are described in the following section.

6.1.1 Domestic Water Supply

Community water supply is the most significant water requirement, accounting for around 5% of overall water use. The volume of water being utilized for domestic needs such as drinking, cooking, bathing, and sanitation as well as businesses, offices, and other commercial uses are significantly less than that used for irrigated agriculture. Estimates show that about 7 km^3 surface water and 18 km^3 ^{of} Groundwater is being used for water supply in urban and rural areas. In India, several water supply and sanitation programs have been launched in recent years under the Ministry of Jal Shakti. The government of India pledged to provide adequate drinking water and sanitation facilities to the population.

The country's population determines the amount of water available per person, and in India, the population is expanding more quickly than the quantity of water available per person. The average annual per capita water availability was estimated to be 1816 cubic meters in 2001 and 1545 cubic meters in 2011, respectively. In 2021 and 2031, this amount is expected to drop to 1486 cubic meters and 1367 cubic meters, respectively.

The core needs of domestic use of water are for drinking, cooking, washing, and bathing. Non-core needs are for toilet flushing, sewer flushing, washing clothes, water for lawns, etc. As per the Working Group (WG, 1999), the demands for domestic uses are given in Table 6.1. Additionally, the Ministry of Housing and Urban Affairs suggests that the standard for urban water supply be 135 liters per capita per day (lpcd). Under the Jal Jeevan Mission, a minimum service delivery of 55 lpcd has been set for rural regions; governments may choose to augment this to a greater level.

Table 6.1: Basic water requirements for human needs

Purpose	**Per capita requirement in rural areas (lpcd)**	**Per capita requirement in urban areas (lpcd)**
Drinking water	3	5
Cooking	5	5
Ablution	10	-
Bathing	15	55
Washing of utensils, clothes & household	22	40
Flushing of toilets/sewer	-	30
Total basic water requirement (BWR)	55	135

Source: WG (1999), Jal Jeevan Mission

6.1.2 Water for Agriculture

Agriculture is the most fundamental form of human activity. The Latin words "Agri" (meaning soil) and "culture," which mean cultivation, are the origins of the English word "agriculture." It is stated that irrigation supports agriculture, and agriculture supports society. Water is a crucial resource in agriculture, playing a vital role in the growth and development of crops. Adequate water supply is essential for ensuring high agricultural productivity. For farming sustainably and productively, water management in agriculture requires a variety of methodologies, including effective irrigation, understanding of crop water requirements, conservation of soil moisture, water quality monitoring, and use of modern technology. Rainfed agriculture refers to farming practices where irrigation facilities are unavailable and the crops are entirely dependent on rainfall.

Irrigation is a crucial practice in agriculture to provide water to crops when natural rainfall is insufficient. There are several types of irrigation methods, each with its advantages and disadvantages. The choice of irrigation method depends on factors such as the type of crop, soil characteristics, water availability, and the climate of the region. Various methods of irrigation to fulfil the requirement of water for agriculture such as surface irrigation, drip irrigation, sprinkler irrigation, and sub-surface irrigation, are employed based on the type of crops, soil conditions, and water availability.

6.1.3 Water for Hydropower

Hydropower, also known as hydroelectric power, mainly generated by harnessing the energy of flowing water to produce electricity. Hydropower plants are often built on rivers or streams where there is a consistent flow of water. The process typically involves the construction of dams or other structures to create reservoirs, which store water at a higher elevation. In the concept of Hydropower generation, water falls under gravity turns the blades of a turbine which is connected to a rotating generator. The rotating generator uses the kinetic energy of the flowing water to generate electricity. The height of the dam determines the potential energy available for electricity generation. When this stored water is released, it flows downhill and drives the turbines connected to the generators, causing them to spin, converting the kinetic energy of the moving water, from mechanical energy into electrical energy.

Types of Hydropower

There are several types of hydropower systems, each with its own characteristics and applications. Some common types of hydropower systems are described in the following sections.

a. **Conventional Hydropower:** This is the most common type of hydropower, involving the construction of dams and reservoirs to store water. The stored water is released from the reservoir, and the flow is used to turn turbines connected to generators, producing electricity. Examples include large hydroelectric dams like the Tehri Hydropower Project (Figure 6.1), Bhakra Nangal Project etc.

Figure 6.1: (a) Tehri Dam; (b) Powerhouse of Tehri Hydropower Project

b. **Run-of-River Hydropower:** Unlike conventional hydropower, run-of-river systems do not require large reservoirs. Instead, they use the natural flow of rivers to generate electricity. A portion of the river flow is diverted through a channel or penstock, and the water passes through turbines to generate electricity. An example includes Chilla Hydropower project in Uttarakhand.

c. **Pumped Storage Hydropower:** Pumped storage involves two reservoirs positioned at different elevations. During periods of low electricity demand, excess electricity is used to pump water from the lower reservoir to the upper reservoir. During periods of high demand, water is released from the upper reservoir to the lower reservoir, passing through turbines to generate electricity.

Advantages of Hydropower Generation

Hydropower generation has several unique benefits:

a. Since water is a renewable energy source, there are no fuel costs associated with it;

b. Hydro plants are well-suited to provide peak energy during the day, saving significant sums of money on installation and operating costs associated with alternative thermal capacity;

c. Compared to 2.5% (excluding the cost of fuel) in the case of coal-based thermal plants, operation and maintenance expenses are extremely low, often around 1% of the capital cost;

d. Compared to coal-based thermal plants, which require 8–10% of the total energy produced, auxiliary usage is just 0.5% to 1%;

e. Hydro units are perfect for handling peak demands because they can be started and shut down within a matter of minutes;

f. There are no issues with air or surface pollution caused by hydro plants;

g. The average cost of generation, depending on the kind of scheme (run-of-the-river or storage-based), is reasonably inexpensive, ranging from Rs 0.50 to Rs 1.00 per kWh;

6.1.4 Water for Electric Production

Water plays a crucial role in various forms of energy production, both directly and indirectly other than hydropower generation. It is important to note that while water is essential for energy production, there are environmental considerations related to water use, such as the impact on aquatic ecosystems, water availability, and potential conflicts between energy and water needs. Sustainable water management practices and the development of water-efficient technologies are critical for addressing these challenges. Additionally, the growing demand for water and energy highlights the importance of integrated water and energy planning to ensure the efficient and responsible use of resources. Some key ways in which water is involved in energy generation other than hydropower are given in the following sections.:

a. **Thermal Power Plants:** Many power plants, especially fossil fuel and nuclear plants, use water for cooling purposes. Water is circulated through the plant to absorb heat produced during the generation of electricity. In this method, Water is drawn from a source, generally used to cool the plant, and then discharged back into a body of water.

b. **Geothermal Power:** Geothermal power plants use the heat from the Earth's interior to generate electricity. In some cases, water is injected into geothermal reservoirs to enhance heat extraction. Steam from underground reservoirs is used to turn turbines and generate electricity.

c. **Biomass Power:** Biomass power generation, which involves burning organic materials for energy, may require water for processing feedstocks and cooling systems. Water is also used in the growth and cultivation of biomass crops.

d. **Nuclear Power:** Nuclear power plants use water for cooling purposes. Water is circulated through the reactor core to absorb heat generated during nuclear fission. The heated water is then used to produce steam, which drives turbines connected to generators.

e. **Solar Power:** Certain types of solar power systems, such as concentrating solar power (CSP), use water for cooling purposes. Water can be used as a heat-transfer fluid in CSP systems to absorb and transfer heat from the solar collectors to a power cycle.

f. **Fuel Extraction and Processing:** Water is used in the extraction and processing of fossil fuels (such as oil and natural gas) and in the production of biofuels.

6.2 FLOOD AND ITS MANAGEMENT

Floods are natural disasters characterized by the inundation of normally dry land, often caused by heavy rainfall, storm surges, melting snow, or the overflow of rivers and other bodies of water. A flood is characterized as a major water flow, especially a body of water that rises, swells, and overflows land that is not typically covered: a deluge, a freshet, and an inundation. It is commonly understood to be a phenomenon linked to an exceptionally high stage or flow covering land or coastal area, with extremely negative consequences. Floods can have severe and widespread impacts on communities, ecosystems, and infrastructure. There are various types of flooding mentioned below:

6.2.1 Types of Floods

a. **Riverine Floods:** Result from the overflow of rivers or streams due to heavy rainfall or snowmelt.

b. **Flash Floods:** Rapid and sudden floods occur within a short period, often in urban or mountainous areas.

c. **Coastal Floods:** Caused by storm surges or high tides, leading to the inundation of coastal areas.

d. **Urban Flooding:** Resulting from poor drainage systems, impermeable surfaces, and intense rainfall in urban areas.

e. **Pluvial Flooding:** This is caused by excessive rainfall that overwhelms drainage systems, leading to surface water flooding.

f. **Floods due to Failure of Dams:** A high flood generated due to heavy precipitation could cause a dam failure. This results in a surge of water that travels at a tremendous speed, destroying property and loss of life.

6.2.2 Causes of Floods

The reasons for flooding are the same in all the major river systems. The following is a description of flood causes.

a. **Heavy Rainfall:** Intense and prolonged rainfall can lead to an excess of water that exceeds the capacity of rivers and drainage systems, causing them to overflow.

b. **Storm Surges:** Coastal areas may experience flooding due to storm surges, which are elevated water levels driven by strong winds associated with tropical storms or hurricanes.

c. **Snowmelt:** Rapid melting of snow, especially during warmer periods, can contribute to increased river flow and potential flooding.

d. **Flash Flooding:** Sudden and intense rainfall, often in mountainous or urban areas, can result in flash floods with rapid water accumulation.

e. **Ice Jams:** In colder climates, ice jams occur when river ice blocks the natural flow, causing water to back up and flood the surrounding areas.

f. **Dam or Levee Failures:** The failure of dams or levees due to structural issues, excessive rainfall, or other factors can lead to downstream flooding.

6.2.3 Impacts of Floods

Since the intangible element of flood losses is a determining factor, it is challenging to offer a quantitative estimate of the losses. Typically, the predicted annual damages are calculated using a probabilistic approach. On average, a few hundred lives are lost, and several million individuals are harmed annually. In addition, the loss of livestock results in significant losses. After all, property valued at several hundred crores is lost; this amount does not include the damages and hardships caused by communications failure, the interruption of vital services, environmental degradation, etc. The main impacts of floods are described in the following sections.

a. **Infrastructure Damage:** Floods can damage roads, bridges, buildings, and other critical infrastructure.

b. **Loss of Lives and Property:** Floods pose a significant risk to human life that can result in the loss of homes, businesses, and personal belongings.

c. **Economic Impact:** The costs associated with rebuilding and recovery, as well as disruptions to businesses, can have a substantial economic impact.

d. **Displacement of Communities:** Floods can force people to evacuate their homes and seek temporary shelter, leading to the displacement of communities.

e. **Environmental Consequences:** Floods can cause soil erosion, disrupt ecosystems, and lead to contamination of water sources.

f. **Health Risks:** Floodwaters may be contaminated with pollutants, posing health risks through waterborne diseases and infections.

6.2.4 Flood Management Measures

There are two types of flood management strategies: (i) short-term and (ii) long-term. The type and extent of flood damage, in addition to regional factors, dictate the appropriate course of action. However, long-term interventions are necessary for short-term efforts to be effective.

a. Short-term measures

- Construction of embankments along the low-lying banks in areas prone to periodic flooding.
- Construction of an elevated platform to serve as a temporary shelter during floods.
- Pumping out of water from flooded towns when gravity discharge of floodwater is impractical.
- Constructions of floodwalls near populated regions in towns, cities, and industrial areas.

b. Long-term measures

- Storage reservoirs will be constructed to control floods downstream by reducing the flow peak.
- In the catchment of the hilly area, integrated watershed management provides a decrease in surface runoff and erosion and an increase in infiltration capacity, all of which lessen the impact of flooding.
- Flood forecasting and warning systems, which can be issued with some advance notice, are based on hydro-geomorphological research that helps to minimize property loss and human casualties by moving people to safer areas.

6.2.5 Flood Control Strategies

The following is a summary of the strategy that can be used to reduce flood-related losses:

a. **Flood Control by Structural Methods:** Building dams, dikes, levees, channel alteration, high flow diversion, and land treatment are the tactics to be used under structural measures of flood control. The main objective is to keep people away from the flood zone.

b. **Flood Forecasting:** Flood forecasting measures are classified as non-structural approaches. Here, the probable damage points are given as a flood forecast. During the possibility of flood damage, the human population, along with animals and moveable assets, relocates to a safer location. Preventing people from the flooded region is the main goal here.

c. **Modify Susceptibility to Flood Damage:** To help people prepare for survival and recovery from floods, individual or group actions are implemented. These actions include flood education and awareness campaigns, flood insurance, taxation, relief efforts, and more.

The Flood control strategies in details are described as follows:

I. Structural Flood Control Measures:

a. **Levees:** Levees (Figure 6.2) can be described as earthen embankments built along riverbanks to prevent river overflow and protect adjacent areas.

Figure 6.2: Levees

b. **Dams and Reservoirs:** Dams (Figure 6.3) regulate water flow, store excess water during heavy rainfall, and release it gradually to prevent downstream flooding.

Figure 6.3: Dam & reservoir

c. **Floodwalls:** Vertical barriers made of concrete or other materials constructed along riverbanks or in urban areas to contain floodwaters (Figure 6.4).

Figure 6.4: Flood wall

d. **Channelization:** Modifying the natural course of rivers by straightening, deepening, or widening channels to improve water flow and reduce the risk of flooding.

e. **Retention Basins:** Constructing basins to temporarily hold excess water during heavy rainfall, preventing downstream flooding.

f. **Flood Barriers:** Temporary structures like sandbags, inflatable dams, or deployable barriers erected to protect specific areas during flood events.

II. Non-Structural Flood Control Measures

a. **Land-Use Planning and Zoning:** Implementing regulations to restrict development in flood-prone areas and promote responsible land use.

b. **Afforestation and Reforestation:** Planting trees and vegetation to stabilize soil, reduce erosion, and absorb excess water.

c. **Stormwater Management:** Designing systems that control the flow of rainwater, incorporating permeable pavements, green roofs, and retention ponds.

d. **Elevated Structures:** Constructing buildings and infrastructure above potential flood levels to prevent damage.

III. Early Warning Systems and Preparedness

a. **Flood Forecasting and Monitoring:** Advanced systems to monitor weather conditions, river levels, and potential flood risks.

b. **Emergency Preparedness:** Developing and practicing emergency response plans, including evacuation procedures and coordination among agencies.

c. **Community Education:** Raising awareness about flood risks, promoting insurance coverage, and educating residents on evacuation procedures.

IV. Natural Flood Control Measures

a. **Wetland Restoration:** Preserving or restoring wetlands, which act as natural buffers and absorb excess water.

b. **Riverbank Stabilization:** Implementing measures to prevent erosion and maintain the stability of riverbanks.

c. **Floodplain Management (Figure 6.5):** Allowing floodplains to function naturally by restricting certain types of development and activities.

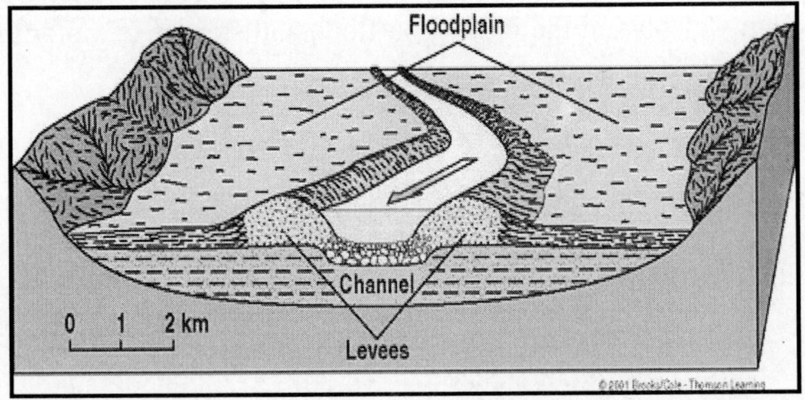

Figure 6.5: Typical sketch of floodplain, levees, and channel

V. International Cooperation

1. **Transboundary Cooperation:** Collaborative efforts between countries to manage and control rivers that cross borders.

2. **Knowledge Sharing:** Sharing expertise, technology, and best practices globally to enhance flood control measures.

The flood control techniques listed above can be applied separately or in combination. The summary of various flood management components and approaches as proposed by the National Commission on Floods (1980) are shown in Figure 6.6.

6.3 ANALYSIS ON SURFACE WATER SUPPLY

Analyzing surface water supply involves a comprehensive assessment of the quantity, quality, and sustainability of water from various surface sources such as rivers, lakes, reservoirs, and ponds. Details of the key aspects involved in the analysis of surface water supply are described below:

6.3.1 Quantity Analysis

a. **Flow Rates:** Measure and monitor the flow rates of surface water sources to understand the volume of water available.

b. **Seasonal Variations:** Assessing how surface water availability changes throughout the year, considering factors like rainfall, snowmelt, and dry seasons.

c. **Water Budgeting:** Calculating the inflow, outflow, and storage of water in surface water bodies to determine the overall water balance.

d. **Drought Analysis:** Evaluating the vulnerability of surface water supply to drought conditions and developing strategies to mitigate potential impacts.

6.3.2 Quality Analysis

a. **Water Quality Testing:** Conduct regular tests for various water quality parameters such as pH, dissolved oxygen, turbidity, nutrients, heavy metals, and pathogens.

b. **Seasonal Variations:** Assessing how surface water availability changes throughout the year, considering factors like rainfall, snowmelt, and dry seasons.

c. **Water Budgeting:** Calculating the inflow, outflow, and storage of water in surface water bodies to determine the overall water balance.

d. **Drought Analysis:** Evaluating the vulnerability of surface water supply to drought conditions and developing strategies to mitigate potential impacts.

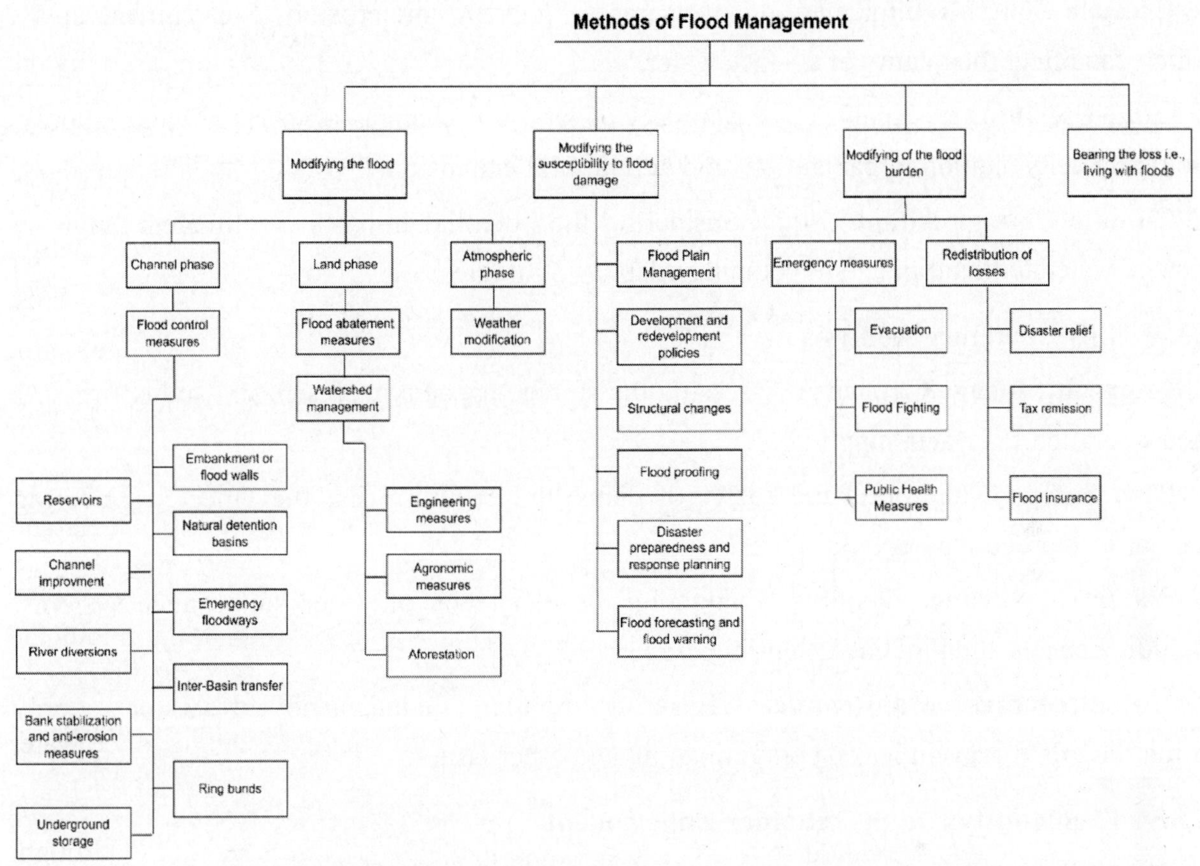

Figure 6.6: Flood management components and approaches (National Commission on Floods, 1980)

e. **Pollution Sources:** Identifying and monitoring potential sources of pollution, including industrial discharges, agricultural runoff, and urban runoff.

f. **Compliance with Standards:** Comparing water quality data with regulatory standards to ensure that the water meets health and environmental requirements.

g. **Water Treatment Needs:** Assess the level of treatment required to meet drinking water standards and identify appropriate treatment technologies.

6.3.3 Sustainability and Management

a. **Aquifer Recharge:** Evaluating natural or artificial methods to recharge aquifers and maintain sustainable water levels.

b. **Erosion Control:** Implementing measures to prevent soil erosion in catchment areas, which can affect the quality of surface water.

c. **Land-Use Impact:** Analyzing the impact of land-use changes on surface water supply, including deforestation, urbanization, and agricultural expansion.

d. **Climate Change Adaptation:** Considering the potential impacts of climate change on surface water availability and developing adaptive strategies.

6.3.4 Infrastructure Analysis

a. **Reservoir Storage Capacity:** Assessing the storage capacity of reservoirs and optimizing their operation for water supply.

b. **Dam Safety:** Monitoring the safety of dams and ensuring that they meet engineering standards to prevent potential failures.

c. **Water Conveyance Systems:** Evaluating the efficiency of water conveyance systems, including canals and pipelines, to minimize losses.

d. **Infrastructure Maintenance:** Ensuring regular maintenance of water supply infrastructure to prevent leaks, contamination, and other issues.

6.3.5 Community and Stakeholder Engagement

a. **Water Demand Analysis:** Analyzing current and future water demand to ensure that supply meets the needs of the community.

b. **Public Awareness:** Collaboration with the community is needed to disseminate awareness regarding sustainable water usage, pollution avoidance, and water conservation.

c. **Stakeholder Collaboration:** Collaborating with government agencies, local communities, and other stakeholders to develop and implement effective water management strategies.

6.3.6 Regulatory Compliance

a. **Permitting and Reporting:** Ensuring compliance with water abstraction permits and reporting requirements set by regulatory authorities.

b. **Environmental Impact Assessment:** Conducting assessments to understand the potential environmental impacts of water abstraction and storage.

6.4 WATER REQUIREMENT OF CROPS

The water requirement of crops, often referred to as crop water demand or irrigation needs, is a critical factor in agricultural management. Every place where irrigation is used has a direct impact on agricultural water requirements. Arid locations, defined as those with an annual rainfall of less than 250–375 mm, are difficult and nearly impossible to produce crops without irrigation; For the production of crops like rice, sugarcane, tobacco, vegetables, and other water-loving plants, irrigation is optional but essential in semi-arid regions (regions with annual rainfall between 375 and 750 mm); in humid regions (regions with annual rainfall above 750 mm), irrigation serves as a buffer against rainfall failure or insufficiency during the crop-growing season However, for optimum crop yield, irrigation is necessary everywhere.

Over a wide area, crops have varying needs for water. Rainfall, crop type, climate, soil properties, growth stage, and local environmental conditions are some of the variables that affect it. Rainfall varies greatly throughout India; in Western Rajasthan, it is only 250 mm, whereas in North Eastern India, it exceeds 5000 mm. In many regions of India, there are also significant differences in soil classifications. Owing to significant variances in rainfall, soil properties, climate, and other factors, a given crop's water needs may fluctuate significantly across the nation.

Optimum use of water: The amount of water needed by a crop to reach its highest yield during growth is known as its optimal requirement. Both irrigation and precipitation-fed water supply contribute to the optimum amount of water.

6.4.1 Factors Affecting Water Requirements of Crops

Here are key considerations related to the water requirement of crops:

a. Crop Types: Different crops have varying water needs. For example, water-intensive crops like rice and sugarcane generally require more water than less thirsty crops such as wheat or millet.

b. Climate: The local climate, including temperature, humidity, and evaporation rates, influences the water requirements of crops. Hot and arid regions often require more irrigation to compensate for higher evaporation rates.

c. Soil Characteristics: Soil type affects water retention and drainage. Sandy soils may require more frequent irrigation, while clayey soils may retain water for longer periods.

d. Growth Stage: The water needs of crops vary during different growth stages. For example, crops often need more water during the flowering and fruiting stages.

e. Irrigation Efficiency: The efficiency of irrigation systems influences the amount of water needed. Efficient systems, such as drip or sprinkler irrigation, can reduce water wastage compared to less efficient methods.

f. Crop Rotation and Intercropping: Crop rotation and intercropping can influence the overall water requirements of a farming system. Some crop combinations may enhance water use efficiency.

g. Topography: Topography is also an important factor which influences the quantity of water required for crops.

h. Rainfall and Natural Water Sources: Regions with ample rainfall may rely less on irrigation, while arid regions may require extensive irrigation. The availability of natural water sources like rivers and lakes also plays an important role in the water requirements for crops.

i. Water Stress Tolerance: Some crops are more tolerant of water stress than others. Drought-resistant varieties may require less water than those sensitive to water shortages.

j. Water Management Practices: Efficient water management practices, such as irrigation scheduling based on crop needs and soil moisture levels, can optimize water use.

k. Socio-Economic Factors: The economic use of water is also influenced by socio-economic considerations. In a particular area, to estimate the optimal amount of water required for various crops, these criteria must be satisfied to the extent possible.

6.5 CROPPING PATTERN

Cropping pattern refers to the distribution of various crops grown in an agricultural area over a specific period. It involves the selection of crops based on factors such as climate, soil type, water availability, market demand, and the region's overall agricultural strategy. The cropping pattern in an area can vary seasonally, annually, or over a more extended period. Here are some key aspects of cropping patterns:

a. Seasonal Cropping Patterns

- **Kharif Season:** Kharif is the monsoon season that starts in June or early July and lasts for a few months. The crops are harvested in September and October, when the southwest monsoon winds blow after being sowed in June or July. This is the main planting season for crops. During the Kharif season, the main crops planted are rice, sorghum, maize, pulses (tur, urad, moong), millets (bajra, jowar), Cotton, jute, and groundnuts, tea, rubber, coffee, oilseeds etc.

- **Rabi Season:** Rabi is the winter season when crops are sown following the monsoon rains and after the kharif season. The crops are seeded in November and harvested in April or May. In the rabi season, crop yield is primarily contingent upon subsurface moisture levels. The major crops grown in rabi season are wheat, Barley, gram and oilseeds like mustard and rapeseed. Pulses are also grown during this season.
- **Zaid Season:** In addition to the two main seasons of Rabi and Kharif, farmers in irrigated areas can plant a third crop between May and July between the Rabi and Kharif seasons. This is a brief season that usually occurs in the summer, referred as *Jayad* (hot weather season). The principal crops of this season are pulses like urad and moong. In addition, during the Jayad season, cucumber, watermelon, muskmelon, and vegetables like tomato and brinjal are also grown.

b. Multiple Cropping: Some regions practice multiple cropping, where more than one crop is cultivated on the same piece of land within a single agricultural year.

c. Crop Rotation: Farmers may adopt crop rotation, changing the type of crop grown in successive seasons to enhance soil fertility and prevent pest and disease build-up.

d. Cash Crops vs. Food Crops: The choice of crops may be influenced by whether they are primarily grown for sale (cash crops) or for local consumption (food crops).

e. Specialization: Certain regions may specialize in the cultivation of specific crops based on agro-climatic conditions. For example, regions with suitable conditions for sugarcane may specialize in sugar production.

f. Government Policies: Government policies, subsidies, and support can influence cropping patterns. For instance, incentives for certain crops may lead to changes in what farmers choose to cultivate.

g. Climate-Resilient Crops: With changing climate patterns, there is a growing emphasis on cultivating crops that are more resilient to climate extremes, such as drought-tolerant varieties.

h. Market Demand: The demand for certain crops in local, national, or international markets can influence cropping decisions. Farmers may choose crops with higher market prices.

i. Crop Diversification: Farmers may diversify their crops to reduce risks associated with dependency on a single crop. This practice also helps in managing soil health and pest control.

6.5.1 Crop and Crop Season in India

India has diverse agro-climatic zones, allowing for the cultivation of a wide variety of crops throughout the year. India's agriculture is characterized by its diversity, and farmers adopt a mix of traditional and modern farming practices. The choice of crops is influenced by factors such as water availability, temperature, and soil fertility. Crop patterns may also vary based on government policies and market demand. The country experiences three main crop seasons: Kharif, Rabi, and Zaid. The main crops and the seasons in which they are usually grown are summarized in section 6.5. In addition to the above-mentioned three main crop seasons, some other non-food crops and their seasons have been described below:

a. Other Non-food crops

- Sugarcane: Grown in various states throughout the year, with a longer growing season.
- Cotton: Primarily grown during the Kharif season in states like Gujarat, Maharashtra, and Andhra Pradesh.
- Tea and Coffee: Plantations in states like Assam, West Bengal, Kerala, Karnataka, and Tamil Nadu.

b. Regional Variations

Different states in India have their own specific crops based on agro-climatic conditions, soil types, and water availability. For example:

- Punjab and Haryana: Known as the "Granary of India" for wheat and rice cultivation during the Rabi and Kharif seasons.
- Andhra Pradesh and Telangana: Major producers of rice, pulses, and cotton.
- Maharashtra: Known for the cultivation of sugarcane, grapes, and oranges.

6.5.2 Duty and Delta

In the context of water resources management, "duty" and "delta" are terms related to the allocation and utilization of water in irrigation systems. These terms are described below in detail:

Duty of water: Duty in irrigation refers to the quantity of water applied to a unit of land over a specific period. It is expressed as the volume of water per unit area of crop or depth of water applied over the cropped area (hectares or acres). Duty of water (D) is expressed as Equation 6.1. Duty is a crucial parameter in irrigation planning and management. It helps to determine the efficiency of water use and ensures that the right amount of water is supplied to crops for optimal growth.

Duty D (mm) = Volume of water applied (m^3) / Irrigated area (hectares) ……(6.1)

a. Factors Influencing Duty

The main factors influencing duty are described below:

- **Crop type**: Different crops have varying water requirements.
- **Climate:** Evapotranspiration rates depend on climate conditions.
- **Soil Type:** Soil characteristics influence water retention and drainage.
- **Stage of Crop Growth:** Water needs vary during different growth stages.

Delta in Irrigation: In the context of irrigation, the term "delta" refers to the change in duty between two points along a water conveyance system (such as a canal). Delta (Δ) is expressed as Equation 6.2. Delta is used to assess the changes in water availability or distribution efficiency along the conveyance network. It helps to identify areas where adjustments or improvements may be needed.

Delta Δ = Duty at Point 2 (downstream point) - Duty at Point 1(upstream point) ......(6.2)

a. Factors Influencing Delta

The main factors influencing delta are described below:

- **Canal Seepage and Conveyance Losses:** Water losses along the canal system can result in variations in duty.
- **Water Flow Control:** Uneven water flow or control issues in the canal may lead to differences in duty.
- **Topography:** Changes in elevation or terrain can affect water distribution.

Example 6.1: If 1000 cubic meters of water is applied to irrigate a 1-hectare field, the duty would be 1000 mm (assuming uniform distribution). Now, if we move along the canal system and find that at a different point, only 900 cubic meters of water is reaching a 1-hectare field, the Delta would be:

Delta = Duty at Point 2 – Duty at Point 1

Delta=900mm-1000mmv= −100mm

This negative delta indicates a reduction in duty, suggesting that adjustments or improvements in water distribution may be needed.

In summary, "duty" quantifies the amount of water applied per unit area in irrigation, while "delta" assesses the changes in duty along a water conveyance system. Both terms are vital in optimizing water use efficiency and ensuring effective irrigation practices.

6.6 QUALITY OF IRRIGATION WATER

The quality of irrigation water is a critical factor in agricultural productivity and soil health. The composition of irrigation water can impact plant growth, soil structure, and the efficiency of water use. Key parameters used to assess the quality of irrigation water include physical, chemical, and biological characteristics. Many countries and organizations have established water quality standards and guidelines for irrigation water to ensure safe and sustainable agricultural practices. Here are some factors and parameters commonly considered when evaluating the quality of irrigation water.

6.6.1 Physical Characteristics

a. **Salinity:** Salinity can be defined as the concentration of dissolved salts in water. High salinity can lead to soil salinization, affecting plant growth and reducing water uptake.

b. **Sediment Load:** Sediment load is known as the amount of suspended particles (sediments) in water. Excessive sediment can clog irrigation systems and affect water infiltration into the soil.

c. **Temperature:** Extreme water temperatures can influence plant growth and microbial activity in the soil.

6.6.2 Chemical Characteristics

a. **pH value:** The pH scale indicates the acidity or alkalinity of water. The range is 0 to 14, whereas 7 is the neutral value. Acidity is indicated by a pH of less than 7, and Alkalinity/ baseness is indicated by a pH of greater than 7. The relative concentration of free hydrogen and hydroxyl ions in water is measured by pH. pH value influences nutrient availability. High alkalinity can affect nutrient availability and soil structure.

b. **Hardness:** The **hardness of water** is due to the presence of soluble bicarbonates, chlorides and sulphates of calcium and magnesium. Hard water is defined as water that does not lather when soap is added. Excessive hardness can affect soil permeability.

c. **Nutrient Content:** Nutrient concentrations in water impact plant growth. Excessive nutrient Contents i.e., Nitrates, Phosphates, Potassium, etc., may lead to water pollution.

d. **Trace Elements:** Excessive levels of trace elements, such as iron, manganese, zinc, etc., in water, can be harmful to plants and stunt their growth.

6.6.3 Biological Characteristics

a. **Microbial Contamination:** Microbial Contamination is defined as contamination of water due to the presence of bacteria, viruses, or other microorganisms. Microbial contamination can pose risks to human health and affect crop safety.

6.6.4 Other Considerations

a. **Heavy Metals:** Accumulation of heavy metals i.e. Lead, Cadmium, Mercury, etc.in soil and crops can have long-term health implications.

b. **Pesticide Residues:** A pesticide is a substance or a mixture of substances used for killing pests: organisms dangerous to cultivated plants or to animals. The term applies to various pesticides such as insecticides, fungicides and herbicides. Pesticide Residues from agricultural chemicals can affect water quality and ecosystem health.

6.6.5 Water Quality Management Strategies

a. **Water Treatment:** On the basis of raw water quality, several treatment processes, like filtration, sedimentation, and chemical treatment, may be adopted.

b. **Precision Irrigation:** Optimizing irrigation practices based on the specific water needs of crops and soil conditions.

c. **Crop Selection:** Choosing crops that are more tolerant to the quality of available irrigation water.

d. **Monitoring and Testing:** Regular testing and monitoring of water quality to detect changes and adjust management practices accordingly.

It is important to note that the suitability of water for irrigation depends not only on individual parameters but also on the interactions between various factors. Sustainable water management practices are crucial for maintaining the quality of irrigation water and ensuring long-term agricultural productivity.

6.7 SOIL-WATER RELATIONSHIP

Particles of varying sizes, both organic and inorganic, make up the porous material known as soil. Compared to fine-textured soil, medium-textured soil with medium-sized particle composition has wider pores, making it a better soil for crop production. For crop growth, soil can serve as a reservoir for storing water that provides moisture to the soil. The water that a plant stores in its root zone can be used to meet its daily needs for water for the crop to grow and develop properly. The soil-water relationship is a critical aspect of agriculture and environmental science, influencing plant growth, nutrient availability, and overall ecosystem health. This relationship is dynamic and involves the interactions between soil and water, affecting the movement, retention, and availability of water in the soil. Here are key aspects of the soil-water relationship.

6.7.1 Soil Properties

a. **Soil Texture:** Soils can be categorized as sand, silt, or clay based on the results of the grain size study. The size of soil particles, water infiltration, drainage, and water-holding capacity are all influenced by the retention percentage of sand, silt, and clay. The United States Department of Agriculture's Triangular Texture Diagram (Figure 6.7) can be used to classify soils based on the percentage of sand, silt, and clay sizes. The flow of soil water, air circulation, and the rate of chemical transformation—all crucial processes for plant life—are significantly impacted by a soil's texture.

b. **Soil Structure:** The arrangement of the soil particles is referred to as the soil structure. It is dependent on the nature and characteristics of the soil water, including its ionic composition, the shape and orientation of the soil particles in relation to one another, their electrical and mineralogical composition, and the forces that interact between the soil water and the soil particles. All aspects of plant growth, including water availability, aeration, nutrient availability, microbial activity, and other critical elements like porosity, permeability, and water retention, are influenced by the structure of the soil.

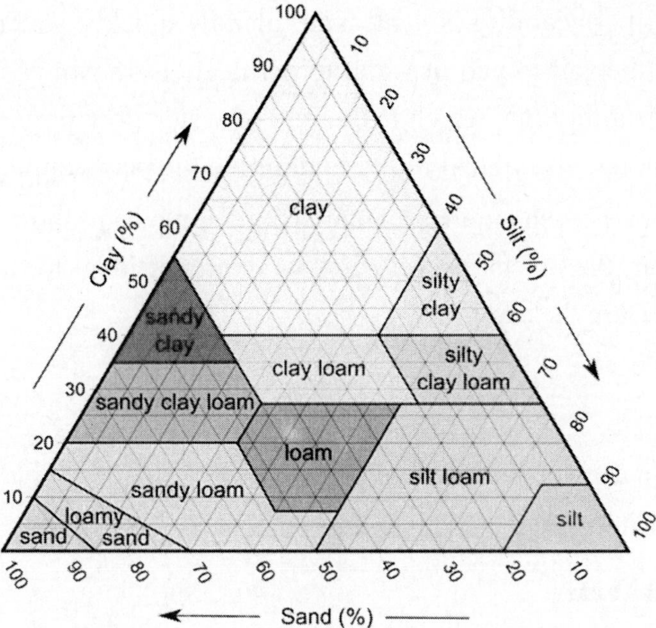

Figure 6.7: Triangular texture diagram

c. **Soil Structure:** The arrangement of the soil particles is referred to as the "soil structure." It is dependent on the nature and characteristics of the soil water, including its ionic composition, the shape and orientation of the soil particles in relation to one another, their electrical and mineralogical composition, and the forces that interact between the soil water and the soil particles. All aspects of plant growth, including water availability, aeration, nutrient availability, microbial activity, and other critical elements like porosity, permeability, and water retention, are influenced by the structure of the soil.

d. **Soil Porosity:** Soil porosity (Equation 6.3) can be referred to as the amount of pores, or open space, between soil particles. The movement of worms, insects, and roots, the dissolving of the parent material of the soil, and/or the expansion of gasses trapped in these spaces by groundwater can all result in the formation of pore spaces. Soil porosity can also be influenced by soil texture. The amount and arrangement of pore spaces in the soil influence water movement and storage.

Porosity (%) = (Volume of Voids / Total Volume) x 100 ......(6.3)

e. **Permeability**: Soil permeability is a measure of how quickly water passes through the soil. The ability of the soil to permit water to permeate is known as its permeability. The coefficient of permeability (k), which is the rate of water flow per unit area of soil under a unit hydraulic gradient, is typically used to describe the permeability of soil.

Darcy's Law: In 1856, a French engineer, Darcy (Equation 6.4) proposed that when the flow through soils is laminar, the discharge velocity (v) is proportional to the hydraulic gradient (i). Thus, according to Darcy's law:

$$\mathbf{v \propto i \text{ or } v = k\,i} \qquad \text{... ....(6.4)}$$

where,

k = constant, known as the coefficient of permeability

Since i is dimensionless, k has the unit of velocity (m/s)

6.7.2 Root Zone Soil Water

The root zone refers to the area of soil around plant roots where they actively absorb water and nutrients. The process by which plants absorb water from the soil through their root systems is called osmosis. It is a critical region for plant growth and development. Soil water within the root zone plays a crucial role in providing the necessary hydration for plants and facilitating the uptake of essential nutrients. Understanding and managing root zone soil water is crucial for sustainable agriculture, landscaping, and overall ecosystem health. It involves considerations of soil type, climate, plant species, and water management practices. Efficient use of water resources in the root zone contributes to the conservation of water and promotes healthier and more productive plant growth. The available moisture capacity of the soil may be calculated as Equation 6.5.

$$\mathbf{AMC = \frac{\gamma d}{100}(\text{Field capacity} - \text{Permanent Wilting point})} \qquad \text{...........(6.5)}$$

where,

γ = Soil density (gm/cm^3)

d = Root zone depth (cm)

Example 6.2: Determine the available moisture capacity of sandy loam soil whose percentage moisture content based of dry weight of soil at field capacity is 12% and at a permanent wilting point is 5%. The bulk density of soil is 1.25 gm/cm^3. The effective depth of root zone of the crop is 0.8 m. If optimal soil moisture is 9% and the water requirement of the crop (consumptive use) is 5 mm/day, compute the theoretical quantity and frequency of irrigation also.

Solution

$$\text{AMC} = \frac{\gamma d}{100}(\text{Field capacity} - \text{Permanent Wilting point})$$

$$= \frac{1.25 \text{ X } 80}{100}(12 - 5)$$

$$= 7 \text{ cm}$$

$$\text{Depth of irrigation required} = \frac{\gamma d}{100}(\text{Field capacity} - \text{Optimum soil moisture})$$

$$\text{Depth of irrigation required} = \frac{1.25 \text{ X } 80}{100}(12 - 9)$$

$$= 3 \text{ cm}$$

Water available for consumption by crop = 3 cm

Water requirement of crop per day = 5 mm/day

$$\text{Number of days or frequency of irrigation} = \frac{3x10}{5}$$

$$= 6 \text{ days}$$

a. Factors affecting root zone soil water

Here are some key points related to root zone soil water:

a. **Water Holding Capacity:** The ability of soil to retain water is known as water holding capacity. Different types of soils have varying water-holding capacities. Sandy soils typically drain quickly, while clayey soils retain water for longer periods.

b. **Root Uptake:** Plant roots absorb water and nutrients from the soil through a process called root uptake. This is a vital physiological process that sustains plant growth, photosynthesis, and overall plant health.

c. **Irrigation:** Managing soil water content in the root zone is essential for crop production. In agriculture, irrigation is often used to supplement natural rainfall and ensure that plants receive an adequate and consistent water supply.

d. **Soil Moisture Monitoring:** Farmers and gardeners use various methods to monitor soil moisture levels in the root zone. This includes visual inspection, soil moisture sensors, and technology-driven irrigation systems.

e. **Aeration:** Adequate soil water content is necessary for proper aeration in the root zone. Soil that is too dry can lead to poor aeration, while waterlogged soil can deprive roots of oxygen.

f. **Root Zone Management:** Effective management of the root zone involves balancing water availability with the needs of the plants. Overwatering or underwatering can have negative effects on plant health and productivity.

g. **Drought Stress:** Insufficient soil water in the root zone can lead to drought stress in plants, affecting their growth, yield, and overall vitality. Drought-tolerant plants are adapted to thrive in environments with limited water availability.

6.7.3 Infiltration

Infiltration refers to of the movement of water from the surface into soil or the process by which water on the ground surface enters the soil. It is a crucial component of the water cycle and plays a significant role in groundwater recharge. Infiltration rate can be defined as soil characteristics determining the maximum rate at which water infiltrates soil under specific conditions. The rate at which water infiltrates into the soil depends on various factors, including soil type, vegetation cover, slope, and the presence of impervious surfaces etc. Understanding infiltration is important for various applications, including water resource management, agriculture, and environmental conservation. It helps to assess the vulnerability of an area to flooding, contributes to groundwater recharge, and informs decisions related to land use planning and water conservation strategies. The factors affecting the Infiltration are described in the following sections.

a. **Soil Characteristics:** The texture and structure of the soil influence its permeability and, consequently, the rate of infiltration. Sandy soils generally have high infiltration rates because they have larger pores, allowing water to move through more easily. Clayey soils with smaller pores tend to have lower infiltration rates.

b. **Vegetation Cover:** Plants and their root systems affect infiltration. Vegetation can enhance infiltration by reducing surface runoff, promoting water absorption, and improving soil structure. Plant roots create channels in the soil, facilitating the movement of water into deeper layers.

c. **Land Use and Land Cover:** Urbanization and the presence of impervious surfaces, such as roads and buildings, can reduce infiltration. Paved surfaces prevent water from entering the soil, leading to increased surface runoff, and potentially contributing to issues like urban flooding.

d. **Slope:** The slope of the land surface influences the rate of infiltration. On steeper slopes, water may run off more quickly, reducing the time available for infiltration. Gentle slopes generally allow for better infiltration.

e. **Rainfall Intensity:** The intensity and duration of rainfall events impact infiltration. Intense rainfall can lead to surface runoff, especially if the soil cannot absorb water quickly enough. Lighter, steady rainfall often allows for more effective infiltration.

f. **Soil Moisture Content:** The existing moisture content in the soil also affects infiltration. Dry soils can initially absorb more water, while saturated soils may experience reduced infiltration rates.

g. **Infiltration Capacity:** This is the maximum rate at which soil can absorb water. It is determined by soil properties and is often depicted graphically as an infiltration curve, showing how the rate of infiltration changes over time during a rainfall event.

h. **Percolation:** The downward movement of water within the soil profile can be described as percolation. Percolation can be affected by gravity, soil texture, and the presence of macrospores.

i. **Capillary Action:** The movement of water in an upward direction through small soil pores can be described as Capillary action. It can be influenced by soil texture and moisture content.

j. **Field Capacity:** The maximum amount of water the soil can hold against gravity after excess water has drained away can be described as field capacity or water retention capacity of the soil. It may be influenced by soil texture and structure.

k. **Wilting Point:** The moisture content at which plants can no longer extract water from the soil is known as the wilting point. It is determined by soil texture and plant species.

l. **Available Water:** The range of soil moisture between field capacity and the wilting point is the volume of available water in the soil. It represents the water available for plant uptake.

6.8 CONSUMPTIVE USE

Consumptive use refers to the portion of water withdrawn or diverted from its source that is consumed or lost in a process and does not return to the original water source. In other words, it is the amount of water that is removed from a water body or aquifer and is not returned, either because it is evaporated, transpired by plants, incorporated into products, or otherwise not available for immediate use. Understanding consumptive use is crucial in water resource management, especially in regions where water scarcity is a concern. It helps assess the sustainability of water use practices and make informed decisions regarding water allocation, conservation measures, and long-term planning. Efforts to reduce consumptive water use often involve improving water use efficiency, implementing water-saving technologies, and promoting sustainable practices in agriculture, industry, and other sectors. There are three main components of consumptive use:

a. **Evaporation:** Water exposed to the atmosphere can be evaporated, particularly from open water surfaces, such as lakes, rivers, and reservoirs. This water is essentially lost to the original source.

b. **Transpiration:** Plants absorb water through their roots and release it into the atmosphere through a process known as transpiration. This water, once transpired by plants, is considered as consumed and does not return to the original water source.

c. **Incorporation into Products:** When water is used in industrial processes, agriculture, or other activities and becomes part of a product or is otherwise not returned to the water source, it is considered for consumptive use.

The term evapotranspiration or consumptive use expresses the entire amount of water returned to the atmosphere by the combined processes of transpiration and evaporation throughout the growth of the plant. The evapotranspiration or consumptive use can be expressed as hectare cm. Therefore, the consumptive use is defined as the total amount of water supplied to the field by irrigation and rainfall minus the amount of water removed as surface runoff and plus or minus the amount of water absorbed by the soil moisture in the root zone, neglecting of course deep percolation losses. The term evapotranspiration or consumptive use can be expressed as equation 6.6.

$$\mathbf{U = W_t + W_r - W_e + W_m} \qquad \text{...........(6.6)}$$

where,

U = Total Consumptive use

W_t = Irrigation Water Supply

W_r = Rainfall

W_e = Surface Runoff

W_m= Difference in soil moisture content at the root zone at the beginning and end of plant growth.

6.9 IRRIGATION REQUIREMENT

Irrigation requirement refers to the amount of water that needs to be applied to a particular area of land to supplement natural rainfall and meet the water needs of crops or plants. The irrigation requirement is influenced by various factors, and determining it accurately is crucial for efficient water management in agriculture and landscaping. Accurately assessing irrigation requirements helps to prevent over-irrigation, which can lead to water wastage, environmental issues, and increased production costs. Conversely, under-irrigation can result in reduced crop yields and overall plant health. By considering these factors, farmers and land managers can develop effective irrigation plans that balance water conservation with the needs of crops or plants.

6.9.1 Factors Affecting Irrigation Requirement

The key factors that affect irrigation requirements are described below:

a. **Climate:** The local climate, including temperature, humidity, wind, and solar radiation, plays a significant role in determining the water needs of plants. Hot and arid regions typically require more irrigation compared to cooler and humid areas.

b. **Crop Type:** Different crops have varying water requirements at different growth stages. For example, some crops are more drought-tolerant than others, and the type of crop being cultivated influences the irrigation schedule and water needs.

c. **Soil Type:** Soil characteristics, such as texture, structure, and water-holding capacity, affect how water moves through the soil and how much it can be retained for plant use. Ex. Sandy soils may require more frequent irrigation than clayey soils.

d. **Plant Growth Stage:** The water needs of plants vary throughout their growth cycle. For instance, during germination and early growth, plants often require more frequent irrigation, while mature plants may have different water needs.

e. **Irrigation System Efficiency:** The type of irrigation system used, such as drip irrigation, sprinklers, or flood irrigation, affects how efficiently water is delivered to the plants. Modern and efficient irrigation systems can help to reduce water wastage.

f. **Local Water Availability:** The availability of water resources in a specific region, including surface water and groundwater, influences irrigation practices. Water scarcity may require more careful water management and conservation measures.

g. **Weather Patterns:** Seasonal variations and weather patterns, such as droughts or unusually wet periods, can impact irrigation requirements. Adjustments to irrigation schedules may be needed in response to changing weather conditions.

h. **Management Practices:** The practices employed by farmers or land managers, including scheduling, monitoring, and optimizing irrigation, can impact the overall irrigation requirement.

Accurately assessing irrigation requirements helps to prevent over-irrigation, which can lead to water wastage, environmental issues, and increased production costs. Conversely, under-irrigation can result in reduced crop yields and overall plant health. By considering these factors, farmers and land managers can develop effective irrigation plans that balance water conservation with the needs of crops or plants.

6.9.2 Frequency of Irrigation

The frequency of irrigation, or the schedule at which water is applied to plants or crops, is a critical aspect of efficient water management in agriculture and landscaping. The ideal irrigation frequency depends on various factors, and finding the right balance is essential for promoting plant health while avoiding water wastage. Farmers and land managers often use a combination of these factors to develop irrigation schedules that maximize water use efficiency and promote optimal plant growth. It is important to strike a balance between providing enough water to meet plant needs and avoiding over-irrigation, which can lead to water wastage and other environmental concerns.

The frequency of irrigation is calculated by dividing the potential quantity of soil moisture loss by the daily consumptive consumption. The following relation (Equation 6.7) can be used to calculate the depth of watering at each irrigation to raise the moisture content in a soil of depth d to the field capacity w_{fc}. The irrigation efficiency can be taken as E_a.

Depth of water to be applied $= \frac{((w_{fc}-w)\times d)}{E_a}$ **...........(6.7)**

Example 6.3: A crop's consumptive water use at a given stage of growth is 2.6 mm/day. When the amount of water in the soil is 30% of the maximum depth of water available in the root zone, which is 75 mm, calculate the number of days between irrigations and the depth of water to be administered. Assume that 70% of irrigation is efficient.

Solution

$$\text{Frequency of irrigation} = \frac{\text{potential quantity of soil moisture loss}}{\text{daily consumptive consumption}}$$

$$\text{Frequency of irrigation} = \frac{((1 - 0.30) \times 75)}{2.6}$$

$$= 20.19 \text{ days}$$

$$= 20 \text{ days (say)}$$

$$\text{Depth of water to be applied} = \frac{((w_{fc}-w)\times d)}{E_a}$$

$$\text{Depth of water to be applied} = \frac{((1 - 0.30) \times 75)}{0.70}$$

$$= 75 \text{ mm}$$

a. Factors affecting frequency of irrigation: The key factors that influence the frequency of irrigation are described below.

a. **Soil Type:** Different soil types have varying water-holding capacities. Sandy soils drain water more quickly and may require more frequent irrigation, while clayey soils retain water for longer periods. Understanding the soil type helps to determine the appropriate irrigation frequency.

b. **Plant Water Needs:** Different plants have varying water requirements at different stages of growth. The growth phase, maturity, and specific water needs of the crops or plants being cultivated influence how often irrigation is needed.

c. **Climate:** Local climate conditions, including temperature, humidity, wind, and sunlight, affect the rate of evaporation and transpiration. Hot and dry climates typically require more frequent irrigation, while cooler and more humid conditions may necessitate less frequent watering.

d. **Weather Conditions:** Seasonal variations, weather patterns, and precipitation levels impact irrigation frequency. Adjustments to the irrigation schedule may be necessary in response to changes in weather conditions, such as droughts or unusually wet periods.

e. **Irrigation System Efficiency:** The type of irrigation system used influences how water is delivered to plants. Drip irrigation and sprinkler systems are more efficient than traditional flood irrigation, and their design can impact the frequency of irrigation needed.

f. **Water Availability:** The availability of water resources, including surface water and groundwater, in a specific region influences irrigation practices. Water scarcity may necessitate more careful water management and potentially lead to less frequent irrigation.

g. **Root Depth:** Understanding the depth of the plant roots helps determine how deep water needs to penetrate into the soil. Tailoring the irrigation frequency to match root depth ensures that water reaches the active root zone.

h. **Cultural Practices:** Farming or landscaping practices, such as mulching and soil amendments, can affect the moisture retention of the soil and, consequently, the frequency of irrigation required.

i. **Monitoring and Technology:** Regular monitoring of soil moisture levels, weather forecasts, and advancements in irrigation technology, such as soil moisture sensors, can help to optimize irrigation scheduling.

6.9.3 Methods of Irrigation

There are several methods for applying water to the fields in agriculture, and the choice of method depends on factors such as water availability, cost, energy requirements, crop type, soil characteristics, local climate conditions, and available resources. Efficient irrigation practices aim to deliver water to crops in a way that maximizes water use efficiency, minimizes water waste, and promotes optimal plant growth. Advances in technology, such as soil moisture sensors and automated irrigation systems, have further improved the precision and efficiency of water application in agriculture. Some common methods of irrigation to fulfil the requirement of water for agriculture are surface irrigation, drip irrigation, sprinkler irrigation, and sub-surface irrigation. These methods are described in the following sections.

I. Surface Irrigation

Surface irrigation (Figure 6.8) is a method of applying water to crops or plants by allowing it to flow over the soil surface. This method relies on the force of gravity to distribute water across the field. There are several types of surface irrigation methods such as the Furrow Irrigation method, the Basin Irrigation method, and the Border Irrigation method. Each method, with its own characteristics, is described below.

Figure 6.8: Surface irrigation

a. **Furrow Irrigation:** In furrow irrigation, water is diverted along with small channels or furrows between rows of crops. Water is allowed to flow down these furrows, and it infiltrates into the soil to reach the root zone. This method is suitable for row crops and is widely used in various agricultural settings.

b. **Basin Irrigation:** In this method, water is applied to a defined basin or depression around the individual plants or groups of plants. Water is applied directly into these basins, allowing it to infiltrate the soil around the plants. This method is commonly used for tree and vine crops.

c. **Border Irrigation:** In border irrigation, water is applied to long, narrow strips of land with a slight slope, bordered by levees or ridges. Water is released at the upper end of the strip, and it flows slowly across the field, irrigating the entire strip. This method is suitable for crops with uniform water requirements.

- ***Advantages of Surface irrigation methods:*** Surface irrigation methods have following advantages:
    - *Simplicity*: Surface irrigation methods are relatively simple to implement and require minimal equipment.
    - *Low infrastructure cost*: Compared to some other irrigation methods, surface irrigation systems often have lower upfront costs.
    - *Suitable for various crops*: Surface irrigation can be adapted for a wide range of crops and field configurations.
- **Challenges *of Surface irrigation methods:***
    - *Water distribution unevenness*: Achieving uniform water distribution across the field can be challenging, leading to overwatering in some areas and underwatering in others.
    - *Water loss*: Evaporation and runoff can result in water loss, reducing the overall efficiency of surface irrigation.
    - Soil erosion: The movement of water across the soil surface can contribute to soil erosion, particularly on sloping terrain.

Efficient water management practices, proper field design, and appropriate scheduling can help address some of the challenges associated with surface irrigation. Additionally, advancements in technology, such as precision land levelling and automated irrigation systems, have improved the efficiency and effectiveness of surface irrigation methods.

II. Drip or Trickle Irrigation

Drip irrigation or Trickle irrigation (Figure 6.9) is a method of applying water directly to the root zone of plants in a controlled and efficient manner. The term "trickle irrigation" is less common than "drip irrigation," but the two are essentially synonymous and describe the same water-efficient irrigation practice. This irrigation technique involves the slow, precise delivery of water through a network of tubes, pipes, valves, and emitters. Drip irrigation is known for its water efficiency, as it minimizes water wastage, reduces evaporation, and allows for targeted water application. Drip irrigation is widely used in agriculture, horticulture, and landscaping. It offers a sustainable and efficient solution for water management, particularly in regions facing water scarcity or where conservation is a priority. The key components and features of drip/ trickle irrigation are described as follows:

a. **Components of Drip or Trickle Irrigation**

- **Drip Tubing or Pipes:** Drip tubing, often made of flexible polyethene, is used to transport water from the water source to the plants. The tubing can be laid on the soil surface, buried below the ground, or suspended above the plants, depending on the specific application.
- **Emitters:** Emitters are devices attached to the drip tubing that release water in controlled amounts directly to the root zone of plants. Common types of emitters include drip emitters, micro-sprinklers, and soaker hoses.
- **Filters:** Filters are installed in the drip system to prevent clogging of emitters by removing particles and debris from the water. Screen or disc filters are commonly used in drip irrigation systems.
- **Pressure Regulators:** Pressure regulators ensure a consistent and optimal pressure level within the drip system, preventing damage to emitters and ensuring uniform water distribution.
- **Valves:** Valves control the flow of water through the drip system. They are used to start and stop irrigation and can be manually or automatically operated.
- **Backflow Preventers:** Backflow preventers are installed to ensure that water from the irrigation system does not flow back into the main water supply, preventing contamination.
- **Mainline and Sub-main:** The mainline is the primary pipe that delivers water from the water source to the different sections of the irrigation system. Sub-main lines branch off the mainline and distribute water to specific zones or areas within the field.

Figure 6.9: Drip/Trickle irrigation

b. **Advantages of Drip Irrigation:**

- **Water Efficiency:** Drip irrigation is highly efficient, delivering water directly to the root zone, minimizing evaporation and reducing water wastage.
- **Precision Irrigation:** The ability to precisely control the amount of water delivered to each plant allows for tailored irrigation based on plant needs.
- **Weed Control:** By delivering water only to the root zone, drip irrigation helps minimize moisture on the soil surface, reducing weed growth.
- **Nutrient Application (Fertigation):** Drip systems can be integrated with fertigation, allowing for the simultaneous application of water and fertilizers.

c. **Challenges and Considerations:**

- **Initial Cost:** The upfront cost of installing a drip irrigation system can be higher compared to some other irrigation methods.
- **Clogging Risk:** Drip emitters are susceptible to clogging, and regular maintenance is required to prevent issues.
- **Suitability for Different Crops:** Drip irrigation is suitable for a wide range of crops, but some crops may have specific requirements or may not be well-suited for this method.

III. Sprinkler Irrigation

Sprinkler irrigation is a method of applying water to crops or plants in a controlled manner by distributing it in the form of droplets or fine spray through a network of pipes, pumps, and sprinklers. This irrigation technique is widely used in agriculture, landscaping, and sports field maintenance. Sprinkler systems can be fixed or mobile and are suitable for a variety of crops. Sprinkler irrigation is versatile and can be adapted to various crops, including field crops, orchards, and vegetable gardens. The choice of sprinkler system type depends on factors such as field size, crop type, and local climate conditions. Efficient water management practices, proper system design, and regular maintenance are essential for maximizing the benefits of sprinkler irrigation.

Originally, sprinkler irrigation (Figure 6.10) was mostly used in hills for plantation crops in the North Eastern States and the Western Ghats, such as tea, coffee, and other crops. Following the emergence of widespread issues related to salinity, waterlogging, and lack of water in certain areas, the government encouraged farmers to switch to sprinkler irrigation, frequently by offering subsidies. Nowadays, sprinkler systems with spinning heads and perforated pipes are both in use in India. Indian farmers are also starting to use rain gun sprinkler systems. States like Rajasthan, Madhya Pradesh, and Haryana use sprinklers a lot. Apart from rice, these work incredibly well with all closely spaced crops, including cereals, oil seeds, pulses, and other income crops. The key components and features of sprinkler irrigation are given below:

Figure 6.10: Sprinkler irrigation

a. **Components of Sprinkler System**

- **Pipes and Pumping System:** Water is transported from its source (such as a well or reservoir) to the field through a system of pipes. A pump is often used to pressurize the water for efficient distribution.
- **Sprinklers:** These are devices attached to the pipes that release water in the form of droplets or spray. Sprinklers are available in various types, including impact, rotary, stationary, and oscillating, each suitable for different applications.

b. **Types of Sprinkler Systems**

- **Fixed or Stationary Sprinklers:** These are mounted in a fixed position and irrigate a specific area without moving. They are often used in smaller-scale applications.
- **Rotary Sprinklers:** These rotate in a circular or semi-circular pattern, covering a larger area. Rotary sprinklers are suitable for larger fields or lawns.
- **Oscillating Sprinklers:** These move back and forth, creating a fan-shaped pattern. Oscillating sprinklers are often used for lawns and gardens.
- **Impact Sprinklers:** These have a rotating arm that generates a pulsating spray pattern. Impact sprinklers are commonly used in agriculture for larger crop areas.

c. **Advantages of Sprinkler Irrigation**

- **Uniform Water Distribution:** Sprinkler systems can provide even water coverage across the field, promoting uniform plant growth.
- **Flexibility:** Sprinkler systems can be adjusted for different crops and field shapes.
- **Reduced Soil Erosion:** By delivering water in a controlled manner, sprinklers can help minimize soil erosion compared to certain surface irrigation methods.

d. **Challenges and Considerations**

- **Evaporation and Wind Drift:** Wind and high temperatures can lead to water loss through evaporation and wind drift, affecting the efficiency of sprinkler systems.
- **Initial Cost:** The installation of a sprinkler system can involve higher upfront costs compared to some other irrigation methods.
- **Maintenance:** Regular maintenance is necessary to ensure the proper functioning of sprinklers, including cleaning nozzles and checking for clogs.

e. Center Pivot Irrigation

- A specialized form of sprinkler irrigation, centre pivot systems involve a rotating sprinkler system mounted on wheeled towers that move in a circular pattern. Centre pivot systems are commonly used for large-scale agriculture, providing efficient coverage for circular or square fields.

IV. Sub-surface Irrigation

- Sub-surface irrigation is a method of delivering water directly to the root zone of plants below the soil surface. This method helps to reduce evaporation losses and is suitable for certain crops. Unlike surface irrigation methods, where water is applied to the soil surface and allowed to flow over the ground, subsurface irrigation aims to minimize water loss through evaporation and reduce surface runoff. This method is particularly useful in areas where water conservation and efficiency are top priorities.
- The method is similar to drip irrigation; however, in this method, similar to traditional drip irrigation, subsurface drip irrigation involves burying the drip tubing or pipes below the soil surface. This system is typically installed at a shallow depth to ensure that water reaches the root zone effectively. This method is suitable for row crops, orchards, and vineyards.
- In some cases, subsurface irrigation may involve flooding the soil below the surface through buried pipes or tubes. This method is less common and may be used in specific agricultural applications.
- Subsurface irrigation can be a valuable tool in promoting efficient water use, especially in regions where water resources are limited. Proper design, installation, and maintenance are crucial for maximizing the benefits of subsurface irrigation systems.

a. Advantages of Subsurface Irrigation

- **Water Conservation:** Subsurface irrigation minimizes water loss due to evaporation and surface runoff, making it a more water-efficient method.
- **Reduced Weed Growth:** By delivering water directly to the root zone, subsurface irrigation can help reduce weed growth since surface moisture is minimized.

- **Improved Nutrient Management:** Subsurface irrigation systems can be combined with fertigation (the application of fertilizers through the irrigation system) for efficient nutrient delivery to plants.

b. Challenges and Considerations

- **Initial Cost:** The installation of subsurface irrigation systems can have higher upfront costs compared to some surface irrigation methods.
- **Maintenance:** Proper maintenance is essential to prevent clogging of emitters or tubes and ensure the efficient operation of the system.
- **Suitability for Different Crops:** While subsurface irrigation is effective for many crops, the suitability may vary depending on the type of crop and soil conditions.

V. Flood Irrigation

Water is applied by allowing it to flow over the entire field. Flood irrigation is often used in flat or gently sloping areas.

SUMMARY

The term "water withdrawals and uses" refers to the crucial process of drawing water from its many sources, including rivers, lakes, subterranean aquifers, etc., for the purpose to meet the needs of many industries, including irrigation, hydropower production, home usage, and agriculture. The maintenance of ecosystems and human activity depends on these withdrawals, which are a vital component of water management. This unit describes the demand for water for household consumption, agriculture, particularly the amount needed for Indian crops and crop seasons, energy production, and other applications. Additionally, there has been discussion of the examination of surface and groundwater supplies, including duty and delta, irrigation water supply, soil-water interaction, root zone soil water, infiltration, and consumptive usage. Furthermore, an explanation of irrigation needs, how often they occur, and different irrigation techniques, such as surface, sub-surface, sprinkler, and trickle/drip irrigation, have been discussed.

EXERCISE

Revision Questions

1. What do you understand by the demand for water?
2. Describe the process of water withdrawals and uses?
3. Categorize various sectors for the withdrawal of water?
4. What do you understand by irrigation? Explain different types of irrigation.
5. Explain Hydropower. How is the water converted in Electrical energy?
6. Explain the advantages of Hydropower Generation.
7. Why are floods called natural disasters?
8. Explain the different types of floods.
9. Explain different causes of flood.
10. What are the impacts of floods on life and property?
11. Explain short-term and long-term flood management measures.
12. Explain in brief the different strategies that is used for flood control as well as to reduce flood-related losses.
13. Describe Structural and Non-structural strategies for flood control.
14. Describe key aspects involved in the analysis of surface water supply.
15. What do you understand about the optimum use of water?
16. What are the key factors that affect the Water Requirements of Crops?
17. What do you understand by Cropping Pattern? Describe some key aspects of cropping patterns.
18. What do you understand by multiple cropping?
19. Explain the main crops and crop seasons in India.
20. Define the duty of water. What are the main factors influencing the duty of water?
21. Define the delta of water. What are the main factors influencing the delta of water?
22. What is the relation between duty and delta?
23. Explain key parameters to maintain the quality of irrigation water and to ensure long-term agricultural productivity.

24. Explain the soil-water relationship. Define key aspects of the soil-water relationship.
25. Draw the United States Department of Agriculture's Triangular Texture Diagram. Also, classify the soils based on the percentage of sand, silt, and clay sizes?
26. Explain the characteristics of porosity and permeability of soil.
27. Explain osmosis. How can the available moisture capacity of the soil be calculated?
28. Explain root zone soil water. Define factors affecting root zone soil water.
29. What do you understand by infiltration? Describe several factors affecting the Infiltration.
30. Describe the consumptive use of water. Describe the main components of the consumptive use of water.
31. What do you understand by irrigation requirement? Describe key factors that affect irrigation requirements.
32. Describe the frequency of irrigation. Explain the key factors that affect the frequency of irrigation.
33. Describe several methods of Irrigation.
34. Explain surface irrigation. What are the advantages and challenges of surface irrigation?
35. Explain drip irrigation. What are the components of drip irrigation?
36. Describe the advantages of drip irrigation.
37. What are the components of sprinkler irrigation? Explain the advantages & challenges of sprinkler irrigation.
38. What do you understand by subsurface irrigation? Explain the advantages and challenges of subsurface irrigation.
39. What do you understand by flood irrigation?

Numerical Problems

1. If 5000 cubic meters of water is applied to irrigate a 5-hectare field, what will be the duty of water assuming uniform distribution? Now, if we move along the canal system and find that at a different point, only 3500 cubic meters of water is reaching at a 5-hectare field, what will be the delta?
2. Determine the available moisture capacity of sandy loam soil whose percentage moisture content based of the dry weight of soil at field capacity is 11% and at a permanent wilting

point is 6%. The bulk density of soil is 1.35 gm/cm^3. The effective depth of the root zone of the crop is 0.9 m. If optimal soil moisture is 9% and the water requirement of the crop (consumptive use) is 6 mm/day, compute the theoretical quantity and frequency of irrigation also.

3. A crop's consumptive water use at a given stage of growth is 2.8 mm/day. When the amount of water in the soil is 35% of the maximum depth of water available in the root zone, which is 85 mm, calculate the number of days between irrigations and the depth of water to be administered. Assume that 75% of irrigation is efficient.

Multiple Choice Questions

1. What is the largest global use of freshwater?

A) Industrial B) Agricultural
C) Domestic D) Recreational

2. Which of the following is a primary use of water in households?

A) Irrigation B) Cooling industrial processes
C) Drinking and sanitation D) Hydroelectric power generation

3. What percentage of freshwater is typically used for industrial purposes?

A) 10% B) 20%
C) 30% D) 50%

4. Which sector is the fastest-growing consumer of water globally?

A) Domestic B) Agricultural
C) Industrial D) Energy production

5. What is one of the main uses of water in agriculture?

A) To generate electricity B) To create recreational lakes
C) To irrigate crops D) To cool power plants

6. What is an example of recreational use of water?

A) Drinking water supply B) Fishing and boating
C) Industrial cooling D) Crop irrigation

7. How is water primarily used in energy production?

A) As a cleaning agent

B) For irrigation

C) In hydropower plants to generate electricity

D) For drinking water supply

8. What is the main use of water in the textile industry?

A) Irrigation
B) Cooling processes
C) Dyeing and finishing fabrics
D) Drinking water supply

9. What is the primary purpose of irrigation?

A) To improve soil fertility
B) To provide water to crops
C) To control pests
D) To enhance biodiversity

10. Which type of irrigation delivers water directly to the root zone of plants?

A) Surface irrigation
B) Drip irrigation
C) Flood irrigation
D) Sprinkler irrigation

11. What is a key characteristic of surface irrigation?

A) Water is applied under pressure
B) Water flows over the soil surface to reach plants
C) Water is distributed through a network of pipes
D) Water is applied directly to the leaves

12. Which irrigation method is most water-efficient?

A) Flood irrigation
B) Drip irrigation
C) Furrow irrigation
D) Sprinkler irrigation

13. What is the main advantage of sprinkler irrigation?

A) It requires less labour
B) It can be used on uneven terrain
C) It minimizes water evaporation
D) It is less expensive to install

14. What is flood irrigation?

A) Water is sprayed over crops
B) Water is applied to furrows between rows of crops
C) Water is allowed to flow freely over the field
D) Water is distributed through a network of underground pipes

15. Which type of irrigation is best suited for orchards and vineyards?

A) Flood irrigation
B) Surface irrigation
C) Drip irrigation
D) Sprinkler irrigation

16. What is a common disadvantage of flood irrigation?

A) High installation cost
B) Risk of waterlogging and soil erosion
C) Inefficiency in water usage
D) Difficulty in maintenance

17. What is a flood?

A) An excess of water in a river or stream

B) A temporary overflow of water onto normally dry land

C) A tsunami

D) A drought condition

18. Which of the following is NOT a type of flood?

A) Flash Flood
B) River Flood
C) Coastal Flood
D) Solar Flood

19. What primarily causes flash floods?

A) Heavy rainfall over a long period
B) Sudden, intense rainfall in a short time
C) Snowmelt
D) Tidal waves

20. Which of the following can contribute to flooding?

A) Deforestation
B) Urbanization
C) Poor drainage systems
D) All of the above

21. What is a common method of flood management?

A) Ignoring the problem
B) Constructing levees and floodwalls
C) Planting more trees without planning
D) Building houses in flood-prone areas

22. Which type of flood is most likely to occur after heavy rainfall on already saturated ground?

A) River Flood
B) Coastal Flood
C) Flash Flood
D) Tidal Flood

23. Which of the following is a long-term flood management strategy?

A) Emergency response plans
B) River channelization
C) Evacuation routes
D) Sandbagging

24. What role do wetlands play in flood management?

A) They exacerbate flooding

B) They provide recreational areas only

C) They absorb excess water and reduce runoff

D) They have no impact on flooding

25. Which of the following technologies is often used to predict floods?

A) Satellite imagery B) Radio communication

C) Social media D) Printed newspapers

26. What is the primary purpose of floodplain zoning?

A) To encourage development in flood-prone areas

B) To restrict development in areas susceptible to flooding

C) To allow unrestricted access to rivers

D) To promote tourism

27. What is one major social impact of flooding?

A) Increased employment opportunities B) Displacement of communities

C) Improved public health D) Enhanced educational access

28. Which of the following is an economic consequence of floods?

A) Rise in property values B) Decrease in agricultural productivity

C) Boost in tourism D) Increase in job creation

29. What type of flood control strategy involves altering the physical landscape to manage water flow?

A) Legislative measures B) Structural measures

C) Community education D) Emergency response planning

30. Which flood control method involves building barriers along riverbanks?

A) Floodplain zoning B) Levees

C) Flood forecasting D) Wetland restoration

31. What is a common environmental impact of flooding?

A) Decrease in biodiversity B) Improvement of air quality

C) Restoration of habitats D) Increased soil fertility

32. Which of the following is NOT a flood control strategy?

A) Constructing reservoirs
B) Deforestation
C) Building floodwalls
D) Implementing early warning systems

33. What is the goal of floodplain zoning?

A) To develop flood-prone areas for housing
B) To restrict certain activities in flood-prone areas
C) To promote industrial growth
D) To enhance recreational activities

34. Which flood control strategy focuses on community preparedness and education?

A) River channelization
B) Public awareness campaigns
C) Flood insurance policies
D) Building dams

35. What is one psychological impact of flooding on affected communities?

A) Increased community cohesion
B) Higher rates of anxiety and depression
C) Enhanced trust in government
D) Improved mental health services

36. Which of the following is a non-structural flood management approach?

A) Building levees
B) Creating wetlands
C) Constructing dams
D) Installing flood barriers

37. What is meant by "crop water requirement"?

A) The amount of water needed for crop growth and development
B) The amount of rainfall received during the growing season
C) The total water available in the soil
D) The water needed for pest control

38. What does the term "delta" refer to in irrigation?

A) The area of land cultivated
B) The depth of water applied over a specific area
C) The rate of evaporation from the soil
D) The nutrient content of the soil

39. Which of the following factors does NOT affect crop water requirements?

A) Soil type
B) Crop type
C) Economic conditions
D) Weather conditions

40. What is the "duty" of water in irrigation?

A) The total volume of water used in a season
B) The area irrigated per unit of water supplied
C) The amount of water needed to grow one crop
D) The cost of irrigation per hectare

41. If a crop has a duty of 1.5 hectares per cubic meter per season, how many hectares can be irrigated with 1000 cubic meters of water?

A) 150 hectares
B) 100 hectares
C) 75 hectares
D) 50 hectares

42. Which of the following crops generally requires the highest water input?

A) Wheat
B) Barley
C) Rice
D) Sorghum

43. What is a primary method for calculating crop water requirements?

A) Soil texture analysis
B) Evapotranspiration estimation
C) Crop rotation planning
D) Fertilizer application rates

44. How can water use efficiency be improved in irrigation?

A) Using more water
B) Implementing drip irrigation systems
C) Delaying planting
D) Increasing crop density

45. In irrigation terminology, what does a higher delta indicate?

A) Increased crop yield
B) More water loss
C) Greater depth of water applied
D) Improved soil fertility

46. Which of the following statements is true regarding crop water requirements?

A) They remain constant throughout the growing season.
B) They vary based on growth stage and climate.
C) They are only influenced by soil type.
D) They are irrelevant in drought conditions.

47. What is infiltration in the context of soil and water?

A) The process of water evaporating from soil

B) The movement of water into the soil surface

C) The collection of water in surface runoff

D) The retention of water in plant roots

48. Which factor does NOT affect the rate of infiltration?

A) Soil texture B) Soil structure

C) Air temperature D) Vegetation cover

49. What is the term for the water held in the soil that is available for plant use?

A) Gravitational water B) Capillary water

C) Hygroscopic water D) Runoff water

50. How does soil texture influence water retention?

A) Coarse soils retain more water

B) Fine soils retain less water

C) Coarse soils drain quickly and retain less water

D) All soils retain water equally

51. Which soil property is most important for determining infiltration rates?

A) Soil colour B) Soil pH

C) Soil texture D) Soil organic matter

52. What is the "field capacity" of soil?

A) The maximum amount of water the soil can hold

B) The amount of water that drains from the soil

C) The water content after excess water has drained

D) The amount of water available to plants

53. In the root zone, which type of water is most crucial for plant uptake?

A) Gravitational water B) Capillary water

C) Hygroscopic water D) Groundwater

54. Which of the following conditions can lead to reduced infiltration?

A) Compacted soil
B) Sandy soil
C) Mulched surfaces
D) Well-structured soil

55. What is the primary function of the root zone in relation to soil water?

A) To store excess water for runoff
B) To facilitate drainage and prevent flooding
C) To supply water and nutrients to plants
D) To hold water for evaporation

Answer: 1-B; 2-C; 3-B; 4-C; 5-C; 6-B; 7-C; 8-C; 9-B; 10-B; 11-B; 12-B; 13-B; 14-C; 15-C; 16-B; 17-B; 18-D; 19-B;20-D; 21-B; 22-C; 23-B; 24-C; 25-A; 26-B; 27-B; 28-B; 29-B; 30-B; 31-A; 32-B; 33-B; 34-B; 35-B; 36-B; 37-A; 38-B; 39-C; 40-B; 41-A; 42-C; 43-B; 44-B; 45C; 46-B; 47-B; 48-C; 49-B; 50-C; 51-C; 52-C;53-B;54-A;55-C.

KNOW MORE

For more information related to this topic scan the QR code.

OR

Type this link in your browser

https://drive.google.com/file/d/14mY-eSeigDdYshkf2uyqjfUQFSjubMnJ/view?usp=sharing

BIBLIOGRAPHY

1. Asawa, G.L. (1993), "Irrigation Engineering", Published by Willey Eastern Limited, ISBN 81-224-0457-X.
2. Asawa, G.L. (2005), "Irrigation and Water Resources Engineering, New Age International (P) Ltd., ISBN-13: 978-8122416732
3. Chow. V.T, (Ed), (1964), "Handbook of Applied Hydrology", McGraw-Hill, New York, NY, 1964.
4. Ghosh, S.N. (2014), "Flood control and drainage engineering", Published by CRC Press/Balkema, The Netherlands, ISBN: 978-1-138-02627-8.

5. Jain S. K., Agarwal, P. K. And Singh, V. P. (2007), Hydrology and Water Resources of India, Water Science and Technology Library, Vol 57, Springer, The Netherland, ISBN-10 1-4020-5179-4(HB).
6. Punmia, B.C. and Lal, Pande B.B., (1992), "Irrigation and Water Power Engineering", Twelfth Edition. Laxmi Publications (P) Ltd.
7. RBA (1980), "Report of National Commission of Flood, Govt. of India.
8. Singh, V. P. (1992). Elementary hydrology. Prentice Hall, New Jersy, 973.
9. Subramanya, K. (2015), "Engineering Hydrology", McGraw-Hill Education (India) Private Limited, New Delhi, 2015, ISBN(13): 978-1-25902997-4.
10. Working Group (1999), The national Commission for Integrated Water Resources Development, Ministry of Water Resources, Govt. of India.

7 Distribution Systems

UNIT SPECIFICS

Water distribution through canals is essential to manage water resources and support human activities. Usually, it refers to a system of artificial water channels intended for various purposes, including drainage, flood control, transit, irrigation, and water supply. Constructing and maintaining canals is necessary to transport water from a reservoir or river to farms and crops. Today's canal distribution systems use technology to increase productivity and save water. This could entail the application of sensors, automated control systems, and targeted irrigation techniques. Canals are lined with vegetation or materials like concrete. This unit describes the canal distribution system and its alignment. It further discusses the types of channels such as rigid boundary and alluvial channels. Furthermore, channel design methodologies such as Kennedy's and Lacey's theory, canal water losses, and its design discharge are described. In addition, different types of canal outlets and types of canal lining are discussed. Finally, methods of irrigated land drainage and causes, effects, and remedial measures for waterlogging are also discussed.

RATIONALE

To learn about the canals as a system of water distribution, theories of regime channels, canal design, outlets, water logging and remedial measures.

PRE-REQUISITE

Nil

UNIT OUTCOMES

The list of outcomes of this unit is as follows:

U7-O1: Canals as a system for water distribution, canal alignment and losses

U7-O2: Canal design and theories of regime channels

U7-O3: Canal outlets, water logging, irrigated land drainage and canal lining

Unit Outcomes	Expected Mapping with Course Outcomes *(1- Weak Correlation; 2- Medium correlation; 3- Strong Correlation)*							
	CO-1	CO-2	CO-3	CO-4	CO-5	CO-6	CO-7	CO-8
U7-O1	3	3	3	1	3	3	3	3
U7-O2	3	3	3	1	3	3	1	3
U7-O3	1	1	3	3	3	3	3	3

7.1 INTRODUCTION

Any network intended to provide a particular good or service to multiple end users from a central source is referred to as a distribution system. A network of canals created specifically to move water from a primary source to multiple locations is known as a canal distribution system in the context of water management. These destinations typically include controlled irrigation of agricultural fields, managing and rerouting excess water to prevent flooding, and delivering water to urban areas for domestic or other designated uses. A canal distribution system improves local economies by promoting farming and associated industries, controls surplus water flow to mitigate flooding, and guarantees a steady supply of water for agriculture. Regular maintenance of canals is important to address erosion, obstructions, and structural deterioration since seepage and evaporation can diminish the efficiency of a canal distribution system. Utilizing canal water effectively is also crucial for management. Waterlogging, a situation when the soil becomes so saturated with water that it is unable to retain any more moisture, is caused by the overuse of canal water. This usually happens in places where water collects more quickly than it can be removed. The environment, infrastructure, and agriculture may be significantly impacted by waterlogging.

7.2 CANAL DISTRIBUTION SYSTEMS

A "canal distribution systems" typically refers to a network of artificial water channels designed for various purposes, such as irrigation, transportation, water supply, drainage, or flood control. This system involves the construction and management of canals to deliver water from a water source, such as a river or reservoir, to fields and crops. Canal distribution systems have been used for centuries and play a significant role in managing water resources and supporting human activities in many parts of the world. Canal distribution systems have been essential for supporting agriculture in many regions, enabling farmers to cultivate crops by providing a reliable and controlled water supply. However, challenges related to water scarcity, sustainability, and the need for modernization are prompting ongoing efforts to improve the efficiency and effectiveness of canal irrigation systems.

Modern canal distribution systems often incorporate technologies to enhance efficiency and conserve water. This may include the use of automated control systems, sensors, and precise irrigation methods. Canals may be lined with materials like concrete or lined with vegetation to prevent water seepage. Canal distribution systems can face challenges such as water loss through seepage, evaporation, and inefficient water use. Regular proper maintenance and modernization are essential to ensure the proper functioning of the canal distribution system to address these challenges. In some cases, canal distribution systems are managed and maintained by local communities or water user associations, emphasizing community involvement in water resource management.

7.2.1 Classification of Canals

The system begins with a water source, which could be a river, lake, reservoir, or another water body. A dam, barrage, or weir is built across a river in a river system to direct the water into the main canal. A head regulator manages supplies in the main canal. Depending on the grade and amount of silt, silt exclusion devices typically consist of a sediment ejector, excluder, or both that have been installed within the main canal in the river. The main canal feeds water to the field through a canal system that includes branch canals, distributaries, and minors. Water is eventually supplied to the field through irrigation outlets by distributors or minors. For convenience, various branches, distributaries, and minors are categorized based on their discharge capabilities. Figure 7.1 shows the schematic layout of a typical canal system. Some essential characteristics and elements of canal distribution systems are given below:

a. **Main Canal:** The main canal is a primary water channel that diverts water from the water source. It is designed to transport a significant volume of water over long distances. It does not provide any irrigation directly.

b. **Branch Canals:** Branch canals are smaller channels that take off from the main canal. They distribute water to specific regions or agricultural fields. The head capacity of branch canal may not be less than 30 cumec.

c. **Distributary Channels:** Distributary channels, also known as furrows, are the channels that take off from a main canal or a branch canal or from another distributary having head capacity under 30 cumec and above 2.5 cumec.

d. **Minors:** Small Distributary channels, having head capacity under 2.5 cumec are known as minors. These channels take off from a branch canal or from other main distributaries.

e. **Field Channels:** Field channels are the smallest channels that deliver water directly to individual fields, generally constructed by cultivator themselves. They are designed to distribute water evenly across the cultivated area from the outlet provided in distributary channel or minors.

f. **Regulation Structures:** To control the flow of water and manage its distribution, regulation structures such as gates, weirs, outlets, and sluices are often installed at key points along the canals.

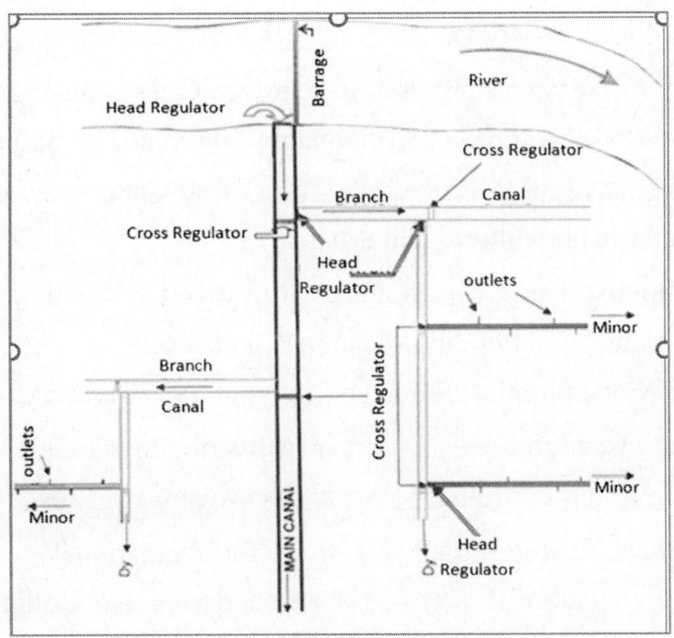

Figure 7.1: A Schematic layout of a typical canal system

7.2.2 Categorisation of Canal Systems

Canals can be categorized in different categories based on their uses, such as irrigation, navigation, flood control, recreation, etc. Canal systems vary widely in scale, design, and function, and their development is influenced by geographical, climatic, and socio-economic factors. Proper management and maintenance are crucial for the efficient and sustainable operation of canal systems. Some common categories of canals based on their uses are described below:

a. **Irrigation Canals:** These canals are used to supply water to agricultural fields for irrigation. The main components of these canals include main canals, branch canals, distributaries, minors, and field channels. For example, Upper Ganga Canal, Middle Ganga Canal, Lower Ganga Canal, The Indus Valley Canal System etc.

b. **Navigation Canals:** These canals are used to facilitate the transportation of goods and people by boats or ships. These canals have locks and gates to regulate water levels. For example: The Suez Canal, Panama Canal, Erie Canal etc.

c. **Water Supply Canals:** These canals are used to transport water from a source to areas for domestic, industrial, or municipal use. The components of these canals have main conduits and distribution networks.

d. **Drainage Canals:** These canals are used to remove excess water from low-lying areas to prevent waterlogging and improve soil drainage. The major components of these canals include drainage canals and pumping stations. These canal systems have designed to manage excess water in agricultural and urban areas.

e. **Flood Control Canals:** These canals are used to divert or contain excess water during periods of heavy rainfall or flooding. Channels and levees are the main components to control water flow. For examples: Flood control systems in river basins.

f. **Hydroelectric Canals:** In these canals water is diverted to generate hydroelectric power. The Canals provide water to turbines in hydroelectric power plants.

g. **Recreational Canals:** These canals are used for recreational activities i.e. boating, fishing, and other recreational activities. These canals are designed for leisure and aesthetics. For example; Venice Canals, San Antonio River Walk.

h. **Urban Storm-water Canals:** Canals and drainage systems are made to manage storm-water runoff in urban areas to prevent flooding. Urban storm-water management systems are examples of these canals.

7.3 COMMAND AREA

The term "command area" typically refers to the area of land that is irrigated by a particular water management system, such as a canal or irrigation project by gravity. The command area is the land that gets benefit from the water resources supplied by the system. Gathering data about the command area is the initial stage in project planning when irrigation is extended to a new area.

7.3.1 Gross Command Area (GCA)

The term "Gross Command Area" typically refers to the total geographical area that is intended to be served or irrigated by a water management project, such as a canal or irrigation system by gravity. It represents the overall extent of land that is supposed to get benefit from the water resources supplied by the project. Gross Commanded Area (Equation 7.1) is the entire area over which a canal system's water can flow gravitationally. The boundaries of a vast command are typically delineated by drainages on both sides, wherein carrying the canal to the subsequent "doab" may not be financially feasible if the intervening drain is sizable.

7.3.2 Cultivable Command Area (CCA)

The cultivable command area includes every piece of land, which can be used for cultivation. Therefore, to get the cultivable command area, pastures and underdeveloped fallow lands are included, but Abadi areas, ponds, usar land, and reserve woods are excluded.

Gross Command Area (GCA) = Cultivable Command Area (CCA) + Uncultivable area ....(7.1)

The process of creating land use maps with a minimum scale of 1: 15000 yields the area statistics. Sharja sheets that are available at the Tehsil headquarters are particularly helpful for this purpose, and they are used extensively. The sheets typically depict field cover area, fallow land, highways, drainage lines, settlement borders, buildings, places of worship, burial ground, spot level and other features at a scale of 16 feet to a mile.

7.4 ALIGNMENTS OF CANALS

The term "alignments of a canal" refers to the planned route or path that a canal follows from its source to its destination. The alignment is a crucial aspect of canal design and construction, and it involves determining the path that the canal will take, considering factors such as topography, geology, land use, and water availability. Canal alignments are typically determined through a comprehensive engineering and planning process, involving surveys, feasibility studies, and consultations with relevant stakeholders. Modern technologies, such as Geographic Information System (GIS) mapping and modelling, are often used to optimize canal alignments based on various factors. The goal is to create a sustainable and efficient canal system that meets the water distribution needs while minimizing negative impacts on the environment and communities. The various approaches can be used to align irrigation channels are (i) as a watershed channel; (ii) as a contour channel; or (iii) as a side slope channel. Figure 7.2 represents the various types of canal alignments. Key considerations and factors in determining the alignments of a canal are described in the following sections.

a. **Topography:** The natural contours of the land, including slopes and elevation changes, play a significant role in determining the alignment. Canals often follow the natural topography to minimize excavation and earthmoving.

b. **Hydraulic Considerations:** The hydraulic characteristics of the area, including the flow of water and potential waterlogging, influence the canal alignment. Engineers consider factors such as water velocity, sediment transport, and the need for regulation structures.

c. **Land Use and Ownership:** Existing land use and ownership patterns affect the feasibility of canal alignments. The canal route must navigate through available land and accommodate existing land uses, avoiding conflicts with private properties or sensitive areas.

d. **Water Source:** The location of the water source, whether it is a river, reservoir, or another water body, influences the starting point and alignment of the canal.

e. **Accessibility:** The alignment should consider accessibility for construction, maintenance, and operation. It should be easily reachable for maintenance crews and equipment.

f. **Economic Considerations:** Economic factors, including construction costs, material availability, and the overall project budget, can influence the choice of canal alignment.

g. **Environmental Impact:** Environmental considerations are crucial in modern canal design. Engineers aim to minimize the environmental impact by avoiding sensitive ecosystems, protecting biodiversity, and incorporating sustainable practices.

h. **Urban Development:** In urban areas, the alignment must consider existing infrastructure, buildings, and other urban features. Canal routes may need to be integrated into the urban landscape.

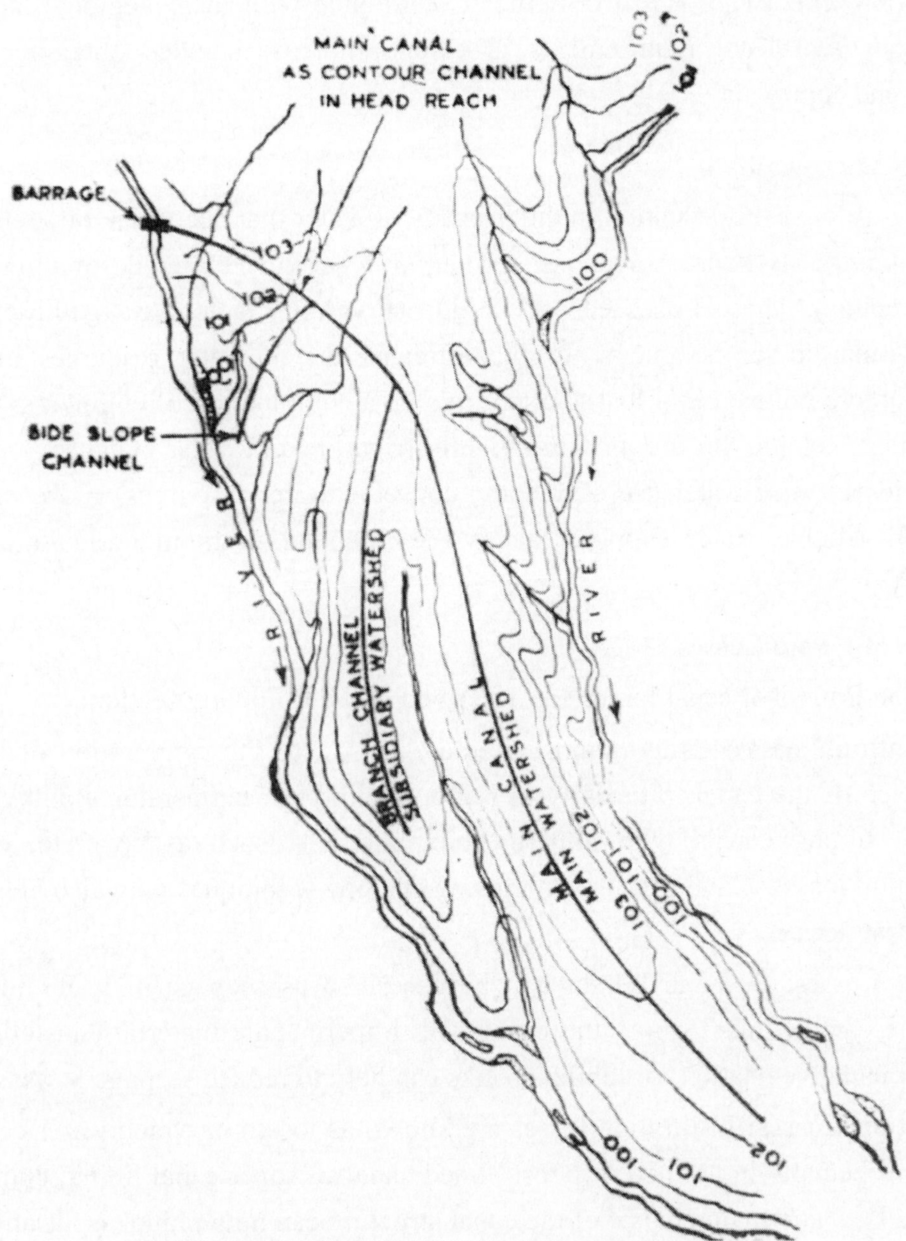

Figure 7.2: Various types of canal alignments.

i. **Cultural and Social Aspects:** Consideration of cultural and social factors is important, especially in areas where canals may impact communities. Community input and involvement in the planning process are often valuable.

j. **Regulatory and Legal Requirements:** Compliance with local, regional, and national regulations and legal requirements is essential. This includes obtaining necessary permits and approvals.

7.5 CANAL LOSSES

"Canal losses" refer to the reduction in the quantity of water that occurs as it travels through a canal system from its source (such as a river or reservoir) to its destination (fields or distribution points). These losses can occur due to various factors along the canal route. Minimizing canal losses is crucial for the efficient use of water resources in irrigation systems. Efforts to reduce canal losses often involve a combination of engineering solutions, proper maintenance, and the adoption of modern irrigation and water management practices. The goal is to optimize water use efficiency, conserve water resources, and ensure that an adequate and reliable water supply reaches the intended users in agricultural or other applications.

7.5.1 Causes of Canal Losses

Some common Causes of canal losses are described in the following section:

a. **Evaporation Losses:** Evaporation losses are known as the evaporation of water from the surface of the canal, especially in regions with high temperatures and exposure to sunlight. Lining canals with impermeable materials, such as concrete, can reduce evaporation losses. Covering canals or using shading techniques may also help to reduce evaporation losses.

b. **Seepage Losses:** Seepage Losses can be described as seepage of water into the soil along the canal banks. Lining the canal with impermeable materials, installing cut-off walls, or using vegetation to stabilize banks can help to reduce seepage losses.

c. **Infiltration Losses:** Infiltration losses are known as losses in which water can infiltrate the soil, especially in unlined or poorly lined canals. Proper canal lining, compaction of the canal bed, and maintenance of the canal structure can help reduce infiltration losses.

d. **Conveyance Losses:** Friction and turbulence in the water flow can lead to losses in the conveyance of water through the canal. Proper design to minimize turbulence, regular maintenance, and ensuring smooth canal beds can reduce conveyance losses.

e. **Structural Losses:** Leaks or structural deficiencies in the canal infrastructure can contribute to water losses. These losses are known as structural losses. Regular inspections, maintenance, and timely repairs of canal structures (such as gates, weirs, and regulators) are essential to minimize structural losses.

f. **Operational Losses:** Inefficient water management practices, such as over-irrigation or improper scheduling, can result in unnecessary water losses. Implementing efficient water management practices, including precision in irrigation techniques and use of modern technology, can reduce operational losses.

g. **Unaccounted-for Water:** Water that is lost but not properly measured or accounted for in the canal system is known as unaccounted- for water. Improved measurement and monitoring systems, such as flow meters and telemetry, can help track water use and identify areas of unaccounted-for water.

7.5.2 Factors Affecting the Canal Losses

The causes of canal losses depend on the several factors, as described below:

a. The permeability of the strata through which the canal runs affect the canal losses.

b. The local drainage system and the level of the groundwater table;

c. The conditions of the canal also affect the canal losses. A canal filled with fine silt experiences less seepage loss than one that has just been built.

d. From the canal in cutting, less water is lost in comparison to the canal which is in filling.

e. Amount of silt, carried by the canal also affect the canal losses. If canal carries more silt the canal is in the loess.

f. Velocities of water also affect the canal losses. If the Velocity of water is more, the less will be the percentage of loss..

g. Channels that run sporadically experience more losses than ones that run continuously.

h. Canal cross Section through which the canal runs also affects the canal losses. The loss due to absorption is directly correlated with the wetted perimeter and the depth of water.

7.6 DISCHARGE IN CANALS

Canal discharge refers to the flow or volume of water passing through a canal over a specific period. Canals are artificial waterways designed to convey water for various purposes, such as irrigation, drainage, navigation, or water supply. The discharge in a canal is typically measured in cubic meters per second (m^3/s) or cubic feet per second (f^3/s), similar to river discharge. Monitoring and managing canal discharge are essential for efficient water resource management. It helps ensure an adequate water supply for agriculture, supports navigation, controls flooding, and facilitates other human activities dependent on water transport. Like river discharge, canal discharge is a critical parameter for water resource engineers and managers to optimize water distribution and usage in a given region.

7.6.1 Factors affecting Canal discharge

Several factors influence canal discharge:

a. **Inflow sources:** Canals receive water from various sources, including rivers, reservoirs, or other water bodies. The volume of water entering the canal from these sources contributes to the overall discharge.

b. **Irrigation demands:** In agricultural regions, canals are often used for irrigation. The discharge in the canal must be sufficient to meet the water needs of the crops.

c. **Human interventions:** Similar to rivers, human activities such as dam operations, water withdrawals, and control structures can affect the canal discharge. Regulation structures like gates and weirs may be used to manage and control the flow in the canal.

d. **Precipitation and runoff:** Rainfall in the canal's watershed or catchment area can impact the water inflow, affecting canal discharge. Additionally, runoff from adjacent areas may contribute to the canal's flow.

7.6.2 Head discharge and cultivable commanded area (CCA)

The head discharge of the main canal is chosen to maximize the use of available water based on the known availability of river discharge and the amount of water needed for a specific cropping plan. To calculate the discharge that would be accessible at the outlets, the channel system's expected losses are estimated. Consequently, the cultivable command area is calculated and plotted on the map based on the available discharge and duty of water. Higher areas that are not irrigated by gravity are also noted and included on the map. An irrigation project's most crucial component is the fixing of head discharge and C.C.A. To meet the demands of both the techno-economic and socio-economic domains, meticulous planning is necessary.

7.7 DESIGN DISCHARGE

The term "design discharge" typically refers to the planned or engineered rate of water flow through a water conveyance system, such as a canal, river, or pipeline. The design discharge is a critical parameter in the planning and engineering of water infrastructure, and it is determined based on various factors and considerations. Accurate determination of design discharge is crucial for the proper functioning and safety of water infrastructure. Engineers must balance the need for conveying water efficiently with the need to manage potential risks associated with extreme flow conditions.

Definition: *Design discharge is the maximum expected flow rate that a water conveyance system is designed to accommodate. It is expressed in terms of volume per unit of time, often measured in cubic meters per second (m^3/s).*

7.7.1 Factors Influencing Design Discharge

a. **Hydrological Data:** Engineers analyze historical and statistical hydrological data, including rainfall patterns, river flow data, and watershed characteristics, to estimate the potential water flow in the system.

b. **Land Use and Development:** The design discharge considers the impact of land use changes and development on the water flow patterns. Urbanization, deforestation, and changes in land cover can influence runoff and flow rates.

c. **Climate Conditions:** Climate factors, such as precipitation, temperature, wind velocity, humidity, sun shine hours, etc., are considered to estimate the potential for weather events that may affect water flow.

d. **Regulatory Requirements:** Design discharge may be subject to regulatory requirements and standards set by governmental agencies to ensure the safety and sustainability of water infrastructure.

7.7.2 Applications of Design Discharge

a. **Irrigation Canals:** In irrigation systems, the design discharge determines the amount of water that can be safely conveyed to irrigate agricultural fields.

b. **River Channels:** For river channels, the design discharge is essential in designing bridges, culverts, and other hydraulic structures to handle the maximum expected flow.

c. **Storm-water Management:** In urban areas, Stormwater management systems are designed with a specific discharge capacity to handle rainfall and prevent flooding.

d. **Water Supply Systems:** Design discharge is considered in the planning of water supply systems to ensure an adequate supply of water for domestic, industrial, or municipal use.

e. **Safety Considerations:** Design discharge is calculated to ensure that the water conveyance system can safely handle the peak flow conditions without causing flooding, erosion, or structural failure.

f. **Hydraulic Modelling:** Engineers often use hydraulic modelling and simulation tools to analyze different scenarios and determine the appropriate design discharge for a given water system.

g. **Long-Term Planning:** Design discharge is part of the long-term planning for water infrastructure, considering factors such as population growth, changes in land use, and potential climate variations.

7.8 DESIGN DISCHARGE ESTIMATION

The design discharge estimation involves the process of determining the maximum expected flow rate that a water conveyance system, such as a canal, river, or pipeline, is designed to handle. The process is critical for the planning, design, and engineering of water infrastructure to ensure the system's capacity meets the demands under various conditions. The estimation of design discharge is a complex process that requires a thorough understanding of hydrological principles, local conditions, and potential variations in climate and land use. It is crucial for engineers and planners to use reliable data and modelling techniques to ensure the safety and effectiveness of water conveyance systems. Here are key steps and considerations in the estimation of design discharge:

a. **Hydrological Data Collection:** Usually, it begins with collecting historical and statistical hydrological data for the area of interest. This includes rainfall data, river flow measurements, and information about the watershed characteristics.

b. **Frequency Analysis:** Use of statistical methods, such as frequency analysis, to analyze historical data and estimate the return period of extreme events (e.g., floods). Common probability distributions like Gumbel, Log-Pearson Type III, or others are often used.

c. **Peak Flow Estimation:** Based on the frequency analysis, estimate the peak flow for different return periods (e.g., 2-year, 10-year, 50-year, 100-year events). These events represent the extreme conditions that the design discharge must accommodate.

d. **Climate Change Considerations:** Consideration of the potential impact of climate change on precipitation patterns and extreme weather events. Climate change projections may influence the estimation of design discharge, requiring adjustments for potential future changes.

e. **Land Use and Development Impact:** Assessment of the impact of current and future land use changes on the hydrological characteristics of the area. Urbanization, deforestation, and changes in land cover can affect runoff patterns.

f. **Regulatory Standards:** Checking and adjuration to any regulatory standards or guidelines set by relevant authorities. Regulatory requirements may specify the design discharge criteria based on safety and environmental considerations.

g. **Hydraulic Modelling:** Utilization of hydraulic modelling and simulation tools to analyze the behavior of the water conveyance system under different flow conditions. Hydraulic models can help to simulate the impact of various scenarios on the system's performance.

h. **Safety Margins:** Introduction of safety margins to account for uncertainties in the estimation process. Safety margins help ensure that the system can handle unexpected conditions or variations from the predicted hydrological patterns.

i. **Long-Term Planning:** Consideration of long-term factors such as population growth, urbanization trends, and potential changes in water demand. Design discharge should not only meet current needs but also allow for future expansion.

j. **Interdisciplinary Collaboration:** Engagement in interdisciplinary collaboration involving hydrologists, hydraulic engineers, meteorologists, and other relevant experts. This ensures a comprehensive approach to estimating design discharge.

k. **Documentation and Reporting:** Documentation of the methodology, assumptions, and data used in the estimation process. Clearly report the estimated design discharge values and associated uncertainties.

7.8.1 Methods for Design Discharge Estimation

Several methods are used to estimate the design discharge for various engineering and water management purposes. The choice of method depends on factors such as the characteristics of the watershed, available data, and the level of precision required. Here are some commonly used methods for estimating design discharge: When selecting a method, it is crucial to consider the specific requirements of the project, data availability, and the desired level of accuracy. In practice, engineers often use a combination of methods or models to account for various factors influencing the design discharge.

a. **Rational Method**

Rational Method is a simple and widely used approach for estimating peak runoff based on the runoff coefficient, rainfall intensity, and watershed area. It is suitable for small to medium-sized watersheds with relatively short time of concentration. The Rational Method is a widely used empirical equation for estimating peak runoff or design discharge from a watershed or catchment area. It is commonly applied in hydrology and civil engineering for designing Stormwater drainage systems, culverts, and other hydraulic structures. The Rational Method is relatively simple but is suitable for small to medium-sized urban or rural watersheds where the time of concentration is relatively short. The Rational Method provides a quick and straightforward approach for estimating peak runoff, but it has limitations. It assumes a uniform rainfall intensity over the entire watershed, neglects the time distribution of rainfall, and may not be suitable for large watersheds with longer time of concentration. For more complex and accurate analyses, hydrologists may use more sophisticated methods such as the SCS (Soil Conservation Service) Curve Number method or distributed hydrological models. The formula for rational method is given below as Equation 7.2.

$$Q = C\,i\,A \qquad \text{......(7.2)}$$

where,

Q=Design discharge (cubic feet per second or cubic meters per second),

C = Runoff coefficient (dimensionless),

i = Rainfall intensity (inches per hour or millimeters per hour), and

A = Area of the watershed (square feet or square meters).

Here is a breakdown of the components:

- **Runoff Coefficient (*C*):** The runoff coefficient represents the portion of rainfall that becomes runoff. It ranges from 0 to 1, where 0 indicates no runoff (all rainfall infiltrates into the soil), and 1 indicates all rainfall becomes runoff. The coefficient depends on factors such as land use, soil type, slope, and vegetation cover. It is often determined empirically based on local conditions or from published tables.
- **Rainfall Intensity (*i*):** This is the rate of rainfall, usually expressed in inches per hour or millimeters per hour. It can be derived from rainfall frequency analysis or obtained from local meteorological records.
- **Watershed Area (*A*):** This is the total area of the watershed or catchment that contributes to the runoff. It is measured in square feet or square meters.

Example 7.1: In each watershed, Rainfall Intensity (i): 5cm per hour; Area of the Watershed (A): 100 square km; Runoff Coefficient (C): 0.5 (represents soil and land use characteristics). Calculated design discharge (Q) using Rational Method.

Solution: Rainfall Intensity in meter: 5 cm/hour

$= 5/100x3600$ m/s

$=1.388x10^{-5}$m/s

Design Discharge (Q):

The Rational Method formula is $Q=CiA$,

$Q= 0.5 \text{ x } 1.388x10^{-5}\text{x } 100 \text{ x}10^{6}$

$Q= 694 \text{ m}^{3}/\text{sec}$

Adjust for Return Period:

- The value obtained from the Rational Method is often based on a specific return period, such as a 2-year or 10-year storm. If no return period is specified, this is considered a general design discharge.
- For example, if the design discharge is needed for a 10-year storm, adjustments would be made based on regional frequency analysis.

This simplified example illustrates the steps involved in estimating the design discharge using the rational method. It is important to note that real-world applications involve more sophisticated methods, consideration of hydrological data, and collaboration with experts in hydrology and hydraulic engineering.

b. **SCS (Soil Conservation Service) Curve Number Method**

This method, developed by the Natural Resources Conservation Service (formerly Soil Conservation Service) to aid in the planning and design of water conservation and management practices, considers factors such as land use, soil type, and hydrologic condition to estimate runoff. The method is suitable for a range of watershed sizes and provides a more detailed analysis compared to the Rational Method. The SCS (Soil Conservation Service) Curve Number (CN) method is a widely used hydrological technique for estimating direct runoff or peak discharge from rainfall events. The SCS-CN method is particularly applicable too small to medium-sized watersheds. The SCS CN method is often used for event-based rainfall-runoff modelling. It provides a simple and efficient way to estimate direct runoff for different storm events. However, it has limitations, such as assuming uniform rainfall distribution, neglecting snowmelt, and not considering channel routing. For more detailed and comprehensive analyses, hydrologists may use more advanced models. The method involves the following steps:

- **Land Use and Cover Classification:** Classify the land use and land cover within the watershed. The SCS provides tables assigning curve numbers based on different types of land cover, such as forests, grasslands, and urban areas.
- **Hydrologic Soil Group Classification:** Classify the soils in the watershed into one of four hydrologic soil groups (A, B, C, or D) based on their runoff potential. Soils with higher infiltration rates are assigned to Group A, while those with lower infiltration rates are assigned to Group D.
- **Antecedent Moisture Conditions:** Determine the antecedent moisture condition (AMC) of the watershed, which represents the soil moisture conditions prior to the rainfall event. The three conditions are Dry (I), Average (II), and Wet (III).
- **Calculate Curve Number:** Using the land use, soil group, and antecedent moisture condition, look up the corresponding curve number in the SCS CN tables or use equations provided by the SCS. The curve number is a dimensionless number ranging from 30 to 100, representing the runoff potential of the watershed.
- **Estimate Runoff:** The formula for estimating direct runoff (Q) is given as Equation 7.3.

$$Q = \frac{((P-0.2S))^2}{(P+0.8S)} \quad \text{.............(7.3)}$$

where,

P= Rainfall excess, and

S= Potential maximum retention after runoff begins. The initial abstraction is assumed to be 0.2 times the potential maximum retention.

c. **Unit Hydrograph Method**

The Unit Hydrograph method involves using a hypothetical unit hydrograph, which represents the response of a watershed to a unit of excess rainfall. By convolution, the unit hydrograph can be transformed to estimate the discharge for any given rainfall event. The Unit Hydrograph (UH) method is a hydrological technique used to estimate the runoff response of a watershed to a unit of excess rainfall. It is commonly employed for designing Stormwater management systems, flood forecasting, and other water resource planning applications. The method involves developing a unit hydrograph, which represents the temporal distribution of runoff resulting from a unit of effective rainfall spread over a specific duration. This method is often used in larger watersheds. The Unit Hydrograph method assumes of linearity, which means that the shape of the unit hydrograph remains constant regardless of the magnitude of the storm event. While this assumption simplifies the analysis, it may not always hold true in all hydrological conditions. More sophisticated hydrological models, such as the SCS Curve Number method or distributed hydrological models, may be used for more complex and accurate analyses in larger watersheds. Here are the key steps involved in the Unit Hydrograph method:

- **Rainfall Distribution:** Determination of the effective rainfall for a given storm event. Effective rainfall is the portion of total rainfall that contributes to runoff after losses due to infiltration, interception, and other abstractions have been considered.
- **Unit Hydrograph Extraction:** Extraction or derivation of the unit hydrograph from historical rainfall and runoff data for the watershed. The unit hydrograph is a graphical representation of the response of the watershed to a unit of effective rainfall. It shows the relationship between time and discharge.

- **Duration Adjustment:** Adjustment of the duration of the unit hydrograph to match the duration of the effective rainfall for the specific storm event. This involves stretching or compressing the time axis of the unit hydrograph.
- **Intensity-Duration-Frequency (IDF) Curves:** Usage of rainfall intensity-duration-frequency curves to determine the intensity of the rainfall for different durations and return periods. This information helps in selecting appropriate rainfall events for analysis.
- **Convolution:** Convolution of the adjusted unit hydrograph with the effective rainfall hyetograph (rainfall over time). Convolution is a mathematical operation that combines the unit hydrograph and the effective rainfall to produce the resulting hydrograph.
- **Estimate Design Discharge:** The resulting hydrograph represents the expected runoff for the specific storm event. Design discharge can be estimated by identifying the peak flow in the hydrograph.

d. Empirical Rainfall-Runoff Models

Empirical models use statistical relationships based on historical data to estimate runoff. These models may consider factors such as antecedent soil moisture, rainfall characteristics, and land use. Examples include the Clark, Nash, and SCS-CN methods. Empirical rainfall-runoff models are statistical models that establish a relationship between rainfall inputs and resulting runoff based on observed data. These models are empirical in nature, meaning they are developed through the analysis of real-world data rather than being derived from physical principles. Empirical models are often used when detailed information about the underlying hydrological processes is limited or when simplicity is required. Empirical rainfall-runoff models are valuable when detailed hydrological data are scarce or when there is a need for a simplified representation of the watershed response. However, they may have limitations in capturing the complexity of hydrological processes, especially in larger and more diverse watersheds. In such cases, more comprehensive and physically based hydrological models may be employed for a more accurate representation of the watershed behaviors. Here are some common types of empirical rainfall-runoff models:

- **Clark's Model:** Clark's model is one of the earliest empirical models and is based on a linear relationship between runoff and effective rainfall. It involves a unit hydrograph approach and assumes a linear reservoir for runoff production. The model parameters are determined from observed data.

- **Nash Model:** The Nash model is an improvement upon Clark's model and considers the nonlinear relationship between rainfall and runoff. It introduces a parameter that accounts for the time delay in runoff response. The Nash model is often used for small to medium-sized watersheds.
- **SCS-CN (Soil Conservation Service Curve Number) Models:** While the SCS-CN method is often used for direct runoff estimation, it can also be considered an empirical model. It uses curve numbers derived from observed data to estimate runoff based on factors such as land use, soil type, and antecedent moisture conditions.
- **Instantaneous Unit Hydrograph (IUH) Models:** IUH models estimate runoff directly from rainfall by using unit hydrographs that vary with time. These models consider the temporal distribution of runoff and are often applied in urban hydrology studies.
- **Gamma Model:** The gamma model relates rainfall and runoff through a gamma probability distribution. This model has been used in various regions and is based on statistical properties of rainfall-runoff events.
- **Power Law Models:** Power law models express a power relationship between rainfall and runoff. These models assume that runoff is a power function of rainfall, and the model parameters are estimated from the observed data.

e. **Probabilistic Methods**

Probabilistic methods, such as frequency analysis, involve assessing the probability of different rainfall events occurring. These methods are often used to estimate design discharges for specific return periods, considering the likelihood of extreme events. Probabilistic methods in hydrology involve assessing the probability of different hydrological events, such as floods or extreme rainfall, occurring over a given period. These methods are particularly useful for estimating design discharges for various return periods. Probabilistic methods allow engineers and hydrologists to incorporate uncertainty into their analyses and design processes. They help in making informed decisions by providing a range of possible outcomes and their associated probabilities. This is particularly important in designing infrastructure that needs to withstand extreme events while considering the inherent variability in the hydrological system. Here are some common probabilistic methods used in hydrology:

- **Frequency Analysis:** Frequency analysis involves the statistical analysis of historical hydrological data to estimate the probability of different magnitudes of events occurring. It helps in determining the return period of a specific discharge value, which is essential for designing structures like dams or bridges.
- **Probability Distribution Models:** Hydrologists often use probability distribution models to represent the statistical distribution of hydrological variables, such as annual maximum discharge or rainfall. Common distributions include the Gumbel, Lognormal, and Pearson Type III distributions.
- **Return Periods:** Return periods express the average recurrence interval of an event of a specific magnitude. For example, a 100-year flood has a 1% chance of occurring in any given year. Return periods are often determined using frequency analysis and probability distributions.
- **Intensity-Duration-Frequency (IDF) Curves:** IDF curves provide information about the relationship between rainfall intensity, duration, and frequency. These curves help to estimate the likelihood of different rainfall events, which is crucial for designing the Stormwater management systems.
- **Copula Models:** Copulas are mathematical tools used to model the joint distribution of multiple variables. In hydrology, copula models are applied to understand the dependence structure between different hydrological variables, such as rainfall and runoff.
- **Monte Carlo Simulation:** Monte Carlo simulation involves generating a large number of random samples from probability distributions to simulate different scenarios. This method is often used to assess the uncertainty associated with hydrological predictions and design discharges.

f. **Distributed Hydrological Models**

These models are designed to simulate the spatial and temporal distribution of various hydrological variables, such as precipitation, evapotranspiration, soil moisture, and runoff, within a watershed. Distributed hydrological models are sophisticated mathematical representations of the hydrological processes that occur within a watershed. Unlike lumped models that consider the entire watershed as a single unit, distributed models divide the watershed into multiple spatial units, each with its own set of characteristics. Examples include the Soil and Water Assessment Tool (SWAT) and the Hydrologic Engineering

Centre-Hydrologic Modelling System (HEC-HMS). Distributed models are more complex but can provide detailed insights into watershed behavior. Distributed hydrological models are valuable tools for detailed and comprehensive hydrological assessments, especially in larger and more complex watersheds. They provide a more realistic representation of the spatial variability of hydrological processes, enabling better-informed decision-making in water resources management. Here are key features and components of distributed hydrological models:

- **Spatial Representation:** The spatial distribution of land cover, soil types, and other relevant features is considered for each cell.
- **Physical Processes:** Distributed models simulate various physical processes, including rainfall, infiltration, surface runoff, subsurface flow, groundwater flow, and stream flow. These processes are often represented by mathematical equations derived from physical principles.
- **Hydrological Connectivity:** The models consider the connectivity between adjacent cells and their influence on water movement. This connectivity is crucial for capturing the flow pathways, especially in terms of overland flow, interflow, and groundwater flow.
- **Land Surface Processes:** Distributed models account for land surface processes such as vegetation dynamics, snow accumulation and melt, and surface energy balance. These components are essential for understanding the impact of land cover on hydrological processes.
- **Model Calibration and Validation:** Distributed models require calibration to adjust the model parameters to match observed hydrological behavior. Calibration involves adjusting the model to replicate observed streamflow and other hydrological variables. Validation ensures that the model performs well under different conditions.
- **Remote Sensing and GIS Integration:** Distributed models often incorporate remote sensing data and Geographic Information System (GIS) information to obtain accurate and up-to-date spatial data for land cover, topography, and other relevant parameters.
- **Simulation of Extreme Events:** Distributed models are particularly useful for simulating extreme events, such as floods or droughts, as they capture the spatial variability in rainfall and terrain that influences the runoff.

- **Model Outputs:** The models provide outputs such as spatial distribution of soil moisture, evapotranspiration, and streamflow. These outputs can be used for various purposes, including water resource management, flood forecasting, and land-use planning.

7.8.2 Natural Channels

When discussing natural channels for the flow of water, it typically refers to rivers, streams, or watercourses that follow their natural course without significant human modifications. These natural channels play a crucial role in the hydrological cycle and support various ecological functions. Preserving and restoring natural channels is crucial for maintaining the ecological integrity of the river systems and ensuring sustainable water management practices. Human activities, such as channelization, urbanization, and dam construction, can significantly impact these natural channels and alter their functions. Conservation efforts often focus on maintaining or restoring the natural flow and geomorphic characteristics of watercourses to support both environmental and human needs. Here are some key features and aspects of natural water channels.

a. **River Systems:** Natural channels often include river systems with headwaters, tributaries, and main stems. Rivers naturally meander, creating bends and curves, and they may have varying widths and depths.

b. **Bankfull Stage:** The bankfull stage is the level at which a river or stream is just about to overflow its banks. It represents the point where water begins to inundate the floodplain, and it is a key parameter in understanding the natural flow regime.

c. **Fluvial Processes:** Natural channels are shaped and maintained by fluvial processes such as erosion, sediment transport, and deposition. These processes contribute to the formation of riverbeds, bars, and other geomorphic features.

d. **Riparian Zones:** The areas along the banks of natural channels, known as riparian zones, are important for biodiversity. They provide habitat for various plant and animal species, offer floodplain connectivity, and help filter pollutants.

e. **Ecosystem Services:** Natural water channels provide essential ecosystem services, including water purification, groundwater recharge, and the transport of nutrients. They also play a role in flood control by absorbing excess water during high-flow events.

f. **Meanders and Oxbows:** Meanders are bends or curves in a river's course, and oxbow lakes are formed when meanders are cut off from the main channel. These features are part of the dynamic nature of natural watercourses.

g. **Hydrological Variability:** Natural channels exhibit natural variability in flow, influenced by seasonal changes, precipitation patterns, and other environmental factors. This variability is important for maintaining the health of aquatic ecosystems.

Preserving and restoring natural channels is crucial for maintaining the ecological integrity of river systems and ensuring sustainable water management practices. Human activities, such as channelization, urbanization, and dam construction, can significantly impact these natural channels and alter their functions. Conservation efforts often focus on maintaining or restoring the natural flow and geomorphic characteristics of watercourses to support both environmental and human needs.

7.8.3 Rigid boundary channels

Rigid boundary channels refer to channels or conduits with structural boundaries that maintain a fixed shape and do not deform significantly under the influence of flowing water. Unlike natural channels with erodible banks or flexible linings, rigid boundary channels have constructed or engineered boundaries that resist deformation and maintain a stable cross-sectional shape. Rigid boundary channels are a common feature in urban infrastructure and engineered water management systems where predictable flow conditions, stability, and durability are crucial. The design and construction of these channels involve a combination of hydraulic engineering principles, material science, and structural design considerations. Key features and considerations related to rigid boundary channels include:

a. **Material of Construction:** Rigid boundary channels are typically constructed using materials such as concrete, steel, or other rigid materials. These materials provide stability and prevent significant deformation.

b. **Lining Materials:** Concrete is a common material used for lining rigid boundary channels. The concrete lining can be applied to both the bed and the sides of the channel, creating a durable and erosion-resistant structure.

c. **Applications:** Rigid boundary channels are often employed in engineered water conveyance systems, irrigation canals, Stormwater channels, and urban drainage systems where stable and well-defined channel geometry is essential.

d. **Stability and Erosion Resistance:** The primary advantage of rigid boundary channels is their stability and resistance to erosion. They are designed to withstand the hydraulic forces of flowing water and resist the erosive effects that might occur in natural or unlined channels.

e. **Efficient Water Conveyance:** Rigid boundary channels are designed to provide efficient and controlled water conveyance. The fixed geometry allows for predictable flow patterns and minimizes energy losses associated with channel deformation.

f. **Maintenance:** Rigid boundary channels often require less maintenance compared to natural or flexible-lined channels. The durable materials used in construction, contribute to longer service life and reduced susceptibility to erosion.

g. **Hydraulic Design:** The hydraulic design of rigid boundary channels involves considerations of cross-sectional geometry, slope, and flow rates to optimize water conveyance efficiency while ensuring structural stability.

h. **Culverts and Stormwater Channels:** Rigid boundary channels are commonly used in the construction of culverts and Stormwater drainage systems in urban areas. These structures are designed to efficiently transport Stormwater while preventing erosion and maintaining structural integrity.

i. **Environmental Impact:** The environmental impact of rigid boundary channels should be carefully considered. While they provide stability and durability, the alteration of natural drainage patterns may have ecological implications, and efforts may be made to incorporate features that enhance habitat diversity.

j. **Design Standards and Codes:** Engineers follow design standards and codes when designing rigid boundary channels to ensure safety, stability, and compliance with regulations. Local, regional, or national guidelines may dictate specific design criteria.

7.8.4 Alluvial channels

Alluvial channels are river channels characterized by the transportation and deposition of sediments, typically consisting of sand, gravel, silt, and clay. These channels form in response to the dynamic interactions between flowing water and the sediment load it carries. Alluvial channels are common in fluvial environments and undergo continuous adjustments based on factors such as flow characteristics, sediment supply, and base level changes.

Understanding the behavior and characteristics of alluvial channels is crucial for various applications, including river management, water resource planning, and environmental conservation. Hydraulic engineers and geo-morphologists study these channels to predict changes, assess sediment transport dynamics, and implement effective management strategies. Key characteristics of alluvial channels include:

a. **Dynamic Nature:** Alluvial channels are dynamic and constantly changing due to the continuous processes of erosion, transport, and deposition of sediment. Fluctuations in water flow, sediment load, and base level contribute to these changes.

b. **Sediment Transport:** The channels transport a mix of sediments, ranging from coarse particles like sand and gravel to finer particles like silt and clay. The nature of sediment transport is influenced by flow velocity, channel slope, and the size and quantity of sediments available.

c. **Channel Morphology:** Alluvial channels often exhibit diverse morphologies, including braided, meandering, or straight patterns. The channel morphology is influenced by sediment supply, slope, and the balance between erosion and deposition.

d. **Braided Channels:** In areas with abundant coarse sediments and variable flow conditions, alluvial channels may form braided patterns characterized by multiple interconnected channels separated by sediment bars.

e. **Meandering Channels:** Meandering channels have a sinuous, winding course and are often associated with finer sediments. The migration of meanders contributes to the lateral movement of the channel.

f. **Floodplains:** Alluvial channels typically have adjacent floodplains where sediment deposition occurs during periods of flooding. Floodplains play a crucial role in storing sediment and providing habitat diversity.

g. **Natural Levees:** During floods, sediment deposition along the channel banks can lead to the formation of natural levees, raised landforms parallel to the channel. These levees help to confine water within the channel during subsequent floods.

h. **Avulsion and Shifts:** Alluvial channels may undergo avulsion, which is the sudden shift of the main channel to a new location on the floodplain. Avulsion is often associated with changes in river course due to sedimentation and channel dynamics.

i. **Sediment Sorting:** The sorting of sediments within the channel occurs based on size and weight. Finer sediments may be transported farther downstream, while coarser sediments may be deposited closer to the source.

j. **Hydraulic Geometry:** Alluvial channels adhere to hydraulic geometry principles, where channel dimensions, such as width, depth, and slope, adjust in response to variations in discharge and sediment load.

7.9 OPEN CHANNEL FLOW HYDRAULICS

a. Chezy's equation

The design of irrigation channels must provide consistent and even flow. The flow condition is said to be steady and uniform, when the discharge is constant along a specific length of the channel under consideration and the flow depth remains constant. The Equation (7.4) proposed by French engineer, Chezy in 1769, is to describe for flow of water in channels.

$$V = C\sqrt{RS} \qquad \text{.............(7.4)}$$

where,

V = Mean velocity of flow in m/sec;

C = Chezy's coefficient depends upon the surface and shape of the channel;

R = Mean hydraulic depth in meters;

S = channel's slope

b. Ganguillet and Kutter's equation

Various authors have periodically conducted statistical analyses to estimate the value of Chez y's Coefficient, and the values of Chezy's Coefficient from the equation proposed by Ganguillet and Kutter gave satisfactory results. The Ganguillet and Kutter equation is given as Equation 7.5a.

$$C = \frac{23+\frac{1}{n}+\frac{0.00155}{S}}{1+\left(23+\frac{0.00155}{S}\right)\frac{n}{\sqrt{R}}} \qquad \text{.............(7.5a)}$$

Substituting C in Chezy's equation, the Equation (7.5a) can be written as Equation (7.5b).

$$V = \frac{23+\frac{1}{n}+\frac{0.00155}{S}}{1+\left(23+\frac{0.00155}{S}\right)\frac{n}{\sqrt{R}}}\sqrt{RS} \qquad \text{.............(7.5b)}$$

c. Manning's equation

The equation proposed by Ganguillet and Kutter is quite cumbersome. This equation was further modified by Manning, who also gave similar results to those obtained from the equation proposed by Ganguillet and Kutter. The manning's equation can be expressed as Equation (7.6a):

$$C = \frac{R^{1/6}}{n} \quad \text{.............(7.6a)}$$

Substituting the value of C from Chezy's equation, Manning's equation can be rewritten as Equation (7.6b).

$$V = \frac{1}{n} R^{2/3} S^{1/2} \quad \text{...........(7.6b)}$$

where n = roughness or Rugosity coefficient.

The physical roughness of the channel's sides and bottom determines the value of n, which is affected by several elements including (i), channel curvature, (ii) changes in cross-sectional size and shape, (iii) obstructions including debris roots, structures, and (iv) vegetation etc. Table 7.1 lists the values of n that I.S.1 recommends for channels that have been excavated. The projected future channel condition determines the value of n that should be used. The number of weeds, issues with silting or scouring, and maintenance standards are the key factors influencing the condition of the channel. The recommended values of n for alluvial rivers by Buckley are listed in Table 7.2.

Table 7.1: Values of n for excavated open channels

SN		Types of Channels	Values on n
a.		Earth, straight and uniform	
	i	Clean, straight and uniform	0.016-0.020
	ii	Clean, after weathering	0.018-0.025
	iii	With short grass, few weeds	0.022-0.033
b.		Rock cuts	
	i	Smooth and uniform	0.025-0.040
	ii	Tagged and irregular	0.035-0.050

Table 7.2: Values of Rugosity coefficient recommended by Buckley for alluvial channels

SN	Condition of Channel	Value of n
1	Very good	0.0225
2	Good	0.0250
3	Indifferent	0.0275
4	Bad	0.0300

Example 7.2: Find out the normal water depth and velocity in channel carrying a discharge of 12 cumec and having bed width of 8 m. Use roughness coefficient n= 0.0250 and bed slope S = 0.0018.

Solution:

Assuming side slope as 1:1, Channel cross-sectional area (A) = (B+D) x D}

Wetted Perimeter (p) = $(B + 2D\sqrt{2})$

where,

B = Bed width of the channel. and

D = depth of water in the channel

Hence, Hydraulic mean depth R = A/P

By Manning's Formula, $V = \frac{1}{n} R^{2/3} S^{1/2}$

$$V = \frac{1}{0.025}\left[\frac{(B+D) \times D\}}{(B+2D\sqrt{2})}\right]^{2/3} 0.0018^{1/2}$$

$$V = 1.697\, x \left[\frac{(B+D) \times D\}}{(B+2D\sqrt{2})}\right]^{2/3} \quad \text{.................(i)}$$

Now Q = A* V

$$Q = (B+D) \times D\} \times 1.697 \left[\frac{(B+D) \times D\}}{(B+2D\sqrt{2})}\right]^{2/3}$$

$$Q = 1.697 \frac{((B+D) \times D)^{5/3}}{(B+2D\sqrt{2})^{2/3}} \quad \text{................(ii)}$$

Solving Equation (ii) by trial and error method, so that value of Q = 12. We get D = 0.93 m

$$Q = 1.697 \frac{((8+0.93) \text{ x } 0.93)^{5/3}}{(8+2\text{x}0.93\text{x}\sqrt{2})^{2/3}}$$

$$= 11.96\ cumec$$

$$= \text{Say 12 cumec}$$

Hence, the water depth (D) = 93 cm

Substituting values of B and D in the Eqn. (i) we get

$$V = 1.697\ x \left[\frac{(\text{B+D) x D}\}}{(B+2D\sqrt{2})}\right]^{2/3}$$

$$V = 1.697\ x \left[\frac{(8+0.93)\text{x}0.93\}}{(8+2\text{x}0.93\text{x}\sqrt{2})}\right]^{2/3}$$

V= 1.44 m/sec.

Example 7.3: Design a channel with side slope 1:1 to carry 16 cumec discharge at a slope = 0.0016. Maximum permissible velocity in channel is 2.0 m/sec and n = 0.020.

Solution:

Assuming the side slope as 1:1

Area of Channel (A) = (B+D) x D}

Wetted Perimeter (p) = $(B + 2D\sqrt{2})$

Where,

B = Bed width of the channel. and

D = depth of water in the channel

Hence, Hydraulic mean depth

R= A/P

$$\text{R} = \frac{(\text{B+D) x D}\}}{(B+2D\sqrt{2})}$$

Substituting these values in Manning's Formula

$$V = \frac{1}{n} R^{2/3} S^{1/2}$$

$$V = \frac{1}{0.02} \left[\frac{(\text{B+D) x D}\}}{(B+2D\sqrt{2})}\right]^{2/3} 0.0016^{1/2}$$

$$2.0 = \frac{1}{0.02}\left[\frac{(B+D)\ x\ D\}}{(B+2D\sqrt{2})}\right]^{2/3} 0.0016^{1/2}$$

$$1.0 = \left[\frac{(B+D)\ x\ D\}}{(B+2D\sqrt{2})}\right]^{2/3} \qquad \text{..............(i)}$$

Now, $Q = A^* V$

$A = Q / V$

$(B+D)\ x\ D\} = 16/2 = 8$

$$(BD+D^2) = 8 \qquad \text{..............(ii)}$$

Substituting this value in equation (i), we have

$$\left[\frac{8}{(B+2D\sqrt{2})}\right]^{2/3} = 1.0$$

$$\frac{8}{(B+2D\sqrt{2})} = 1.0$$

$$(B + 2D\sqrt{2}) = 8$$

$$B = 8.00 - 2.82\ D$$

Substituting these values of B in equation (ii), and solving we get

$$(BD+D^2) = 8$$

$$D\ (8.00 - 2.82\ D) + D^2 = 8$$

Or D = 1.53 m and B= 3.69 meter

7.10 KENNEDY'S SILT THEORY FOR STEADY REGIME CHANNELS

The charges and silt grade found in the upper Bari Doab Canal and its distributaries served as the foundation for R.G. Kennedy's idea. Based on his observations, he concluded that, in a stable regime, there is only one velocity for non-silting and non-scouring channels. This velocity is known as the "critical velocity," represented by V_o, and it depends on the water depth in the channel. Kennedy found the equation (Equation 7.7a) by charting his measurements of depth and velocity in the steady regime reaches. Kennedy found the equation by charting his measurements of depth and velocity in the steady regime reaches.

$$\boldsymbol{V_0 = 0.546D^{0.64}} \qquad \text{.............(7.7a)}$$

where,

D = Depth of water in meter

V_0 = Velocity of water m/sec

This formula only applied to the silt grade found in the upper Bari-Doab Canal. The 'critical velocity ratio' (C.V.R.), represented by m, was included to the equation as an additional element to account for different grades of silt. The equation may thus be re-written as Equation (7.7b).

$$V = 0.546\, mD^{0.64} \qquad \text{.............(7.7b)}$$

where, $m = V/V_0$ = C.V.R.

Table 7.3 lists the values of m for the various types of silt.

Table 7.3: Values of C.V.R. for different types of soils

Type of silt	Value of m
Light sandy silt in the rivers of Northern India	1.00
Somewhat coarser silt or debris of hard soils	1.10
Sandy, loamy silt	1.20
Rather, coarser silt or debris of hard soils	1.30
Silt of the River Indus in Sindh (Pakistan)	0.70

Kennedy's equation does not account for the channel's breadth, shape, or slope. Assuming trial values for these parameters is important before moving further with the design. The calculated velocity should fulfil Kennedy's equation and provide the necessary discharge for the estimated segment. The channel's mean velocity must not be lower than the critical velocity.

7.10.1 Channel design by Kennedy's theory

a. For designing a channel on Kennedy's theory, the following data is required:

(i) Channel's design discharge, Q (cumec),

(ii) The Rugosity coefficient, n

(iii) The slope, S

(iv) The ratio of critical velocity, $m = V/V_0$

b. The equations to be applied are:

(i) Q = A* V

(ii) Kutter's equation = $V = \frac{23+\frac{1}{n}-\frac{0.00155}{S}}{1+\left(23+\frac{0.00155}{S}\right)\frac{n}{\sqrt{R}}}\sqrt{RS}$

(iii) Kennedy's equation = $V = 0.546\ mD^{0.64}$

Procedure:

(a) Assume a trial value of Depth (D). Find out the value the critical velocity needed (V), for this trial depth, by substituting in Kennedy's equation given as (iii).

(b) Using given value of channel's design discharge (Q) and Velocity (V), Obtain value of A using equation (i) above, A = Q/V,.

(c) Find out the value of Bed width B, using side slope of channel, obtained value A and trail value depth D. If slide slope of channel is not known, it can be assumed as ½:1.

(d) Find out the value of Hydraulic mean depth R, using obtained value A, Obtained value of bed width B and trail value of depth D. Using equation R=A/P; where P= wetted perimeter.

(e) Substitute obtained value R along with the other known values of Rugosity coefficient, n, and Slope S, in the Kutter's equation and obtain actual velocity V.

(f) If the velocity worked out from Kutter's equation and obtained from Kennedy's equation are almost same, the assumed depth is correct. If not, repeat the calculations with changed values of D, till the both values of Velocity V are almost same.

Example 7.4: Design an irrigation channel to carry 50 cumec at a slope of 1/5000 with Kutter's n = 0.0250 and m = 0.90.

Solution:

Assume depth = 2.50 m

Kennedy's V $= 0.546 \times 0.90 \times 2.50^{0.64}$

= 0.88 m/sec.

A = Q/V = 50.0/0.88 = 56.81 m^2

For the purpose of discharging capacity, the side slopes of an irrigation channel may be considered as $\frac{1}{2}:1$

Hence $A = BD + D^2/2 = 56.81$ sq. meter

Substituting $D = 2.50$ m, we have

$2.50\ B + 2.50^2/2 = 56.81$

$2.50\ B = 56.81 - 3.125 = 53.685$

$B = 21.474$ m

$P = B + 2D\sqrt{2}$

$= 21.474 + 2\sqrt{2}\ x\ 2.50$

$= 28.54$ m

Hydraulic mean depth $R = A/P$

$R = 56.81/28.54$

$= 1.99$ m

Chezy's C, according to Kutter's equation $= \dfrac{23+\frac{1}{n}+\frac{0.00155}{S}}{1+\left(23+\frac{0.00155}{S}\right)\frac{n}{\sqrt{R}}}$

$$= \frac{23+\frac{1}{0.0250}+\frac{0.00155}{1/5000}}{1+\left(23+\frac{0.00155}{1/5000}\right)\frac{0.0250}{\sqrt{1.99}}}$$

$$= \frac{23+40+7.75}{1+(23+7.75)0.0177}$$

$$= \frac{70.75}{1.544} = 45.82$$

Actual velocity of flow = $V = C\sqrt{RS}$

$$V = 45.82\sqrt{1.99x1/5000}$$

$$V = 0.914 \text{ m/sec}$$

The actual velocity of flow comes out to be nearly the same as the critical velocity of 0.88 m/sec. Hence the assumed channel dimensions are in order.

Bed width = 21.474 m; Depth D = 2.50 m.

7.11 LACEY'S THEORY

Lacey's theory likely refers to the work of Luna Bergere Lacey, an American hydraulic engineer who made significant contributions to the understanding of sediment transport in rivers and channels. While Lacey's work primarily focused on sediment transport in river channels rather than the design of regime channels, we can draw some general principles from his research that might inform the design of regime channels as follows:

(i) Just as Lacey studied the optimal flow of water in rivers; in designing regime channels, one might aim to optimize the flow of information and communication between states.

(ii) Lacey's equilibrium slope concept suggests a balance between sediment transport and deposition.

(iii) Lacey's work recognized that river channels are dynamic and subject to change over time. Similarly, in designing regime channels, flexibility and adaptability are crucial.

(iv) Channels should be designed to accommodate shifting geopolitical landscapes, evolving interests among states, and changes in the issues or challenges facing the international community.

(v) Lacey's theories considered how the shape and geometry of a river channel influence sediment transport.

(vi) Lacey's equilibrium slope concept aimed to achieve stability in river channels over time.

7.11.1 Lacey's silt theory

"Lacey's silt theory" likely refers to the developed theories and formulas by Lacey to predict the movement of sediment, including silt, in river systems. Lacey's research focused on how sediment particles of various sizes move and settle within river channels. He developed equations and theories that described the relationships between flow velocity, channel geometry, and sediment transport. One of his notable contributions was the development of the "equilibrium slope" concept, which predicts the slope of a river channel necessary to maintain a balance between sediment transport and deposition over time.

In essence, Lacey's silt theory provided valuable insights into the dynamics of sediment transport in river systems, which have practical applications in engineering projects related to river management, flood control, and navigation. His work had a lasting impact on the field of hydraulic engineering and continues to be studied and applied by researchers and practitioners today.

7.11.2 Channel design using Lacey's theory

Lacey's equations primarily pertain to the study of sediment transport in river channels rather than the design of regime channels in international relations. However, we can draw some parallels and use Lacey's principles metaphorically to illustrate the design of regime channels. Here is an attempt to provide a metaphorical interpretation of Lacey's equations in the context of regime channel design: For designing the channel based on Lacey's theory, the following items must be known.

(i) Discharge Q cumec

(ii) Silt factor 'f'

The following equations are to be used for designing the channel:

(i) $V = \left(\frac{Q\ x\ f^2}{140}\right)^{1/6}$ ....(i)

(ii) $R = \frac{5}{2} \times \left[\frac{V^2}{f}\right]$ ....(ii)

(iii) $A = Q \times V$ ....(iii)

(iv) $P = 4.75\sqrt{Q}$ ....(iv)

(v) $S = \frac{f^{5/3}}{3340 \times Q^{1/6}}$ ....(v)

Procedure:

(a) Using the given discharge (Q) and silt factor (f), find out the velocity from equation (i),

(b) Obtain the hydraulic mean depth (R) fusing equation (ii) above.

(c) Work out the channel's area and wetted perimeter using equations (iii) and (iv) above.

(d) Using the Channel's area A, wetted perimeter (P) and hydraulic mean depth (R); obtain the bed width and depth of the channel, assuming the channel side slope $\frac{1}{2}:1$.

(e) The longitudinal slope may be obtained using equation (v).

Example 7.5: Design a regime channel using by Lacey's theory. The available discharge of the regime channel may be taken as 50 cumec and silt factor = 0.80. The corresponding side slope of the channel can be assumed as $\frac{1}{2}:1$.

Solution:

Given, Channel discharge (Q) = 50 cumec

Silt factor (f) = 0.80

$V = \left(\frac{Q\ x\ f^2}{140}\right)^{1/6}$

$V = \left(\frac{50\ x\ 0.80^2}{140}\right)^{1/6}$

V = 0.782 m/sec

$R = \frac{5}{2} \times \left[\frac{V^2}{f}\right]$

$R = \frac{5}{2} \times \left[\frac{0.782^2}{0.80}\right]$

R = 1.91 m

$P = 4.75\sqrt{Q}$

$P = 4.75\sqrt{50}$

P = 33.58 m

$A = Q \times V$

$A = 50 \times 0.782$

A = 39.1 m

For a trapezoidal channel with $\frac{1}{2}$: 1 side slopes:

P = (B + $\sqrt{5}D$)

(B + $\sqrt{5}D$) = 33.58, and

A = BD+ $\frac{D^2}{2}$ **............(i)**

BD+ $\frac{D^2}{2}$ = 39.1 **.............(ii)**

Solving equation (i) & (ii) we get

B = 30.90 m and D = 1.245 m

Now, Silt factor $S = \dfrac{f^{5/3}}{3340 \times Q^{1/6}}$

$$S = \frac{0.80^{5/3}}{3340 \times 50^{1/6}} = 1/9303$$

7.11.3 Factors for channel design consideration

The design of channels, whether for irrigation, drainage, or Stormwater management, involves a systematic process that considers various hydraulic, geometric, and environmental factors. Proper channel design is essential for efficient water conveyance, erosion control, and overall functionality. Here are the key steps and considerations in the design of channels that needs to be followed:

a. **Hydraulic Design:**
   - **Flow Requirements:** Determine the expected flow rates, or design discharge, based on factors such as water demand (for irrigation), Stormwater runoff, or drainage requirements.
   - **Velocity and Manning's Equation:** Calculate the required channel velocity using Manning's equation, considering the channel slope, roughness coefficient, and flow rate. Adjust the channel dimensions to achieve the desired velocity.
   - **Cross-Sectional Shape:** Select an appropriate cross-sectional shape for the channel, such as trapezoidal, rectangular, or triangular, based on hydraulic efficiency and site-specific conditions.

b. **Geometric Design:**
   - **Channel Dimensions:** Determine the width, depth, and side slope of the channel based on hydraulic and geometric considerations. Ensure that the dimensions provide adequate conveyance capacity and stability.
   - **Bed Slope:** Establish the longitudinal slope (bed slope) of the channel to achieve the desired flow velocity and facilitate sediment transport. The slope is often influenced by the topography of the area.

c. **Erosion Control:**
   - **Bank Protection:** Incorporate measures to prevent erosion of channel banks. This may involve adding riprap, vegetation, or other erosion control structures.

- **Velocity Control:** Design the channel to control flow velocities and prevent excessive erosion. Structures like weirs, drop structures, or energy dissipaters may be used.

d. **Sediment Transport:**

- **Velocity and Sediment Load:** Consider the ability of the channel to transport sediment without excessive deposition. Adjust channel dimensions and slope to accommodate sediment transport requirements.

e. **Materials and Linings:**

- **Lining Material:** Select appropriate materials for the channel lining, considering factors such as soil type, erosion potential, and water quality. Common materials include concrete, asphalt, clay, or synthetic liners like geo-membranes.
- **Vegetative Cover:** In some cases, vegetation may be used to stabilize the channel banks and enhance erosion control.

f. **Structural Elements:**

- **Culverts and Bridges:** Design and incorporate culverts or bridges where necessary to facilitate water flow across roads, railways, or other obstacles.
- **Weirs and Drop Structures:** Include structures like weirs and drop structures to control flow velocities, reduce erosion, and manage elevation changes.

g. **Water Quality Considerations:**

- **Vegetative Buffers:** Design vegetative buffers along the channel to filter pollutants and protect water quality.
- **Sediment Basins:** Consider the need for sediment basins or settling ponds to capture sediment before it enters the channel.

h. **Regulatory Compliance:**

- **Permitting and Regulations:** Ensure that the channel design complies with local, regional, and national regulations. Obtain the necessary permits for construction and modifications.

i. **Maintenance Considerations:**

- **Accessibility:** Design the channel for ease of maintenance. Access points, service roads, and maintenance considerations should be incorporated into the design.
- **Sediment Removal:** Plan for periodic sediment removal to maintain the channel's conveyance capacity.

j. **Environmental Impact Assessment:**
   - **Ecological Impact:** Assess and minimize the potential impact of the channel design on the surrounding environment, including aquatic habitats, flora, and fauna.

k. **Hydrological Analysis:**
   - **Rainfall and Runoff:** Conduct a hydrological analysis to estimate rainfall patterns, runoff characteristics, and peak flow rates. Use this information for sizing the channel and designing appropriate structures.

l. **Community Engagement:**
   - **Stakeholder Involvement:** Involve local communities and stakeholders in the design process to gather input, address concerns, and enhance the project's overall success.

Throughout the design process, it is crucial to use engineering principles, consider site-specific conditions, and integrate environmental sustainability practices. Collaborating with hydrologists, hydraulic engineers, environmental scientists, and other relevant experts can contribute to a comprehensive and effective channel design.

7.12 CANAL OUTLETS

Canal outlets refer to structures or mechanisms designed to regulate the flow of water from a canal to its intended destination. These outlets are crucial in managing water distribution for agricultural, industrial, or domestic purposes. The choice of canal outlet depends on factors such as the volume of water to be discharged, the desired flow rate, and the specific requirements of the downstream area.

Here are some common types of canal outlets:

a. **Sluice Gates:** Sluice gates are adjustable gates or barriers that can be raised or lowered to control the flow of water. They are often used in larger canals and are effective in managing water levels.

b. **Check Structures:** Check structures are built across canals to control the flow of water. They can include various features such as gates, weirs, or other flow control mechanisms.

c. **Siphons:** Siphons are structures that allow water to flow over an obstruction or under an obstacle. They are often used to carry water across a depression or under a road, enabling the canal to continue its course.

d. **Culverts:** Culverts are pipes or structures that allow water to pass under a road, embankment, or other obstacle. They are commonly used in canal systems to maintain a continuous flow.

e. **Flumes:** Flumes are open channels designed to carry water over a short distance. They can be used as outlets to direct water to specific areas or as a means of controlling the flow.

f. **Weirs:** Weirs are structures built across a river or canal to control the flow of water. They often involve a low dam or wall with a notch to regulate water levels.

g. **Regulators:** Regulators are structures that help in dividing water among different distributaries. They are used to control the distribution of water in irrigation systems.

7.12.1 Factors Affecting Canal Outlets

The choice of a canal outlet depends on factors such as the type of canal, the terrain, the intended use of the water, and the local environmental conditions. Properly designed and maintained canal outlets are essential for efficient water management in agriculture, and several irrigation factors influence the selection and performance of canal outlets. The choice of a canal outlet is crucial for effective water management and distribution. Here are some key factors that can affect canal outlets:

a. **Flow Rate and Discharge Requirements:** The volume of water that needs to be discharged from the canal influences the type of outlet selected. Different outlets have varying capacities, and the design must accommodate the expected flow rates.

b. **Topography and Terrain:** The physical characteristics of the landscape, such as slopes, elevations, and soil types, can impact the design and location of canal outlets. The outlet must be situated in a way that allows for efficient water flow and minimizes erosion.

c. **Water Quality:** The quality of the water in the canal, including sediment levels and debris, can affect the choice of outlet. Some outlets, like sluice gates and screens, are better at handling debris, while others may require additional filtration systems.

d. **Purpose of Water Use:** The intended use of the water (agriculture, industrial, or domestic) can influence the type of outlet chosen. Different outlets may be more suitable for delivering water to specific applications.

e. **Local Climate and Weather Conditions:** Climate factors such as rainfall patterns, temperature variations, and the possibility of freezing conditions can impact the performance of canal outlets. Proper design and materials should account for these environmental factors.

f. **Maintenance Requirements:** The ease of maintenance and the frequency of required upkeep are essential considerations. Some outlets may be more prone to clogging or damage, requiring more frequent inspection and maintenance.

g. **Regulation and Control Needs:** The level of control needed over water flow, including the ability to adjust flow rates, is a significant factor. Sluice gates, weirs, and other control structures offer varying degrees of regulation.

h. **Cost and Budget Constraints:** The financial resources available for the canal outlet's construction, installation, and maintenance play a major role in the selection process. Different types of outlets have varying costs associated with their design and implementation.

i. **Environmental Impact:** The ecological impact of canal outlets on the surrounding environment, including aquatic ecosystems and wildlife, is a consideration. Designs that minimize negative effects on the environment may be preferred.

j. **Legal and Regulatory Compliance:** Compliance with local laws, regulations, and permits is essential. Authorities may have specific requirements regarding the design, construction, and operation of canal outlets to ensure water resource management and environmental protection.

Considering these factors in the planning and design stages helps to ensure the successful implementation of canal outlets that meet the specific needs of the water distribution system and the surrounding environment.

7.12.2 Modular Canal Outlets

"Modular" typically refers to a system or design that consists of separate components or modules that can be easily connected or combined to create a larger structure. In the context of canal outlets, it might imply a design where various components are assembled in a modular fashion to create an outlet structure. In general, canal outlets can be designed with modular features to enhance flexibility, ease of installation, and adaptability to different conditions. Some features that could be considered modular in canal outlet design include:

a. **Adjustable Components:** Outlets that have adjustable elements, such as gates or weirs, allowing for customization based on water flow requirements.

b. **Modular Screens or Filters:** Filters or screens that can be modular in design, making it easier to replace or clean specific components without affecting the entire system.

c. **Interchangeable Parts:** Components that can be easily replaced or upgraded without requiring extensive changes to the entire canal outlet structure.

d. **Scalability:** Modular designs can often be easily scaled up or down to accommodate varying water flow rates or different canal sizes.

e. **Easy Maintenance:** Components designed for easy access and maintenance, ensuring that individual parts can be addressed without disrupting the entire canal outlet.

7.12.3 Non-modular canal outlets

The term "non-modular canal outlets" refers to canal outlet structures that are not designed with modular components. In this context, "non-modular" suggests that the outlet is constructed as a single, integrated system without easily separable or interchangeable parts. Here are some characteristics of non-modular canal outlets:

a. **Integrated Design:** Non-modular canal outlets are designed and constructed as a single, cohesive structure. The components are typically interconnected and not intended to be easily separated.

b. **Limited Customization:** Unlike modular designs that allow for customization through the addition or removal of specific modules, non-modular outlets may offer limited options for customization once installed.

c. **Fixed Components:** The various elements of the canal outlet, such as gates, weirs, or control structures, are often fixed in place and not easily adjustable or replaceable.

d. **Single-Purpose Structures:** Non-modular canal outlets are typically designed for specific flow rates, water levels, and purposes without the flexibility to adapt to changing conditions.

e. **Complex Maintenance:** Maintenance of non-modular canal outlets may involve more intricate processes, as repairs or replacements may require addressing the entire structure rather than individual components.

f. **Less Scalability:** Non-modular designs may be less scalable in terms of accommodating changes in water flow rates or canal system modifications.

It is important to note that the choice between modular and non-modular canal outlets depends on various factors, including the specific requirements of the water distribution system, environmental conditions, and the level of flexibility needed in managing water flow.

Engineers and designers carefully consider these factors when planning and implementing canal outlet structures to ensure efficient water management.

7.12.4 Semi-modular canal outlets

The term "semi-modular canal outlets" suggests a design that incorporates both modular and non-modular features. While there is not a standard or widely recognized definition for "semi-modular" in the context of canal outlets, the term generally implies a combination of both modular and integrated elements in the design. Here are some characteristics that might be associated with semi-modular canal outlets:

a. **Combination of Modular and Fixed Components:** Semi-modular canal outlets may include some components that are designed to be modular, allowing for flexibility and customization, while other components are integrated and less easily separable.

b. **Adjustable Features:** Some elements of the canal outlet structure might be designed to be adjustable or interchangeable, providing a degree of flexibility in managing water flow.

c. **Limited Customization:** While offering more flexibility than non-modular designs, semi-modular canal outlets may not provide the same level of customization as fully modular systems.

d. **Integrated Control Structures:** Certain critical components, such as control gates or weirs, may be integrated into the structure and not easily replaced or modified.

e. **Adaptability to Changing Conditions:** Semi-modular designs may strike a balance between adaptability and stability, allowing for adjustments to be made based on changing water flow requirements or other conditions.

f. **Ease of Maintenance:** Maintenance of semi-modular canal outlets may involve a combination of straightforward tasks for modular components and more complex procedures for integrated elements.

It is important to note that the specific features and characteristics of semi-modular canal outlets can vary based on the engineering and design choices made for a particular project. The goal is often to find a balance between the benefits of modularity, such as ease of maintenance and adaptability, and the stability of an integrated structure. The choice of a semi-modular design will depend on the specific needs and constraints of the water distribution system in question.

7.13 WATER LOGGING

Water logging refers to the saturation of soil with water to the extent that it hinders or prevents plant growth. This occurs when the soil's ability to absorb and drain water is exceeded, leading to an accumulation of excess water on the surface or in the root zone of plants. Water logging can have detrimental effects on both agricultural and urban areas. Here are some key points related to water logging:

a. **Causes:**
   - **Excessive rainfall:** Heavy and prolonged rainfall can lead to water logging, especially in areas with poor drainage systems.
   - **Poor drainage:** Inadequate drainage systems or blocked drainage channels can contribute to water logging.
   - **High water table:** In some areas, the water table may be naturally high, making the soil more prone to water logging.
   - **Compacted soil:** Soil compaction reduces its ability to absorb water, increasing the likelihood of water logging.

b. **Effects:**
   - **Reduced oxygen availability:** Waterlogged soil restricts the flow of air to plant roots, leading to oxygen deficiency. This can negatively impact root growth and overall plant health.
   - **Nutrient leaching:** Water logging can result in the leaching of nutrients from the soil, making them less available to plants.
   - **Increased susceptibility to diseases:** Waterlogged conditions create a favorable environment for certain plant diseases.
   - **Crop damage:** Agricultural crops may suffer yield losses or even complete crop failure due to water logging.

c. **Prevention and Mitigation:**
   - **Improved drainage:** Installing or maintaining proper drainage systems helps prevent waterlogging.
   - **Land shaping:** Properly shaping the land can promote water runoff and prevent water accumulation.

- **Crop selection:** Choosing crops that are more tolerant to waterlogged conditions can mitigate the impact.
- **Soil management:** Implementing practices like organic matter addition and avoiding soil compaction can improve soil structure and drainage.

d. **Urban Water logging:**

- In urban areas, water logging can lead to flooding of streets, basements, and other low-lying areas.
- Inadequate stormwater drainage systems and improper urban planning can contribute to urban water logging.

e. **Climate Change Impact:**

- Changes in precipitation patterns and more intense rainfall events due to climate change may exacerbate water logging issues in certain regions.

Effective management of water logging involves a combination of proper land use planning, sustainable agricultural practices, and infrastructure development to ensure efficient water drainage.

7.14 LINING OF CANALS

Canal lining refers to the process of adding a protective layer to the bed and/or sides of a canal to reduce seepage and enhance the conveyance efficiency of the canal system. Lining can help conserve water resources by minimizing losses due to seepage into the surrounding soil. The choice of lining material and method depends on factors such as the local soil conditions, climate, and the specific requirements of the canal. Here are some common methods of canal lining:

a. **Earthen Lining:** The canal bed and sides are compacted to form a relatively impermeable layer using locally available soil.

- **Application:** Suitable for canals with cohesive soils that have good compaction characteristics. It is a cost-effective method but may require periodic maintenance.

b. **Clay Lining:**

- **Description:** A layer of clay is applied to the canal bed and sides to form a seal against seepage. The clay layer may be compacted to enhance impermeability.

- **Application:** Effective in areas where suitable clay is available. Proper compaction is essential for the success of clay lining.

c. **Concrete Lining:**

- **Description:** Concrete is applied to the canal bed and sides to form a durable and impermeable lining. It can be precast or applied in situ.
- **Application:** Suitable for canals with high seepage losses and where a durable and long-lasting lining is required. Common in urban areas and large irrigation projects.

d. **Asphaltic Concrete Lining:**

- **Description:** A layer of asphaltic concrete, which consists of asphalt and aggregates, is applied to the canal to provide an impermeable coating.
- **Application:** Effective in reducing seepage losses. Commonly used in conjunction with other lining methods or materials.

e. **Geo-membrane Lining:**

- **Description:** Synthetic liners, such as geo-membranes made of materials like HDPE (High-Density Polyethylene) or PVC (Polyvinyl Chloride), are used to create an impermeable barrier.
- **Application:** Suitable for canals with highly permeable soils. Geo-membranes are resistant to chemical degradation and provide a reliable lining solution.

f. **Flexible Membrane Lining:**

- **Description:** Flexible membranes made of materials like EPDM (Ethylene Propylene Diene Monomer) or PVC are used as liners to reduce seepage.
- **Application:** Effective in areas with expansive soils or where flexibility is required. They can conform to the soil surface and resist cracking.

g. **Grouted Lining:**

- **Description:** Cement or chemical grout is injected into the soil to create an impermeable barrier against seepage.
- **Application:** Used in conjunction with other lining methods or in situations where localized treatment is needed.

h. **Rock Lining:**

- **Description:** A layer of rock or other durable materials is placed along the canal bed and sides to reduce erosion and seepage.
- **Application:** Common in open-channel sections of canals where protection against erosion is a primary concern.

The choice of lining method depends on factors such as the local soil conditions, the level of seepage control required, the project budget, and the expected lifespan of the canal. It is essential to consider the long-term maintenance requirements and the environmental impact of the chosen lining method. Additionally, a proper engineering analysis and design are crucial for the success of the canal lining project.

The drainage of irrigated lands is essential to maintain optimal soil conditions for plant growth and prevent issues such as water logging, salinity, and root damage. Here are the key reasons why drainage is necessary for irrigated lands, along with various methods used to achieve effective drainage:

7.14.1 Necessity of Drainage in Irrigated Lands

a. **Preventing Water logging:**

- **Excess Water Removal:** Irrigated lands may receive more water than the soil can absorb or drain naturally. Proper drainage prevents water logging, which can lead to oxygen deprivation and hinder root development.

b. **Avoiding Soil Salinity:**

- **Leaching of Salts:** Irrigation water can contain dissolved salts. Without proper drainage, these salts may accumulate in the soil, leading to salinity problems that adversely affect plant growth.

c. **Improving Soil Aeration:**

- **Oxygen Infiltration:** Effective drainage enhances soil aeration by allowing the infiltration of oxygen into the root zone. Adequate oxygen levels are crucial for the respiration of plant roots.

d. **Enhancing Nutrient Uptake:**

- **Nutrient Availability:** Proper drainage helps to maintain optimal nutrient levels in the soil by preventing nutrient leaching and ensuring that essential elements are available for plant uptake.

e. **Preventing Erosion:**

- **Surface Runoff Control:** Drainage systems help to control surface runoff, preventing soil erosion and preserving the topsoil structure.

f. **Facilitating Timely Planting and Harvesting:**

- **Workability of Soil:** Well-drained soils are more workable, allowing for timely planting and harvesting operations. In poorly drained soils, fieldwork may be delayed due to wet conditions.

7.14.2 Methods of Drainage in Irrigated Lands:

a. **Surface Drainage:**

- **Contour Ploughing:** Ploughing along the contour lines helps slow down water runoff, preventing soil erosion.
- **Open Ditches:** Constructing open ditches or channels helps to collect and direct excess water away from the fields.

b. **Subsurface Drainage:**

- **Tile Drainage:** Installing perforated pipes or tiles underground to collect and transport excess water to drainage outlets.
- **French Drains:** Buried pipes filled with gravel or rocks that collect and redirect subsurface water.

c. **Land Grading and Levelling:**

- **Precision Grading:** Levelling the land to ensure a uniform slope for efficient water drainage.

d. **Vegetative Measures:**

- **Cover Crops:** Planting cover crops helps stabilize the soil structure, reduce surface runoff, and enhance water infiltration.
- **Buffer Strips:** Planting vegetation along waterways to trap sediments and filter runoff.

e. **Water Control Structures:**

- **Weirs and Check Dams:** Structures that regulate water flow in ditches and canals.
- **Culverts and Bridges:** Allow water to pass under roads and pathways without causing obstruction.

f. **Irrigation Management:**

- **Controlled Irrigation:** Implementing precise irrigation schedules to avoid overwatering and minimize the risk of waterlogging.

g. **Regular Maintenance:**

- **Clearing Obstructions:** Periodically inspecting and clearing drainage channels and structures to ensure unimpeded water flow.

h. **Water-Quality Monitoring:**

- **Testing Water Quality:** Regularly monitoring the quality of irrigation water to detect and address salinity issues.

A comprehensive approach that combines various drainage methods and considers the specific characteristics of the land is essential for effective drainage in irrigated areas. The goal is to create a balanced and sustainable water management system that supports healthy crop growth.

SUMMARY

Canal distribution systems are crucial for maintaining human activity and managing water resources in many parts of the world. It typically describes a network of artificial waterways used for drainage, flood control, transportation, irrigation, and water supply, among other purposes. Canal distribution systems require the construction and upkeep of canals to transport water from a river or reservoir to farms and crops. Technology is widely used in today's canal distribution systems to boost efficiency and conserve water. The use of sensors, automated control systems, and specialized irrigation techniques might be required to achieve this. Materials such as concrete or plants may be utilized to line canals. Water losses from the canal distribution system, estimation of its design discharge, types of channels (such as rigid boundary channels, alluvial channels, etc.), and channel design methodologies (such as Kennedy's and Lacey's theory) have all been addressed during this unit. Furthermore, various kinds of canal linings and outflows have been documented. Lastly, techniques for draining irrigated land have been discussed, along with the causes, consequences, and corrective actions for waterlogging.

EXERCISE

Revision Questions

1. Define the canal distribution systems.
2. What are the uses of a canal system? With the help of a diagram classify a canal system.
3. Provide different category of canals based on their uses.
4. What do you understand by alignment of canals? Describe key considerations and factors in determining the alignments of a canal.
5. With the help of a neat sketch define various types of canal alignments.
6. Give a short note on (a) Command area; (b) gross command area; and (iii) cultivable command area.
7. Describe the causes of canal losses.
8. What are the factors that affects the canal losses?
9. What do you understand by discharge of canal? Describe various factors affecting discharge in a canal system.
10. Define term design discharge. Describe the factors affecting design discharge.
11. Describe the key considerations in the estimation of design discharge.
12. Describe the rational method for design discharge of a water body.
13. Describe SCS (Soil Conservation Service) Curve Number Method for design discharge of a water body.
14. Describe the Unit hydrograph method for design discharge of a canal.
15. Describe in brief different empirical models use statistical relationships based on historical data to estimate runoff.
16. What are the differences between a natural channel and a canal system?
17. Define rigid boundary channels. Describe the key features and considerations related to rigid boundary channels.
18. What do you understand by the alluvial channel?
19. Define the following equations. (i) Chezy's equation; (ii) Ganguillet and Kutter's equation; (iii) Manning's equation.
20. Define the following (i) Kennedy's silt theory; (ii) Lacey's theory.

21. Describe key factors for design of channel.
22. Describe canal outlets. Describe various factors affecting the canal outlets.
23. Define the term, water logging. Describe various factors affecting water logging.
24. What do you understand by lining of canals? Describe various types of canal lining.

Numerical Problems

1. In a given watershed, Rainfall Intensity (i): 6cm per hour; Area of the Watershed (A): 120 square km; Runoff Coefficient (C): 0.5 (represents soil and land use characteristics). Calculated design discharge (Q) using the Rational Method.
2. Find out normal water depth and velocity in channel carrying a discharge of 11 cumec and having bed width of 7 m. Use roughness coefficient n= 0.0225 and bed slope S = 0.0017.
3. Design as channel with side sloped 1:1 to carry 15 cumec at a slope = 0.0018. Maximum permissible velocity in channel is 1.9 m/sec and n = 0.022.
4. Design an irrigation channel to carry 45 cumec at a slope of 1/5500 with Kutter's n = 0.0250 and m = 0.95.
5. Design a regime channel using Lacey's theory. The available discharge of the regime channel may be taken as 55 cumec and silt factor = 0.85. The corresponding side slope of the channel can be assumed as $\frac{1}{2}:1$.

Multiple Choice Questions

1. What is the primary purpose of a canal distribution system?

A) To transport water for irrigation
B) To generate hydroelectric power
C) To provide recreational boating opportunities
D) To manage stormwater runoff

2. Which of the following is a common feature of a canal distribution system?

A) Dams
B) Spillways
C) Sluice gates
D) All of the above

3. What is a "headworks" in a canal distribution system?

A) The area where water is stored
B) The structure that regulates water flow into the canal

C) The main canal itself

D) A type of irrigation technique

4. Which factor is most important in the design of a canal distribution system?

A) Soil type
B) Crop water requirements
C) Climate conditions
D) All of the above

5. What is a "reach" in the context of canal distribution systems?

A) The total length of a canal
B) A segment of the canal between two control structures
C) The area where water is diverted from a river
D) The point where water is used for irrigation

6. How is water typically conveyed in a canal distribution system?

A) Through underground pipes
B) In open channels
C) Using pumps only
D) Through water tanks

7. What is a potential drawback of canal distribution systems?

A) High efficiency in water delivery
B) Evaporation losses
C) Low construction costs
D) Minimal environmental impact

8. What is the term for the process of measuring and managing water flow in a canal?

A) Hydrology
B) Water allocation
C) Water management
D) Flow regulation

9. Which type of canal is primarily used for navigation?

A) Irrigation canal
B) Drainage canal
C) Navigation canal
D) Flood control canal

10. What type of canal is designed specifically to divert water for agricultural purposes?

A) Irrigation canal
B) Stormwater canal
C) Flood control canal
D) Recreational canal

11. Which type of canal helps to manage excess water and prevent flooding?

A) Navigation canal
B) Irrigation canal
C) Drainage canal
D) Recreational canal

12. What is a "feeder canal"?

A) A canal that collects stormwater

B) A canal that supplies water to a larger canal system

C) A canal used for recreational purposes

D) A canal that drains excess water

13. Which type of canal is often used to create artificial lakes for recreational activities?

A) Irrigation canal
B) Flood control canal

C) Recreational canal
D) Navigation canal

14. What distinguishes a "conveyance canal" from other types?

A) It is only used for irrigation

B) It primarily transports water over long distances

C) It is designed for small water bodies

D) It has no controlled flow

15. What is the main function of a "drainage canal"?

A) To provide water for irrigation
B) To remove excess water from land

C) To facilitate navigation
D) To store water for later use

16. What is a major category of canals based on their construction?

A) Concrete canals
B) Earthen canals

C) Steel-lined canals
D) All of the above

17. Which type of canal is specifically designed to accommodate large vessels?

A) Irrigation canal
B) Drainage canal

C) Navigation canal
D) Flood control canal

18. Canals can be categorized based on their maintenance requirements. Which of the following is true?

A) Natural canals require more maintenance

B) Artificial canals are typically easier to maintain

C) All canals need regular maintenance

D) Only irrigation canals require maintenance

19. Based on water flow, canals can be categorized into two types. What are they?

A) Open and closed
B) Static and dynamic
C) Natural and artificial
D) Seasonal and perennial

20. What is a "seasonal canal"?

A) A canal that runs dry in the summer
B) A canal that is used only during certain times of the year
C) A permanent canal with fluctuating water levels
D) A canal used for recreational purposes

21. Canals can also be categorized based on their size. Which of the following is a common size category?

A) Micro canals
B) Major canals
C) Minor canals
D) All of the above

22. What is a "command area" in irrigation terminology?

A) The total area that can be cultivated
B) The area served by a specific irrigation project
C) The area affected by soil erosion
D) The area designated for crop rotation

23. Which of the following is a primary objective of defining a command area?

A) To increase the use of chemical fertilizers
B) To manage water resources efficiently
C) To eliminate pests and diseases
D) To promote urban development

24. In a well-managed command area, which factor is most critical for optimizing crop yield?

A) Soil type
B) Water availability
C) Weather patterns
D) Market access

25. What is a common challenge faced in command areas?

A) Excessive rainfall
B) Water scarcity and distribution
C) Overpopulation
D) High land prices

26. Which method is often used to improve irrigation efficiency in command areas?

A) Flood irrigation
B) Drip irrigation
C) Rain-fed agriculture
D) Traditional ploughing

27. What does GCA stand for in irrigation terminology?

A) Gross Cultivated Area
B) Gross Command Area
C) Geographic Crop Area
D) General Crop Assessment

28. What is the primary characteristic of CCA?

A) The total area under water management
B) The area that can be cultivated with irrigation
C) The area affected by soil salinity
D) The area used for non-agricultural purposes

29. Which of the following statements is true regarding the relationship between GCA and CCA?

A) GCA is always equal to CCA
B) CCA is typically less than GCA
C) CCA can be greater than GCA
D) GCA and CCA are unrelated concepts

30. Why is it important to differentiate between GCA and CCA in irrigation planning?

A) To maximize crop prices
B) To assess environmental impact
C) To ensure efficient water resource management
D) To identify pest control methods

31. Which of the following factors can reduce the CCA within a given GCA?

A) Improved irrigation techniques
B) Urban development and infrastructure
C) Crop rotation practices
D) Increased rainfall

32. What is design discharge?

A) The maximum flow that can be accommodated in a system
B) The average flow during peak usage
C) The flow rate estimated for a specific design condition
D) The flow rate during a storm event

33. Which of the following methods is commonly used for estimating design discharge for drainage systems?

A) Rational Method
B) Average Daily Flow Method
C) Peak Hourly Flow Method
D) Continuous Simulation Method

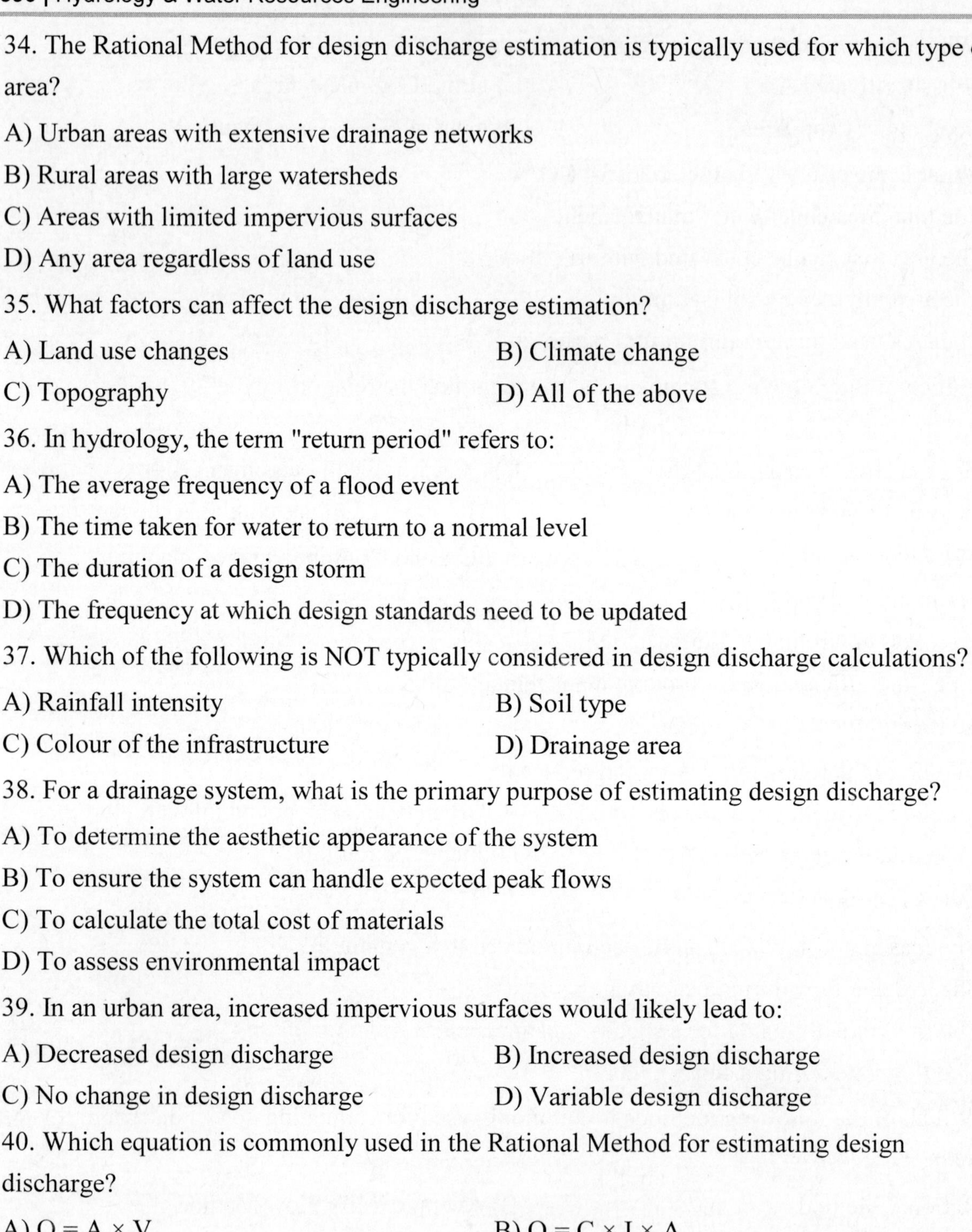

34. The Rational Method for design discharge estimation is typically used for which type of area?

A) Urban areas with extensive drainage networks

B) Rural areas with large watersheds

C) Areas with limited impervious surfaces

D) Any area regardless of land use

35. What factors can affect the design discharge estimation?

A) Land use changes | B) Climate change

C) Topography | D) All of the above

36. In hydrology, the term "return period" refers to:

A) The average frequency of a flood event

B) The time taken for water to return to a normal level

C) The duration of a design storm

D) The frequency at which design standards need to be updated

37. Which of the following is NOT typically considered in design discharge calculations?

A) Rainfall intensity | B) Soil type

C) Colour of the infrastructure | D) Drainage area

38. For a drainage system, what is the primary purpose of estimating design discharge?

A) To determine the aesthetic appearance of the system

B) To ensure the system can handle expected peak flows

C) To calculate the total cost of materials

D) To assess environmental impact

39. In an urban area, increased impervious surfaces would likely lead to:

A) Decreased design discharge | B) Increased design discharge

C) No change in design discharge | D) Variable design discharge

40. Which equation is commonly used in the Rational Method for estimating design discharge?

A) $Q = A \times V$ | B) $Q = C \times I \times A$

C) $Q = P \times H$ | D) $Q = R \times T$

41. Why is it important to consider climate change in design discharge estimation?

A) It affects the aesthetic design of infrastructure

B) It can alter precipitation patterns and increase flood risks

C) It has no impact on design discharge

D) It helps in determining construction materials

42. What defines an alluvial channel?

A) A channel carved into bedrock

B) A channel composed of loose sediment

C) A channel that is always dry

D) A channel formed by glacial activity

43. Which of the following is a characteristic of open channel flow?

A) Flow is confined within pipes

B) Flow depth is greater than the channel width

C) Flow is influenced by gravity and friction

D) Flow is only unidirectional

44. In an alluvial channel, sediment transport occurs primarily due to:

A) Gravity

B) Wind

C) Ice

D) Biological activity

45. What is the main factor influencing the flow velocity in an open channel?

A) Channel depth

B) Channel roughness

C) Bed slope

D) All of the above

46. Which formula is commonly used to estimate the discharge in an open channel flow?

A) Q = A × V

B) Q = C × R^2/3 × S^1/2

C) Q = V × T

D) Q = P × H

47. What does the term "critical flow" refer to in open channel flow?

A) The maximum possible flow rate

B) Flow conditions where specific energy is minimized

C) Flow conditions at which sediment begins to settle

D) Flow that is stable and constant

48. The Manning's equation is used for:

A) Estimating sediment transport

B) Calculating flow velocity in open channels

C) Designing hydraulic structures

D) Determining rainfall intensity

49. Which factor does NOT influence the flow regime in an alluvial channel?

A) Channel shape
B) Flow depth
C) Type of vegetation
D) Air temperature

50. Which type of flow is characterized by a high Reynolds number and turbulent flow conditions?

A) Laminar flow
B) Subcritical flow
C) Supercritical flow
D) Steady flow

51. What is the primary environmental concern with alluvial channels?

A) Sediment accumulation
B) Water temperature fluctuations
C) Erosion and habitat destruction
D) Noise pollution

52. What are canal losses?

A) The total flow rate in the canal
B) Water that is lost due to evaporation and seepage
C) Water that is diverted for irrigation
D) Water that flows back to the river

53. Which of the following is NOT a type of canal loss?

A) Seepage loss
B) Evaporation loss
C) Loss due to sedimentation
D) Loss due to irrigation

54. What is the primary function of a canal outlet?

A) To measure flow velocity
B) To control water levels in a canal
C) To release water from the canal into another system
D) To filter water before release

55. In the SCS-CN method, the Curve Number (CN) is used to estimate:

A) Peak discharge
B) Rainfall intensity
C) Runoff potential from a given area
D) Evapotranspiration

56. What factors influence the Curve Number in the SCS-CN method?

A) Land use and hydrologic condition
B) Soil type and cover
C) Antecedent moisture conditions
D) All of the above

57. Which of the following statements is true about seepage loss in canals?

A) It is always negligible.
B) It can lead to water shortages downstream.
C) It is beneficial for groundwater recharge.
D) It occurs only in lined canals.

58. When using the SCS-CN method, what is typically used to convert rainfall to runoff?

A) A runoff coefficient
B) The curve number
C) The infiltration rate
D) The evaporation rate

59. Which type of outlet is designed to allow for controlled water release at varying levels?

A) Fixed outlet
B) Adjustable outlet
C) Spillway outlet
D) Weir outlet

60. In the context of canal systems, which of the following is a method to reduce evaporation losses?

A) Increasing flow velocity
B) Reducing surface area exposed to air
C) Deepening the canal
D) Using lined channels

61. What is the main advantage of using the SCS-CN method for hydrological analysis?

A) It requires complex data input.
B) It can be applied universally regardless of location.
C) It is a straightforward method for estimating runoff.
D) It does not consider land use changes.

Answer: 1-A; 2-D; 3-B; 4-D; 5-B; 6-B; 7-B; 8-D; 9-C; 10-A; 11-C; 12-B; 13-C; 14-B; 15-B; 16-D; 17-C; 18-C; 19-A;20-B; 21-D; 22-B; 23-B; 24-B; 25-B; 26-B; 27-B; 28-B; 29-B; 30-C; 31-B; 32-C; 33-A; 34-A; 35-D; 36-A; 37-C; 38-B; 39-B; 40-B; 41-B; 42-B; 43-C; 44-A; 45D; 46-B; 47-B; 48-B; 49-D; 50-C; 51-C; 52-B;53-D;54-C;55-C;56-D;57-B;58-B;59-B;60-B;61-C.

KNOW MORE

For more information related to this topic scan the QR code.

OR

Type this link in your browser

https://drive.google.com/file/d/1jLKmyqGKZSXd6gp5PO2CyZlakem12xwE/view?usp=sharing

BIBLIOGRAPHY

1. Asawa, G.L. (1993), "Irrigation Engineering", Published by Willey Eastern Limited, ISBN 81-224-0457-X.
2. Asawa, G.L. (2005), "Irrigation and Water Resources Engineering, New Age International (P) Ltd., ISBN-13: 978-8122416732
3. Chow. V.T, (Ed), (1964), "Handbook of Applied Hydrology", McGraw-Hill, New York, NY, 1964.
4. Jain S. K., Agarwal, P. K. And Singh, V. P. (2007), Hydrology and Water Resources of India, Water Science and Technology Library, Vol 57, Springer, The Netherland, ISBN-10 1-4020-5179-4(HB).
5. Punmia, B.C. and Lal, Pande B.B., (1992), "Irrigation and Water Power Engineering", Twelfth Edition. Laxmi Publications (P) Ltd.
6. Singh, V. P. (1992). Elementary hydrology. Prentice Hall, New Jersy, 973.
7. Subramanya, K. (2015), "Engineering Hydrology", McGraw-Hill Education (India) Private Limited, New Delhi, 2015, ISBN(13): 978-1-25902997-4.
8. Varshney, R.S., Gupta, S.C and Gupta R. L.(2009), "Theory and design of Irrigation structures (Vol-1: Channels and Tubewells)", Published by Nem Chand & Bros, Civil Lines, Roorkee, ISBN 81-85240-73-6.
9. Varshney, R.S., Gupta, S.C and Gupta R. L.(2009), "Theory and design of Irrigation structures (Vol-I1: Canal and Storage works)", Published by Nem Chand & Bros, Civil Lines, Roorkee, ISBN 81-85240-48-5.

Dams and Spillways

UNIT SPECIFICS

Dams and spillways are crucial components of water management systems. Dams are barriers built across rivers or streams to store water, create reservoirs, and control water flow. They help with water supply, flood control, hydroelectric power generation, and recreational activities. There are various types of dams, including gravity, arch, and earthen dams. Spillways are channels or structures designed to safely release excess water from a dam to prevent overflow and potential damage. They are essential for managing the water levels in the reservoir and protecting the dam. This unit discusses the classification, design consideration, causes of failure, estimation, and control of seepage for various types of dams such as embankment dams and arch dams. In addition, this unit deals with components of spillways and types of gates for spillway crests. The types, yield, regulation, and sedimentation of a reservoir are also described in this unit.

RATIONALE

To learn about the dams and spillways, types of dams, control of seepage, spillway, types of reservoirs, reservoir regulations, sedimentation, and selection of suitable sites for dams.

PRE-REQUISITE

Nil

UNIT OUTCOMES

The list of outcomes of this unit is as follows:

U8-O1: Understand the fundamentals of dams and spillways

U8-O2: Understand the types of dams like gravity, embankment, and arch dams, and seepage control from the dams.

U8-O3: Spillways, types, yield, sedimentation, and regulations of reservoirs

U8-O4: Selection of suitable sites for dams

Unit Outcomes	Expected Mapping with Course Outcomes *(1- Weak Correlation; 2- Medium correlation; 3- Strong Correlation)*							
	CO-1	CO-2	CO-3	CO-4	CO-5	CO-6	CO-7	CO-8
U8-O1	3	3	3	1	3	3	3	3
U8-O2	3	3	3	1	3	3	1	3
U8-O3	1	1	3	3	3	3	3	3
U8-O4	3	3	3	3	3	3	3	3

8.1 INTRODUCTION

An extensive wall built across a river or stream to restrict or stop the flow of water is called a dam. The purpose of building dams is to create reservoirs that can hold water for industrial, agricultural, and drinking purposes; producing power by turning turbines with the force of flowing water Managing water levels and flow to support habitats and manage ecosystems; providing possibilities for activities like boating, swimming, and fishing; reducing the likelihood of downstream flooding during heavy rains or snowmelt. There are several different kinds of dams, such as gravity dams, arch dams, and buttress dams, and each is intended to address engineering and environmental issues. For them to be safe and effective, proper design and upkeep are necessary. While constructing a dam, a big natural or man-made lake is created behind the dam is called a reservoir. The primary purposes of reservoirs are to store water for drinking, irrigation, hydroelectric power generation, flood control, and recreational activities. A spillway is a structure designed for safely release of excess water from a reservoir or dam. It helps to manage water levels and prevent overflow or damage to the dam. The primary purpose of a spillway is to control the flow of water when it exceeds the normal capacity of the reservoir.

8.2 EMBANKMENT DAMS

Embankment dams are commonly used for water storage, flood control, and hydropower generation. Embankment dams are structures built using compacted earth, rock, or other suitable materials to form a stable barrier across a river or watercourse. These materials are distinct from one another and have spaces between them. The location, internal friction, and mutual attraction of the particles in these materials give them their strength. These fragmentary components, as opposed to cemented materials, combine to form a somewhat flexible structure that can flex slightly to accommodate the displacement of the foundation without failing.

The following factors make choosing an embankment dam more likely:

a. A substantial layer of soil deposits covering bedrock,
b. Soft or weak bedrock that could not withstand the intense pressures of a concrete dam,
c. Abutments of weak rock or deep soil layers,
d. The presence of a suitable spillway position and
e. The availability of adequate and appropriate soils from neighboring borrows areas or necessary excavation.

8.2.1 Classification

Embankment dams are mainly of two types:

a. Earth-fill or earth dams
b. Rock-fill or earth-rock dams.

In an earth-fill dam, the bulk of the mass is made up of dirt, whereas in a rock-fill dam, it is made up of rock material. The two kinds of embankment dams share similar design concepts. The two types of earth dams that are further separated/classified as:

a. Homogeneous earth dam, and
b. Zoned earth dam.

One type of earth material is used exclusively or almost entirely in the construction of homogeneous earth dams. On the other hand, various materials are used in different sections of the embankment of a zoned earth dam. When only one kind of material is commercially feasible and/or the dam's height is not particularly high, a homogenous earth dam is typically constructed. A drain of some kind made of material more pervious than the embankment soil should always be present in a homogenous earth dam that is higher than around 6 to 8 meters. These drains enhance the stability of the downstream slope by lowering pore pressures in the dam's downstream section. Moreover, the drains also control the seepage water that emerges from the earth to prevent "piping" from developing. A dam of this type is also classified as homogeneous (or occasionally as "modified homogeneous") dams (Figure 8.1). A homogenous earth dam can nonetheless reap some of the benefits of a zonal earth dam by utilizing various building techniques in different sections of the embankment or by selectively placing dirt to create zones with distinct characteristics. The zonal earth dam is the most widely used earth dam because it results in a more stable and cost-effective design for the dam.

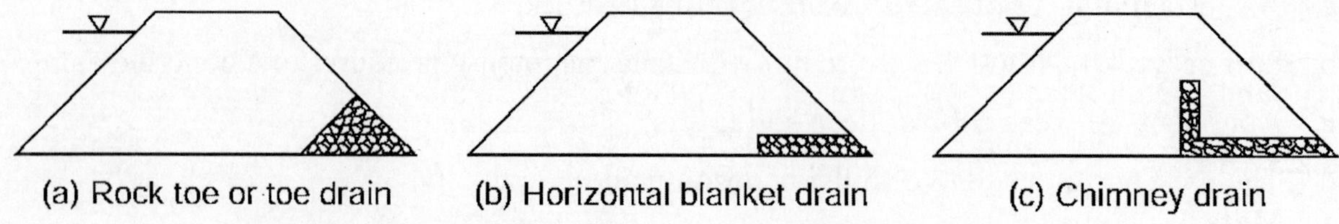

Figure 8.1: Typical Sections of Homogenous Earth Dam

An embankment that uses large-sized rock fragments for stability and an impermeable membrane for waterproofing is called a rock-fill dam (Figure 8.2). The membrane is made of wood, steel, concrete, asphalt, and soil. The impermeable membrane can be positioned as a core inside the embankment or on the dam's upstream face. However, it is better to place the impermeable membrane on the dam's upstream face.

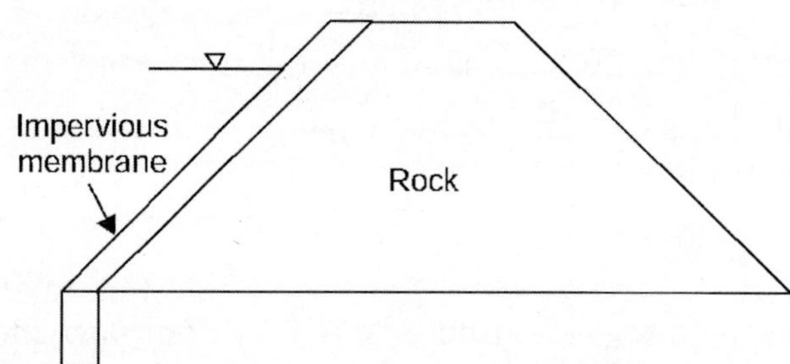

Figure 8.2: Typical Section of a Rock-fill Dam

8.2.2 Design Considerations

An embankment dam's design is derived from both experience and analytical factors. The following are the primary steps involved in designing an embankment dam:

- A careful examination of the abutments and foundation.
- A review of all the materials for building embankments that are readily accessible within a suitable driving distance of the dam site, including their amounts and qualities.
- An examination of every element that could affect the design.

- Choosing the trial designs.
- Evaluation of the trial designs' safety.
- Changing the designs to meet the minimal requirements for stability.
- The creation of thorough cost projections.
- The chosen design appears to provide the optimal balance of construction economy, safety, and convenience.

8.2.3 Factors Influencing the Design of an Embankment Dam

a. Materials Available for Construction

The free availability of construction materials near or adjacent to the embankment dam site is one of its main advantages. The designed embankment can be either a zoned earth dam (when both pervious and impervious soils are available), a rock-fill dam (if the rock is available and impervious material is not), or a homogeneous earth dam (when the soil available is impervious). This depends on the type of material available. For economic reasons, the design may also include the utilization of materials from the necessary excavation (for the construction of the spillway).

b. Foundation Characteristics

Almost any type of foundation can support the construction of an embankment dam. The foundation treatment, which is often the most challenging and significant aspect of designing and building an embankment dam, is primarily influenced by the features of the foundation. Additionally, the embankment's size would be significantly impacted. for instance, a softer foundation, would require an embankment with flatter slopes, a bigger cross-section, a larger freeboard (To minimize the consequences of embankment settlement), considerations for differential settlement cracks, and controls over under seepage to prevent the risk of piping.

c. Climate

Controlling the construction moisture content of fine-grained soils in arid places and working with them during the rainy season are often challenging tasks. Therefore, a sloping core embankment is recommended if the embankment construction must be completed during the rainy season. Similarly, building a small reservoir to store flood water for the purpose of building the dam may take an additional year in arid locations.

d. Shape and Size of Valleys

A vast valley and gently sloping abutments at a dam site might not have an impact on an embankment's design. Narrow valleys and steep abutments, however, might call for extra design considerations. For example, a simpler design requiring few additional construction provisions would be preferred due to the limited working space in a small valley. In the event that building and maintaining haul roads on abutments at varying heights becomes challenging and expensive, it might be necessary to design a rock-fill embankment, which can be built by excavating rock in high lifts from comparatively few haul roads.

e. River Diversion

The difficulty for the designer increases if a river diversion strategy is to be carried out by the contractor or construction engineer. The designer must consider every scenario and modify the design to accommodate each one. However, it can be wise to specify the river diversion system and build the embankment in accordance with it on large rivers. The river in a narrow valley is diverted by a conduit or tunnel. In larger valleys, the river is allowed to flow through the valley's centre while parts of the embankment on the two abutments are built. Known as the "closure" section, this canter portion of the embankment is only built at its end. In order to prevent the dam from overtopping, the closure section must be constructed promptly. As a result, specific design and construction details are required, such as extra filter drains, distinct embankment designs that maximize the material available on the two abutments, compacting the closure section at higher water content, and maintaining a reserve of borrowed material to achieve a rapid construction rate for closure. It would be cost-effective to incorporate coffer dams with significant flows into the dam embankment if they are used for diversion.

f. Probable Wave Action

The length of the reservoir and the wind speed are the most important factors of the intensity of wave action and the quantity of protection required for the embankment's upstream face. The embankment is constantly being driven against by the waves, which results in embankment erosion. Wave protection which is simultaneously most affordable and effective is a layer of riprap constructed from dumped rock.

g. Time Available for Construction

The amount of time allocated for construction determines how an embankment is designed. In the case of high dams, a shorter construction period could cause higher pore pressures, requiring almost flatter slopes. It might not be possible to use all of the material from the necessary excavation during construction, or perhaps just a portion of it could be utilized. Similarly, the duration of time allotted for construction would also have an impact on under seepage measures. It may be beneficial to provide a constructed impermeable barrier in order to save time while handling fine-grained soils, as this can take an enormous amount of time.

h. The function of Reservoir

The permissible water loss from seepage through the foundation and embankment is determined by the reservoir's objective. As a result, depending on whether the embankment component was designed for flood control reservoirs or conservation reservoirs, it may be comparatively more pervious. A flatter upstream slope may be required in hydroelectric projects due to the "sudden drawdown" scenario that will affect the dam's upstream face.

i. Earthquake Activity

The designer might require more conservative design features in seismically prone areas, such as stronger filters, greater capacity downstream drains, thicker cores made of more piping-resistant materials, flatter side slopes, longer construction times, and furthermore.

8.2.4 General Design Criteria for Embankment Dams

Overtopping, piping, and earth movements in some sections of the embankment and its foundation (due to the insufficient shear strength) are the primary causes of the embankment dam's collapse. Among these, overtopping is the most common reason for an embankment dam to fail completely and disastrously. The following safety considerations should be included in the construction of an embankment dam:

a. There is no possibility of an overturn. A spillway with a suitable capacity and enough freeboard should be constructed for this purpose.

b. Horizontal piping is not possible since the seepage line is well within the downstream face.

c. The slopes upstream and downstream are sufficiently level to provide stability of the materials used to construct the embankment under all circumstances, including construction, operation, and rapid drawdown.

d. The foundation's shear stress is less than the foundation material's shear strength. The embankment slopes should be appropriate flat for this purpose.

e. Appropriate protection is provided for the upstream and downstream faces from wave and rainwater action, respectively.

f. There should be no means for water to flow freely through the embankment.

g. Seepage from the foundation should not cause pipework at the dam's downstream toe.

h. The dam's top needs to be high enough to permit the foundation and dam to settle.

i. For all operational and construction conditions (steady seepage and abrupt drawdown), the foundations, abutments, and embankment should be stable.

8.2.5 Estimation and Control of Seepage

I. Seepage Analysis

The estimation of seepage through an embankment dam and its foundation is done using the principle of flow through porous media. The Laplace equation (Equation 8.1) governs two-dimensional seepage, which corresponds to whatever would happen in an embankment dam and its foundation.

$$\frac{\partial^2 h}{\partial x^2} + \frac{\partial^2 h}{\partial y^2} = 0 \qquad \text{.............(8.1)}$$

Here, the seepage head is denoted by h. When incompressible water is entirely saturated in homogeneous, isotropic, and incompressible soil, Equation (8.1) holds true. It is possible to solve Equation (8.1) using graphical, analytical, numerical, or other appropriate approaches, such analogue methods. Equation (8.1) can be solved graphically by sketching a flow net.

Drawing flow nets for sections with single permeability enables the analysis of the majority of seepage issues related to embankment dams. For instance, an investigation of the seepage conditions in the dam's core alone may be sufficient if the outer shells of the dam are more permeable than the core. Nonetheless, it is often necessary to analyze seepage through components with varying permeability in seepage situations. Under these circumstances, while moving from a soil with one permeability to one with a different permeability, a fundamental deflection rule needs to be followed Water seeps through an area with comparatively higher permeability using less energy. Water therefore moves in a way that allows it to remain in the more permeable region for as long as possible when it moves from a region of high permeability to lower permeability. Put another way, water searches out the quickest route to save energy. If all other conditions remain the same, a smaller area of flow cross-section is required in the greater permeability zone, which is another method to understand the seepage behavior in sections with varying permeabilities.

The hydraulic gradient is a measure of the rate of energy loss through seepage when media is porous. For such a reason, one should expect strong hydraulic gradients in low-permeability areas. The deflection of flow lines within boundaries between soils with varying permeabilities are shown in Figure 8.3. The following relationship (Equation 8.2) determines how the flow lines bend:

$$\frac{\tan \beta}{\tan \alpha} = \frac{K_1}{K_2} \quad \text{.............(8.2)}$$

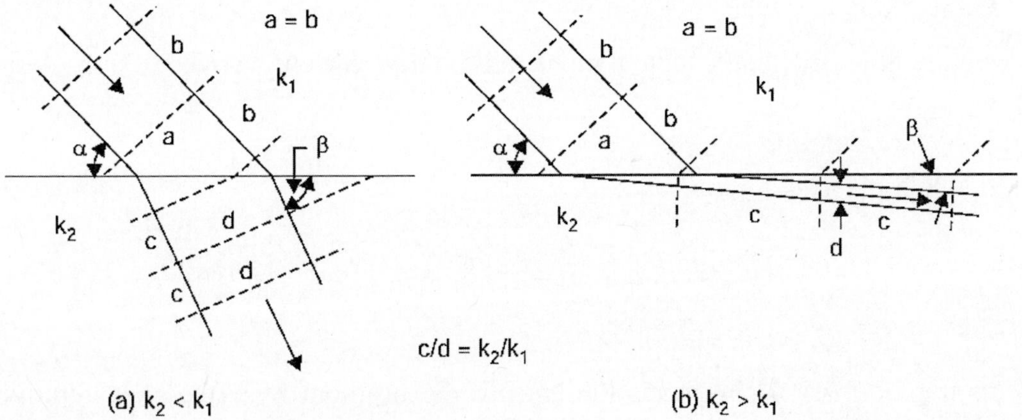

Figure 8.3: Flownet shape change on account of regions of different permeability

Additionally, the regions produced by the equipotential and flow lines intersecting either shorten or elongate in accordance with the following relationship (Equation 8.3).

$$\frac{c}{d} = \frac{K_2}{K_1} \quad \text{.............(8.3)}$$

It is necessary to measure the lengths and widths of the figures frequently while drawing flow nets for sections with varying permeabilities to ensure that Equation (8.3) is satisfied. Compacted embankments and natural soil deposits can occasionally exhibit stratification, making them more permeable horizontally than vertically. Under anisotropic conditions, the velocity components for two-dimensional flow are given by u and v, as shown in the equation 8.4:

$$u = -K_x \frac{\partial h}{\partial x} \text{ and } v = -K_y \frac{\partial h}{\partial y} \quad \text{.............(8.4)}$$

where,

K_x = coefficients of permeability in the x direction

K_y= coefficients of permeability in the y direction,

h = hydraulic head causing the flow,

u = velocity component in the x direction, and

v = velocity component in the y direction

For two-dimensional flow, applying the continuity equation with $\partial w/\partial z = 0$ and combining it with Equation (8.4), yields equation 8.5:

$$K_x \frac{\partial^2 h}{\partial x^2} + K_y \frac{\partial^2 h}{\partial y^2} = 0 \qquad \text{.............(8.5)}$$

In other words, to manage the calculations efficiently, a new variable, x_t is defined as Equation 8.6 and 8.7:

$$\frac{\partial^2 h}{\partial x_t^2} + \frac{\partial^2 h}{\partial y^2} = 0 \qquad \text{.............(8.6)}$$

where,

$$x_t = x\sqrt{\frac{K_y}{K_x}} \qquad \text{.............(8.7)}$$

This transformation simplifies the anisotropic problem by effectively removing the anisotropy through scaling the x-coordinate, thereby reducing it to an isotropic form. Equation (8.8), a well-known Laplace equation utilizing a transformed coordinate system involving x and y, determines flow under conditions of anisotropic seepage. All that is required to create a flow net for the anisotropic condition is to reduce the cross-section's size in the direction of higher permeability. The modified section's flow net is first created, and after that, it is rebuilt on the cross-section that was drawn utilizing the original scale. Rectangles extended in the direction of increased permeability would comprise the flow network on the original cross-section rather than squares.

$$\Delta q = \bar{K}\frac{\Delta h}{b}b = K_x \frac{\Delta h}{b\sqrt{K_x/K_y}} \qquad \text{.............(8.8)}$$

$$\bar{K} = \sqrt{K_x K_y} \qquad \text{.............(8.9)}$$

One can also determine the seepage quantity which is rewritten as equation 8.10:

$$q = Kh\frac{N_f}{N_d} \quad \text{.............(8.10)}$$

By comparing the discharge Δq through any one figure of the modified section with the corresponding figure of the original section (Figure 8.4), one can ascertain the effective permeability K. If the head drop between two consecutive equipotential lines is represented by Δh, then

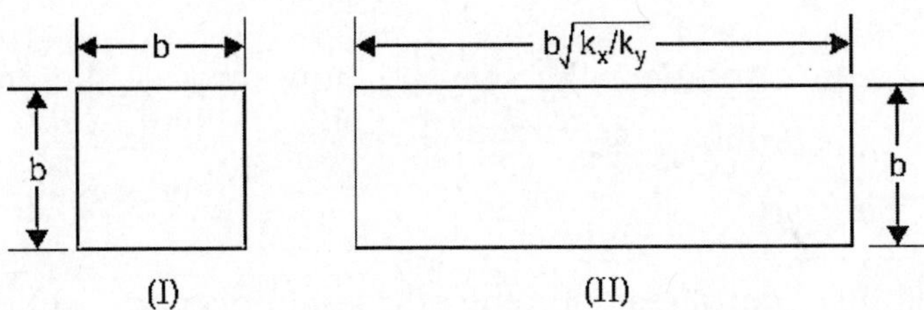

Figure 8.4: Comparison of Flow net in the (I) Transformed and (II) Original Section.

$\overline{K}$ as the geometric mean of the permeabilities K_x, and K_y in an anisotropic medium, ensuring it accommodates directional variability. Therefore, Equation (8.10) can be applied to both isotropic and anisotropic conditions by replacing effective permeability. $\overline{K}$ for K. Note that for both the original and modified sections, the shape factor (i.e., N_f /N_d) remained the same.

The primary challenge in doing a seepage analysis of an embankment dam is from the unknown topmost streamline, sometimes referred as the phreatic or seepage line. The line that has hydrostatic pressure below it and no hydrostatic pressure above it is known as the seepage line. The seepage line is essentially the line of saturation if the embankment comprises of coarse material, and the capillary effects are negligible. However, in the case of a fine-grained soil embankment, there is saturation without hydrostatic pressure, and the capillary fringe above the seepage line experiences very little flow. The piping potential and flow net drawing are aided by the seepage line prediction. The seepage line in a homogeneous earth dam constructed on an impermeable foundation crosses the downstream face above the dam's base unless, naturally, additional drainage techniques are used.

The seepage line and the equipotential lines are required to intersect at identical vertical intervals (Figure 8.5). This requirement allows the seepage line to be graphically determined concurrently with the creation of a flow net. As an alternative, the seepage line can be found using Kozeny's solution, which gives the following expression (Equation 8.11) for the seepage line equation: for an embankment with a parabolic upstream face and a downstream horizontal drain (Figure 8.6),

$$y^2 - y_0^2 + 2xy_0 = 0 \quad \text{.............(8.11)}$$

where,

$$y_0 = 2a_0 = \sqrt{d^2 + h^2} - d \quad \text{.............(8.12)}$$

Additionally, using Kozeny's solution, the seepage flow through the embankment per unit length, q (equation 8.13), is provided as:

$$q = Ky_0 = K\left[\sqrt{d^2 + h^2} - d\right] \quad \text{.............(8.13)}$$

where K is the permeability coefficient, and the other symbols correspond to the explanations given in Figure 8.6.

Note that the seepage line location and the point at which it intersects the downstream face (the downstream drain in Figure 8.6) are independent of the embankment material's coefficient of permeability and solely depend on the dam's cross-section. Furthermore, although many embankment dams feature a horizontal drain downstream, real embankment dams lack parabolic upstream faces. Equation (8.11) shows that the seepage line is a parabola, with minor deviations at the entrance and exit of popular embankment dam types.

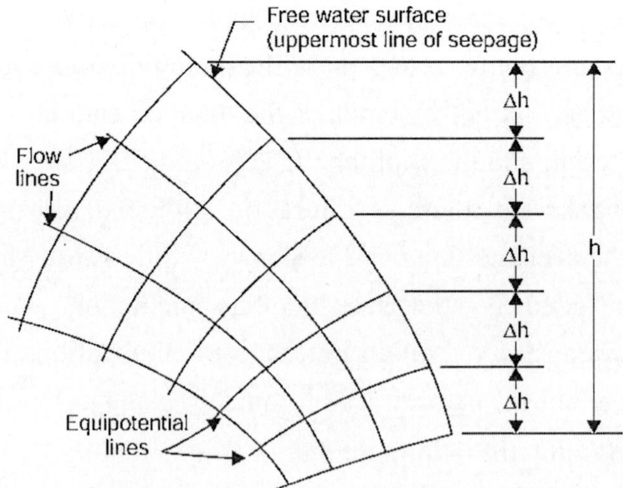

Figure 8.5: General Conditions for Line of Seepage

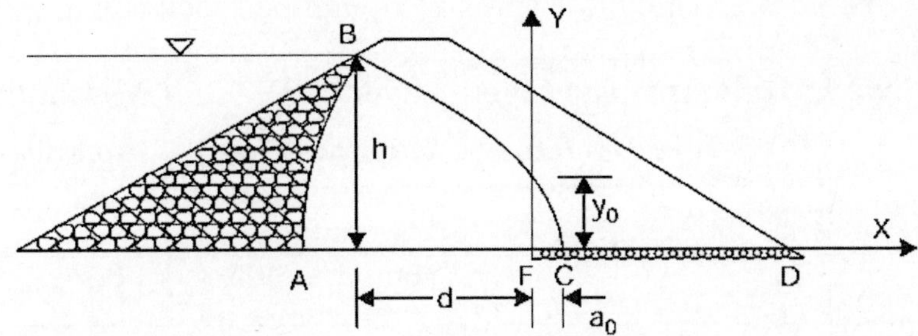

Figure 8.6: Embankment Dam with Kozeny's Conditions

II. Slope Protection

An earth dam that retains a sizable reservoir needs to be shielded from wave action if its upstream slope is made of materials other than rocks or cobbles. Experience from the past has shown that the following techniques can be used to safeguard an upstream slope:

a. riprap made of dumped rock;

b. pitching stones by hand;

c. monolithic RCC slab and

d. asphaltic concrete.

Out of these four techniques, the riprap of dumped rock placed over a layer or layers of finer filter material offers the best wave protection since it is less susceptible to damage from post-construction embankment settlement and is a good dissipator of wave energy. It is important to construct a riprap layer so that: (i) The wave forces do not push the individual rocks out of position, and (ii) the filter beneath the riprap will not be washed out via the gaps in the riprap layer. It should be possible for the filter to stop the underlying embankment material from eroding. Table-1 lists the size and thickness of the riprap made of dumped rock. The minimum thickness for the underlying filter layer suggested by the US Army Corps of Engineers is provided in Table 8.1.

Surface water runoff during rainstorms or wind during windstorms in arid climates may cause significant erosion if the surface of the downstream slope of an earth dam is composed of fine-grained soil. Deep gullies, sometimes reaching a depth of three meters, are caused by surface erosion in the middle section of the dam and at the abutment contacts on the downstream slope.

Table 8.1: Thickness and size of dumped rock riprap

Maximum Wave Height (m)	**Minimum Average Rock Size D_{50} (cm)**	**Layer Thickness (cm)**	**Minimum Filter Layer thickness (cm)**
0-0.6	25	30	15
0.6-1.2	30	45	15
1.2-1.8	38	60	22.5
1.8-2.4	45	75	22.5
2.4-3.0	52	90	30

A healthy grass cover on the downstream slope's surface stabilizes the surface soil and offers the most effective and affordable slope protection. It might not be able to save enough water in really dry regions for the grass cover to continue growing. Under such circumstances, the downstream slope may also be protected by dumping rock riprap, which is often utilized for upstream slopes.

8.3 GRAVITY DAMS

A gravity dam is a solid concrete or masonry structure that, without the use of beams or arches, maintains stability against all applied loads solely via its own weight. These dams typically have a straight plan and a cross-section that resembles a triangle. Gravity dams are typically categorized according to their structural height, which is the difference in elevation between the lowest point in the excavated foundation area, excluding features like narrow fault zones, and the top of the dam, or the roadway's crown, if there is not one (1). In general, gravity dams up to 100 feet (30.48 meters) height are referred to as low dams. Medium-height dams are defined as those height are in between 100 feet (30.48 meters) and 300 feet (91.44 meters). High dams are those that are more than 300 feet (91.44 meters) in height. A gravity dam's downstream face often has a consistent slope that, if it were extended, would meet the vertical upstream face at or close to the reservoir's maximum water level. In addition to being thick enough to support the road or other necessary access, the upper section of the dam is designed to withstand the impact of floating items in the reservoir. In order to successfully withstand the tensile strains caused by the reservoir water loading, a gravity dam's upstream face is often kept vertical. This concentrates most of the dam's weight toward the upstream face. Because the thickness of the dam resists sliding, it can determine the downstream face's slope, which is typically between 0.7 and 0.8 (H):1(V). An upstream batter has the potential to thicken the dam's lower section. If the spillway cannot be placed in the abutment, it may be placed on a section of the dam; in this instance, the dam is changed at the top to provide room for the spillway's crest and at the toe for the energy dissipator. Such overflow sections of gravity dams would have different stability requirements than non-overflow parts.

8.3.1 Forces on a Gravity Dam

Figure 8.7 displays the forces often included in a gravity dam's design. These are listed in the following order:

(i) Dead Load

The weight of the appurtenances, such as piers, gates, and bridges, as well as the weight of the concrete make up the dead load (W_c). It is believed that all dead weight is transferred vertically to the foundation, with no shearing between blocks in between.

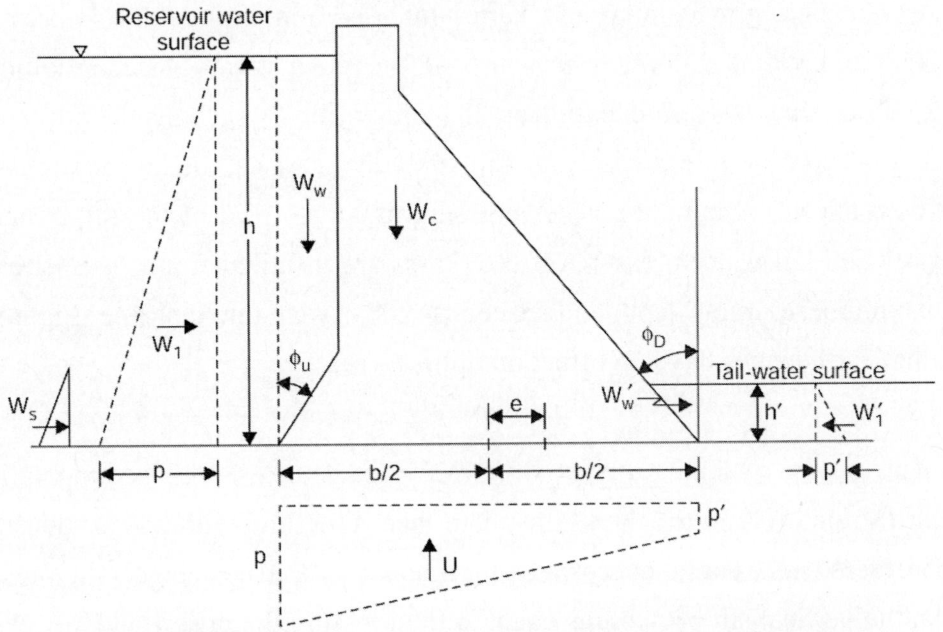

Figure 8.7: Usual Loading Combination for a Gravity Dam

(ii) Reservoir and Tail-water Loads (W_w , W_w', W_1, and W_1')

These are derived from reservoir operation studies' tail-water curves and the reservoir's range of water surface heights. Operating and hydrologic data, including reservoir capacity, storage allotments, stream flow records, flood hydrographs, and reservoir releases for all purposes, form the basis of these studies. Low overflow dams may have a substantial dynamic impact from approach velocity; hence this should be considered. When it comes to applying water pressure, gates and other control elements on the crest are considered a part of the dam. Gravity dams that do not overflow should have their tail water modified for any retrogression. When designing a gravity dam, it is important to consider any rise in tail-water pressure brought on by the downstream bucket's curvature of flow n an overflow type gravity dam.

(iii) Uplift Forces

Internal hydraulic pressures in a dam's pores, fissures, and seams within the dam itself, at the point where the dam and foundation meet, and within the foundation itself cause uplift forces (U). It is assumed that the distribution of internal hydrostatic pressure acts across the entire horizontal section of a gravity dam and varies linearly from full reservoir pressure at the upstream face to zero or tail-water pressure at the downstream face. Additionally, the distribution of pressure is altered based on the dimensions, placement, and separation of internal drains. According to experimental and analytical studies, the average pressure at the drains will be reduced to about tail-water pressure plus one-third of the difference between reservoir water and tail-water pressures if the drains are installed from the upstream face at 5% of the maximum reservoir depth and spaced laterally twice that distance (Figure 8.8). It is considered that earthquakes have no effect on uplift forces.

(iv) Silt Load

Building a dam over a river that carries silt, always leads to reservoir sedimentation, which exerts an extra strain (W_s) on the dam's upstream face. The fluid with a mass density of 1360 kg/m^3 is considered to exert a hydrostatic load equal to the horizontal silt pressure. It is assumed that the vertical silt pressure is equal to the pressure that a soil with a wet density of 1925 kg/m^3 would exert.

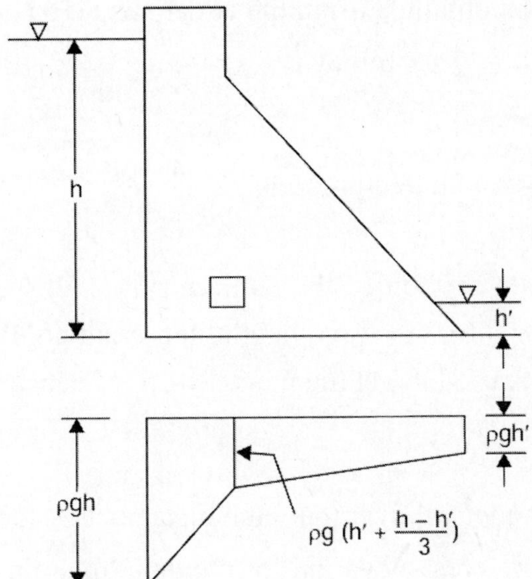

Figure 8.8: Modification in Uplift Force due to Drain

(v) Ice Pressure

A good method of estimating the ice pressures must be used if the designer expects an ice sheet of considerable thickness to form and stay on the reservoir water surface for an extended period of time. If such a method is not available, ice pressure can be estimated as 250 kPa (250 kN/m^2) applied over the expected region where ice will meet the dam's face.

(vi) Wave Pressure

Wave impact is also experienced by the upper section of a dam. Generally, wave pressure against large-scale dams has minimal impact. Wave height h_w and wave pressure are related in the following ways:

(a) The equation 8.14 gives the maximum wave pressure, or p_w (in kilopascals), which happens at 0.125 h_w above the still water level.

$$p_w = 24h_w \qquad \text{.............(8.14)}$$

where, h_w= height of the wave (in metres).

(b) The total wave force P_w (in kilonewtons) is given by equation 8.15.

$$P_w = 20h_w^2 \qquad \text{.............(8.15)}$$

and acts at 0.375 h_w above the still water level in the downstream direction.

(c) The wave height h_w can be obtained using the equations 8.16 and 8.17 as follows:

$$h_w = 0.032\sqrt{VF} + 0.76 - 0.27\,F^{1/4} \text{ For } F < 32 \text{ km} \quad \text{.............(8.16)}$$

$$h_w = 0.032\sqrt{VF} \text{ For } F < 32 \text{ km} \quad \text{.............(8.17)}$$

Here, V = wind velocity (kilometres/hour); and

F = fetch (in kilometres).

The vertical distance between the top of the dam and the still water level is known as the freeboard, and it is determined by the wave height and wind configuration. The formula for Zuider Zee (Equation 8.18) is used to get the wind set-up S (measured in meters).

$$S = \frac{V^2 F}{62{,}000\,D} \quad \text{.............(8.18)}$$

where, D= average depth (in metres) over the fetch distance F.

In order to have a long crest elevation for the dam, the minimum freeboard should be equal to the wind set-up plus 4/3 times the wave height over the usual pool elevation or above the maximum reservoir level corresponding to the design flood. Nonetheless, the freeboard must always be at least 1.0 m above the mean water level that corresponds to the design flood.

(vii) Earthquake

Gravity dams are elastic constructions that seismic disturbances have the potential to cause to resonate. When subjected to the anticipated earthquake, such dams need to be built to maintain their elasticity. The design earthquake should be established considering the following:

a. historical earthquake records to determine the frequency of occurrence versus magnitude,
b. the dam's useful life, and
c. A statistical approach to determine the likelihood of earthquakes of different magnitudes occurring during the dam's lifetime.

Additionally, a gravity dam needs to be built to resist the maximum acceptable earthquake, which is one that is typically larger than any earthquake that has ever been recorded in history. The dam experiences random oscillations from earthquakes, which enhance the strains and pressures from water and silt acting on it. Any direction can experience seismic movement. It is important to apply earthquake loads both vertically and horizontally in the direction that results in the least favorable circumstances. The most unfavorable direction for earthquake movement in a gravity dam when the reservoir is full is upstream (since the inertial forces acting downstream may cause a resultant force to intersect the dam's base outside the middle third of the base, in addition to increasing the water load and, therefore, the increased overturning moment) and downward for vertical earthquake movement because it reduces the weight of the concrete and water above the sloping faces of the dam, reducing the stability of the structure. When the reservoir is empty, additional harmful inertial forces resulting from downstream ground motion may meet the base of the dam outside of its middle third. The following factors determine the influence of earthquake forces:

a. their magnitude, which depends on the earthquake's severity;
b. the structure's mass and elasticity, and
c. whether the earthquake affects the water load.

Understanding earthquake acceleration or intensity, which typically represents relation with acceleration due to gravity (g), is helpful for estimating earthquake load. The seismic coefficient, abbreviated as αh, is the ratio of acceleration caused by earthquakes to that of gravitational forces. The seismic coefficient values for the country's various zones may be obtained from the codes and varies in value for both horizontal and vertical earthquake accelerations. When a structure of mass M moves horizontally during an earthquake with an acceleration of αhg, the horizontal earthquake force acting on the structure, P_e, is given as:

$$P_e = M\,\alpha_h\,g = \frac{W}{g}\alpha_h g = \alpha_h W$$

where,

W = weight of the structure.

In the absence of any other specified value, the value of α_h is often assumed to be 0.1. Similarly, the value of the seismic coefficient in the vertical direction can be assumed to be 0.5. During an earthquake, a force is also applied to the dam's face by the reservoir's water's inertia. Equations 8.19 and 8.20 provide the fluctuation of horizontal hydrodynamic earthquake pressure with depth for dams with vertical or sloping upstream faces:

$$\boldsymbol{P_e = c_1 \alpha_h \rho\, g\, h} \qquad \text{.............(8.19)}$$

$$c_1 = \frac{c_m}{2}\left[\frac{y}{h}\left(2-\frac{y}{h}\right) + \sqrt{\frac{y}{h}\left(2-\frac{y}{h}\right)}\right] \qquad \text{.............(8.20)}$$

where P_e = hydrodynamic earthquake pressure normal to the face,

c_l = dimensionless pressure coefficient,

α_h = horizontal acceleration factor (ratio of horizontal acceleration due to earthquake and the gravitational acceleration),

ρ = mass density of water,

g = acceleration due to gravity,

h = depth of reservoir,

y=vertical distance from the reservoir surface to the elevation under consideration,

c_m = the maximum value of c_l for a given slope (Figure8.9).

Let V_{pe} (or V_{pe}') denote the change in the horizontal component of the reservoir (or tail-water) load on the face above a section as a result of horizontal seismic loads. It is calculated for each elevation increment chosen for the study, and because of the nonlinear response, the totals are derived by summing. Similarly, M_{pe} (or M_{pe}'), which denotes the moment of V_{pe} (or V_{pe}') about the section's centre of gravity, is calculated. The average acceleration factor for each height increment should be used to calculate the inertia forces for the concrete in the dam. The total of all incremental forces above a given elevation and their moment about the elevation's centre of gravity are the inertia forces that should be considered while evaluating a dam elevation. It is possible to compute the horizontal concrete inertia force (V_e) and associated moment (M_e) using Simpson's rule. As an alternative, the following Equations (Equation 8.21 and 8.22) can be used to obtain V_{pe} and M_{pe}.:

$$V_{pe} = 0.726\, p_e y \qquad \text{.............(8.21)}$$

$$M_{pe} = 0.299\, p_e y^2 \qquad \text{.............(8.22)}$$

With appropriate forces, moments, and the vertical acceleration factor αv, one may calculate the implications of vertical accelerations. To calculate the increase (or reduction) brought on by the vertical downward (or upward) accelerations, the forces and moments owing to dead loads and water pressure normal to the dam's faces should be multiplied by the relevant acceleration factors. It is believed that an earthquake has little impact on uplift forces.

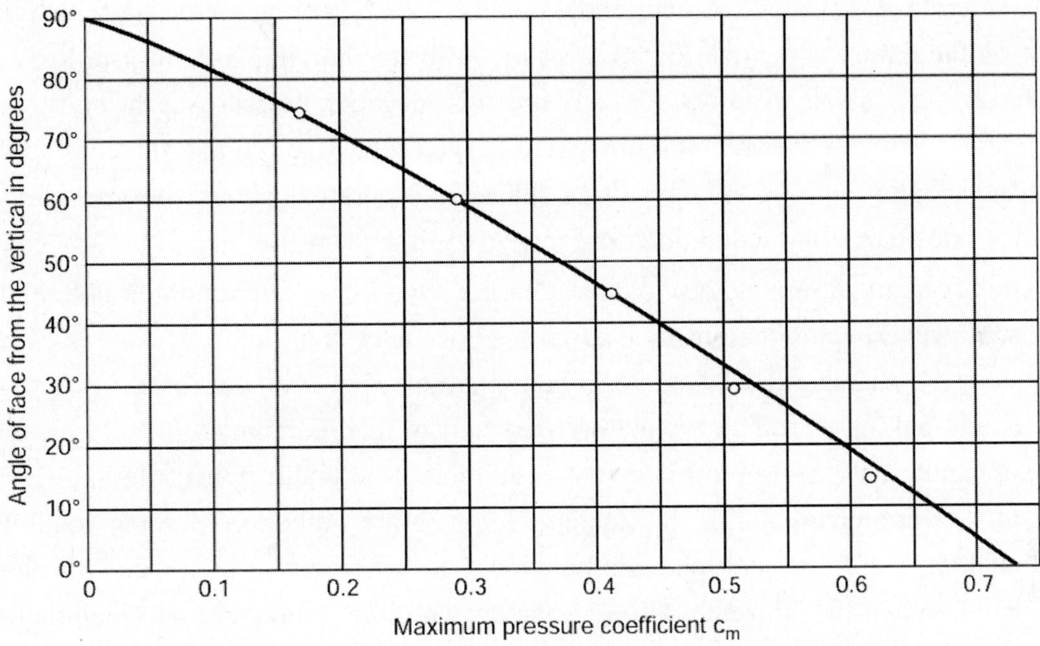

Figure 8.9: Variation of c_m with Upstream Face Inclination

Dams having upstream faces as a combination of vertical and sloping faces are analysed as follows:

a. Examine the dam with the assumption that it had a vertical upstream face all the way around if the height of the vertical section of the upstream face is equal to or higher than half the overall height of the dam.

b. Use the pressure that would result from the upstream face having a constant slope (equal to the slope of the sloping portion of the upstream face) from the water surface elevation to the heel of the dam if the height of the vertical portion of the upstream face of a dam is less than half the total height of the dam.

(viii) Other Miscellaneous Loads

Apart from the forces previously described, there could also be thermal stresses and vertical water loading. Load transmission across joints occurs if the contraction joints are grouted because of the horizontal thrusts brought on by volumetric increases imposed by rising temperatures. The loads at the abutments and the twist effects are increased by this load transfer. The weight of the water on the dam's sloping upstream and downstream faces exerts vertical water loading. Since water tends to reach the spouting velocity, which lowers pressure on the dam, the vertical component of the water moving over the spillway is not considered in the analysis. Additionally, any potential negative pressure on the spillway crest is disregarded. However, any sub-atmospheric pressure that forms on the spillway's downstream sloping surface as a result of inadequate aeration should be considered and imposed as a positive load (acting downstream) on the upstream face.

Gravity dam design needs to consider the worst-case scenario for potential load situations. However, combinations of loads whose occurrence at the same time is extremely unlikely may be disregarded. Most load combinations fall into one of three categories: excessive, unusual, or typical. A typical ordinary sort of load combination could include, for instance, a regular design reservoir height with suitable dead loads, uplift, silt, ice, tail-water, and thermal loads corresponding to standard temperature. When combining loads of the conventional type with unusual load combinations, the maximum design reservoir elevation is considered. When the impacts of the maximum credible earthquake are included to the usual load combination, an extreme load combination is created.

8.3.2 Causes of Failure of a Gravity Dam

A gravity dam overturning could cause it to fail. A gravity dam must have dimensions that allow the combined forces to intersect the base of the dam inside the middle third of the dam in order to be safe from overturning. Assuming that any horizontal portion of a gravity dam, including the base, and the combined value of all the forces operating on the dam above the section. Thus, the dam could collapse if the downstream edge of the segment was not crossed by the dam's line of action. Nonetheless, a gravity dam's section is designed so that the force generated will act only inside its upstream and downstream boundaries, preventing overturning. However, the downstream edge of the section may be destroyed if the resultant's line of action extends far enough beyond the middle third of the horizontal section. This could lead to the resultant passing outside the dam section by reducing the effective breadth and, thus, the section's sliding resistance. Additionally, the resultant develops tensile stresses at the section's upstream edge when it traverses downstream of the middle third of the horizontal section. The dam section may crack as a result of these tensile strains, increasing the uplift pressure. As a result, the stabilizing forces would be decreased. Therefore, various forms of failures, like crushing of toe material, sliding, tension-induced material cracking, and an increase in uplift, may happen before a gravity dam actually overturns. If the following conditions are met, a gravity dam is considered safe against overturning:

a. there is no tension on the upstream face;

b. there is sufficient resistance against sliding, and

c. the concrete or masonry of the dam and its foundation are of an appropriate quality and sufficient strength.

Considering that masonry and concrete are generally weak in tension, a gravity dam's construction should ensure that no tensile stresses exist anywhere throughout the dam section. However, in extremely high gravity dams, if maintaining such a state proves to be challenging, minor tensile stresses up to 50 N/cm^2 may be permitted under the worst loading circumstances.

If the horizontal pressures acting on a dam above any horizontal plane are greater than the resistance to sliding on the plane, the dam may fail as a result of sliding. The frictional resistance and shearing strength of the material along the plane under consideration are responsible for the resistance to sliding. The following Equation 8.23 is an expression for the shear-friction factor of safety, or F_s, which is a measure of stability against sliding or shearing:

$$F_s = \frac{CA + \sum W\mu}{\sum H} \quad \text{.............(8.23)}$$

where,

C = unit cohesion,

A = area of the plane (substituted for the plane's width if one considers the unit length of the dam),

ΣW = total vertical forces acting on the plane,

μ = internal friction coefficient, and

ΣH = driving shear forces/ resultant horizontal forces.

The stability against sliding or shearing at any horizontal component of a dam, at its contact with the foundation, or through the foundation along any plane of weakness can be ascertained utilizing the shear-friction factor of safety. For gravity dams, the lowest permissible values of F_s are 3.0, 2.0, and 1.0 for typical, abnormal, and extreme loading combinations, respectively. Under normal, exceptional, and extreme loading combinations, the value of F_s for any plane of weakness within the foundation should not be less than 4.0, 2.7, and 1.3, respectively.

In general, the acceptable factor of safety against shear and overturning under normal or typical loading conditions is considered 2.0. On the other hand, 1.25 is an equivalent number under extreme loading conditions. The ratio of the coefficient of the static friction to the selected safety factor is the allowable value of the sliding factor, which is equal to the total of the horizontal and vertical forces.

For typical, uncommon, and extreme load combinations, respectively, the maximum permissible compressive stress for concrete in a gravity dam should be less than the concrete's stated compressive strength divided by 3.0, 2.0, and 1.0. For both common and uncommon load combinations, the compressive stress should not be greater than 1035 N/cm^2 and 1550 N/cm^2, respectively.

For typical, uncommon, and extreme load combinations, respectively, the maximum permissible compressive stress in the foundation should be less than the compressive strength of the foundation divided by 4.0, 2.7, and 1.3. To account for uncertainty when evaluating the foundation qualities, these factor of safety values are larger than those for concrete.

8.3.3 Stress Analysis of Gravity Dams

Depending on the dam's configuration, block continuity, and level of refinement needed, one of the following techniques can be used to do stress studies on gravity dams.

a. The gravity method,
b. The trial-load method, and
c. The finite element method

Straight gravity dams, whose transverse contraction joints are neither keyed nor grouted, are designed using the gravity technique for analysis. Regardless, how they are grouted, a gravity dam's transverse contraction joints become keyed, making it a three-dimensional challenge for which the trial load approach should be applied. This approach assumes that the dam is made up of three systems: twisted elements, horizontal beams, and vertical cantilevers. It is believed that each of these systems exists independently of the others and takes up the complete volume of the construction. These systems share the loads on the dam in a way that results in equal rotations and deflections at conjugate sites.

Experiments are implemented to achieve purpose. However, for a preliminary investigation of keyed and grouted dams, the gravity method may be implemented. The finite element method, which has been developed recently, can be used to challenges in two dimensions as well as three dimensions. Only the gravity approach has been covered in this book.

Gravity Method

In the general scenario of a gravity section, where the blocks are not made monolithic by keying and grouting the joints between them, the gravity technique for the stress analysis can be applied. Each of these gravity section blocks functions separately, and the cantilever's weight resists the load as it is delivered to the foundation. The gravity technique analysis depends on the following assumptions:

a. The material used to construct the dam is homogenous, isotropic, and uniformly elastic.
b. Due to the water loads on the reservoir's walls and foundation, there are no differential movements at the dam site.
c. All loads are transferred to the foundation by the gravity action of parallel and vertical cantilevers; neither side's neighboring cantilever elements provide any support.
d. From the upstream face to the downstream face, normal stresses on horizontal planes vary linearly.
e. From the dam's upstream face to its downstream face, horizontal shear stresses vary parabolically across horizontal planes.

With the exception of horizontal planes close to the dam's base, where the effects of foundation yielding alter the stress distributions within the structure, the assumptions at serial numbers (d) and (e) above are precisely accurate. However, these impacts are typically negligible in medium- or low-height dams. However, in high dam situations, these impacts can be substantial, in that case, stresses close to the base should be examined using additional appropriate stress analysis techniques.

However, in high dam situations, these impacts can be substantial, in which case stresses close to the base should be examined using additional appropriate stress analysis techniques. As seen in Figure 8.10, ΣW and ΣH stand for the total of the consequent vertical and horizontal forces operating on a gravity dam's horizontal plane (seen as section PQ). The centroid of the plane under discussion is represented by O, and the consequent R of ΣW and

ΣH intersects the section PQ at O′. The eccentricity of loading, e, is the distance between O and O′. When e is not equal to zero, indicating that the loading on the plane is eccentric., the normal stress σ_{yx} at any point (on the section PQ) x distant from the centroid O is given as equation 8.24.

$$\sigma_{yx} = \frac{\Sigma W}{A} \pm \frac{(\Sigma W)e}{I} x \qquad \text{..............(8.24)}$$

In this case, A stands for the plane PQ's area, and I is the plane PQ's moment of inertia about an axis that passes through its centroid and passes parallel to the dam's length. It should be noticed that the nature of the bending stress (= (ΣW) e x/I) depends on the location of O′ with respect to O, while the direct stress (= ΣW/A) at every point of the section PQ is always compressive. Any point between O and Q and any point between O and P will experience tensile bending stress and compressive bending stress, respectively, if O′ is located between O and Q. As a result, when the reservoir is full, all points between O and Q in Fig. (8.10) should be represented by the positive sign, and all locations between O and P by the negative sign. Similar to this, one should use the positive sign for all points between O and P and the negative sign for all sites between O and Q when the reservoir is empty (in which case ΣH may represent an earthquake force acting in the upstream direction) and O′ sits between O and P.

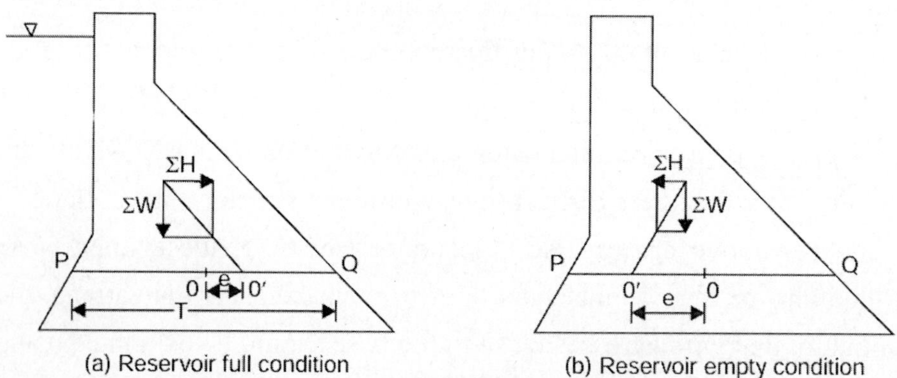

Figure 8.10: Resultant Force on Gravity Dam

One can write A = T and I = T3/12 by taking the unit length of the dam and the horizontal distance (T) between the upstream edge P and the downstream edge Q of the plane PQ. Equation (8.24) thus reduces to:

$$\sigma_{yx} = \frac{\Sigma W}{T}\left(1 \pm \frac{12ex}{T^2}\right) \qquad \text{............. (8.25)}$$

This equation (8.25) can also be used to find the normal stress at the dam's base, BB′ (Figure 8.11). Equation (8.25) for the base of the dam simplifies to equation 8.26, if the width of the base BB′ is b.

$$\sigma_{yx} = \frac{\Sigma W}{b}\left(1 \pm \frac{12ex}{b^2}\right) \qquad \text{............(8.26)}$$

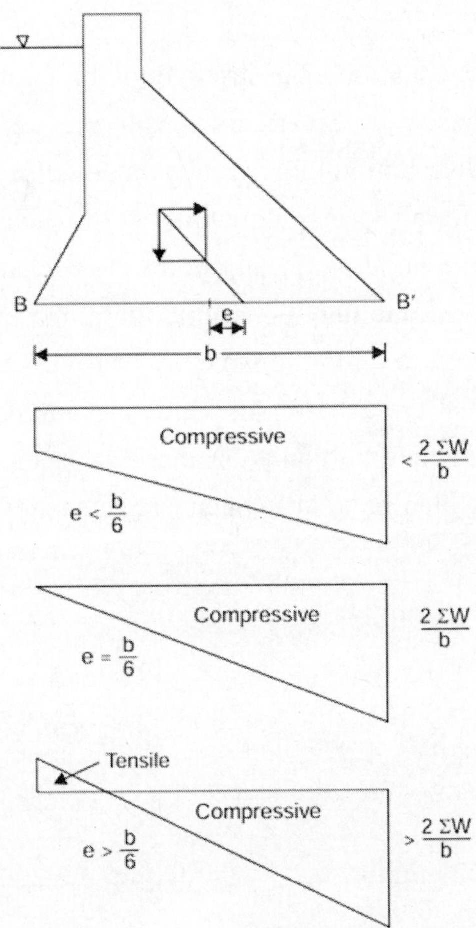

Figure 8.11: Normal Stresses on the Base of a Gravity Dam

x = b/2 holds true for both the dam's heel (B) and toe (B′). Therefore, the typical stresses at the dam's heel (σ_{yU}) and toe (σ_{yD}) are given in equation 8.27 to 8.30 as follows:

When the reservoir is full,

$$\sigma_{yD} = \frac{\Sigma W}{b}\left(1 + \frac{6e}{b}\right) \qquad \text{............(8.27)}$$

$$\sigma_{yU} = \frac{\Sigma W}{b}\left(1 - \frac{6e}{b}\right) \quad \text{.............(8.28)}$$

When the reservoir is empty,

$$\sigma_{yD} = \frac{\Sigma W}{b}\left(1 - \frac{6e}{b}\right) \quad \text{.............(8.29)}$$

$$\sigma_{yU} = \frac{\Sigma W}{b}\left(1 + \frac{6e}{b}\right) \quad \text{.............(8.30)}$$

According to these equations, the stress along the base is compressive if e is less than or equal to b/6, and the base may experience tensile pressures if e is larger than b/6. Figure 8.12 displays the stress distributions for various values of e when the reservoir is full. This implies that the resultant for all loading situations must reach the base inside the middle third of the base if there is to be no strain at any point in the dam's base. By using the principal planes and primary stresses, one may determine the range of stresses operating at a certain site and use that information to build the structure based on extreme values. A primary plane is one that experiences just typical stresses. Such a plane does not experience shear loads. Because water pressure applies normally on these surfaces, the upstream and downstream faces of a gravity dam, which have tail water, are therefore the primary planes because no other force acts on them.

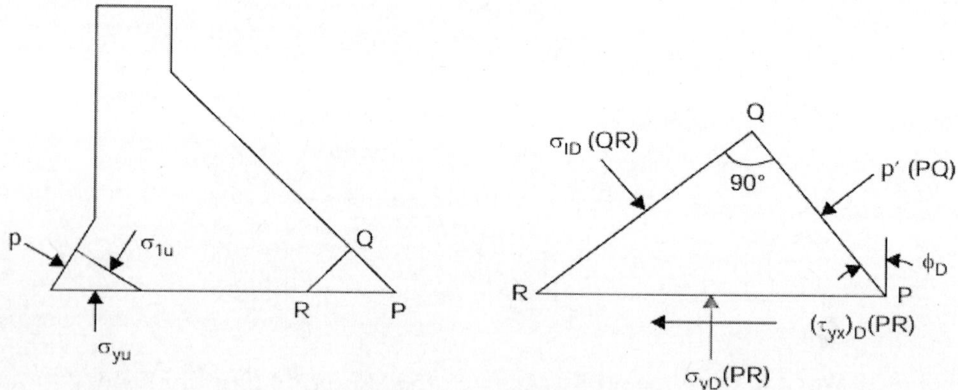

Figure 8.12: Principal Stresses in a Gravity Dam

Moreover, the primary planes of a structure are mutually perpendicular at every location. Therefore, the upstream and downstream faces of a gravity dam would be at right angles to other major planes. The plane QR at the toe of a gravity dam forms an infinite triangle element PQR at a right angle to the downstream face PQ. Therefore, PQ and QR are the principal planes, and PR is a component of the dam's base. The major stresses operating on the principal planes PQ and QR are σ1D and p′ (tail-water pressure), respectively. σ_{yD} and $(\tau_{yx})D$ are the normal and tangential stresses acting on PR, respectively. It is possible to consider all of the stresses as taking place at a point due to the tiny size of the element. Given the element, PQR's equilibrium, the sum of all the vertical forces should be zero. When the dam's unit length is considered, then

$$\sigma_{1D}\,(QR)\cos\varphi_D + p'\,(PQ)\sin\varphi_D - \sigma_{yD}\,(PR) = 0$$

$$\sigma_{1D}\,(PR)\cos^2\varphi_D + p'\,(PR)\sin^2\varphi_D - \sigma_{yD}\,(PR) = 0$$

$$\therefore \sigma_{1D} = \sigma_{yD}\sec^2\varphi_D - p'\tan^2\varphi_D \qquad \textbf{............(8.31)}$$

Thus, the primary stress σ_{1D} at the dam's toe can be obtained from Equation (8.31) by knowing p′ and σ_{yD} [from Equation (8.27)]. Typically, p′ is very small compared to σ_{1D} or zero (no tail-water). As a result, the major primary stress is σ_{1D}, and the minor principal stress is p′. Upon p′ being equal to zero, Equation (8.31) can be re-written as equation 8.32.

$$\sigma_{1D} = \sigma_{yD}\sec^2\varphi_D \qquad \textbf{............(8.32)}$$

The effective minor primary stress becomes p′ – pe′ when the hydrodynamic pressure pe′ induced by the earthquake acceleration (towards the reservoir) is considered, and Equation (8.31) can be rewritten as equation 8.33.

$$\sigma_{1D} = \sigma_{yD}\sec^2\varphi_D - (p' - pe')\tan^2\varphi_D \qquad \textbf{............(8.33)}$$

In the absence of tailwater, p′ and pe′ are both zero, and σ_{1D} is determined using Equation (8.32). Similarly, the expression for σ_{1U} can be obtained by considering an insignificant element at the heel of the dam (Fig. 8.12) from equation 8.34 as follows:

$$\sigma_{1U} = \sigma_{yU}\sec^2\varphi_U - (p + p_e)\tan^2\varphi_U \qquad \textbf{............(8.34)}$$

In the case of an empty reservoir, p = pe = 0, therefore,

$$\sigma_{1U} = \sigma_{yU}\sec^2\varphi_U \qquad \textbf{...........(8.35)}$$

The intensity of water pressure p is frequently higher than the average stress σ_{1U} when the reservoir is full. Consequently, p represents the main primary stress and σ_{1U} the minor principal stress at the heel. Since $\varphi_U = 0$ for the vertical upstream face, $\sigma_{1U} = \sigma_{yU}$. This is obtained by once more estimating the forces acting in the horizontal direction on the infinitesimal element PQR and equating their algebraic sum to zero for the equilibrium condition.

$\therefore (\tau_{yx})_D\,(PR) + p'\,(PQ)\cos\varphi_D - \sigma_{1D}\,(QR)\sin\varphi_D = 0$

$(\tau_{yx})_D = (\sigma_{1D} - p')\sin\varphi_D\cos\varphi_D = (\sigma_{yD}\sec 2\,\varphi_D - p'\tan 2\,\varphi_D - p')\sin\varphi_D\cos\varphi_D$

$(\tau_{yx})_D = (\sigma_{yD} - p')\tan\varphi_D$ **............(8.36)**

In the same way, considering the element's equilibrium at the dam's heel,

$(\tau_{yx})_U = -(\sigma_{yU} - p)\tan\varphi_U$ **............(8.37)**

When the effects of seismic acceleration are considered, equations 8.36 and 8.37 reduce to

$(\tau_{yx})_D = [\sigma_{yD} - (p' - p_e')]\tan\varphi_D$ **............(8.38)**

and

$(\tau_{yx})_U = -[\sigma_{yU} - (p + p_e)]\tan\varphi_U$ **............(8.39)**

Similarly, by considering only the forces acting above the section, one may determine the main and shear stresses at the downstream and upstream sides of the dam at any horizontal section.

8.3.4 Elementary Profile of a Gravity Dam

The right-angled triangular section (Figure 8.13), with its apex as the maximum water level in the reservoir with a wide base where the water pressure is maximum, is termed the elementary profile of a gravity dam. This section is subject to its self-weight W, force due to water pressure P, and uplift force U. The only force operating on the dam in the empty reservoir situation is its own weight, which will reach the base at b/3 from the dam's heel, and thus, it meets the stability criteria with no tension. The elementary profile's base width is chosen by determining which of the two base widths is higher in order to satisfy the no tension and no sliding conditions, as stated below. For the elementary profile depicted in Fig. 8.13, if the downstream middle third point is passed through by the resultant R of all three forces W_c (= 0.5 s ρgbh), W_1 (= 0.5 ρgh2), and U (= 0.5 ρghbc′), (where s = concrete's specific gravity and c′ = correction factor for uplift force), one gets:

$$(0.5\,s\,\rho gbh)\frac{b}{3} - (0.5\,\rho gh^2)\frac{h}{3} - (0.5\,\rho ghbc')\frac{b}{3} = 0$$

$$b^2(s - c') = h^2$$

or, $b = \frac{h}{\sqrt{s-c'}}$

for c' = 1, $b = \frac{h}{\sqrt{s-1}}$

in case the uplift is ignored c' = 0 and $b = \frac{h}{\sqrt{s}}$

Result for no-sliding requirement = $\mu\ (W_c - U) = P = W_1$

where, μ = shear-friction coefficient.

or, $\mu(0.5\ s\ \rho gbh - 0.5\ \rho ghbc') = 0.5\ \rho gh^2$

or, $b = \frac{h}{\mu(s-c')}$

for c' = 1, $b = \frac{h}{\mu(s-1)}$

in case of no uplift c' = 0 and $b = \frac{h}{\mu s}$

It is evident that, the elementary profile of a gravity dam must have a minimum base width equal to the greater of the base widths established from the no-sliding and no-tension criteria to satisfy the stability criterion.

Again, for an elementary profile, $\Sigma W = (W_c - U)$

$$\Sigma W = \frac{1}{2} b\rho gh(s - c')$$

$$\sigma_{yx} = \frac{\Sigma W}{b}\left(1 \pm \frac{12ex}{b^2}\right)$$

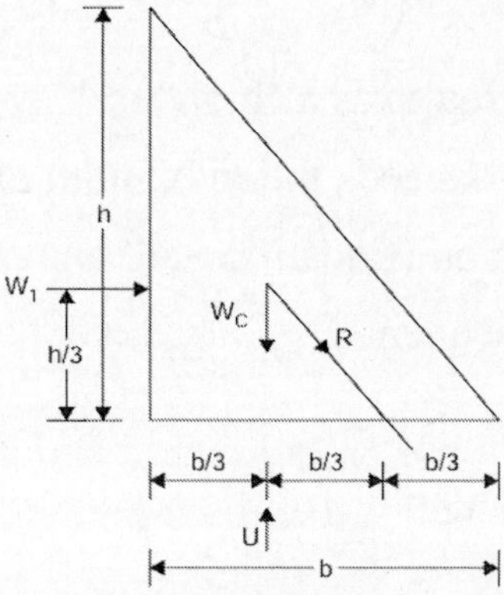

Figure 8.13: Elementary profile of a Gravity Dam

For no tension in the dam, e = b/6

Therefore, at the toe of the dam (*i.e.*, x=b/2)

$$\sigma_{yD} = \frac{2\sum W}{b}$$

$$\sigma_{yD} = \rho gh(s - c')$$

and at the heel of the dam (i.e., x= -b/2) :$\sigma_{yU} = 0$;

Accordingly, the principal stress $\sigma_{1D} = \sigma_{yD} \sec^2\varphi_D$

$$\sigma_{1D} = \rho gh(s - c')[1 + (b/h)^2]$$

$$\sigma_{1D} = \rho gh((s - c')\left[1 + \frac{1}{(s - c')}\right]$$

$$\sigma_{1D} = \rho gh((s - c' + 1)$$

Similarly,

$(\tau_{yx})_D = \sigma_{yD} \tan\varphi_D = \rho gh(s - c')\frac{b}{h} = \rho gh(s - c')\frac{1}{\sqrt{s-c'}}$

$$= \rho gh\sqrt{s - c'})$$

The principal and shear stress at the heel are, obviously, zero. Accordingly, when the reservoir is empty, $\Sigma W = 0.5\, s\, \rho gbh$

$$\sigma_{yD} = 0$$

$$\sigma_{1U} = \sigma_{yU} = \frac{2\sum W}{b} = \rho ghs$$

A gravity dam may be referred to as a "high" or "low" dam, depending on whether the compressive stress at the toe (σ_{1D}) exceeds the maximum allowable stress (σ_m) for the dam's material. Based on this, the expression (Equation 8.40 and 8.41) for σ_{1D} is equivalent with σ_m to yield the limiting height h_l. Consequently,

$$\sigma_m = \rho gh_l(s - c' + 1) \qquad \text{............(8.40)}$$

$$h_l = \frac{\sigma_m}{\rho g(s-c'+1)} \qquad \text{............(8.41)}$$

A gravity dam is considered as low dam if its height is less than h_l; otherwise, it is considered as high dam.

8.3.5 Design of a Gravity Dam

A basic profile must be adjusted before being used in real-world situations. It is merely an ideal profile. A finite crest width, an appropriate freeboard, a batter in the lower portion of the upstream face, and a flatter downstream face are some of the modifications that would be made. A gravity dam's design entails assuming its approximate profile and then segmenting it into several zones using horizontal planes so that stability analysis may be done at the level of each dividing plane. Three-dimensional or two-dimensional analysis is possible for design of gravity dam.

8.4 ARCH DAMS

An arch dam (Figure 8.14) is a dam with a curved design that uses arch action to transfer most of its water load horizontally to the abutments. The amount of curvature is the main determinant of this portion of the water load. Cantilever action transfers the remaining water load to the foundation. Strong canyon side walls are primarily necessary for supporting the arch forces and the thrust created by the water load carried by the arch motion. It is not anticipated that the weight of arch dams will significantly contribute to the resistance against external loads. Uplift on the base is therefore not a crucial design consideration.

8.4.1 Advantages of Arch Dams

a. Arch dams work best in gorges when the length is relatively small compared to the height.

b. The section of an arch dam is significantly smaller compared to the size of a matching gravity dam for a given height. As a result, an arch dam requires less material and is less expensive.

c. The uplift pressure concerns are negligible due to the considerably smaller base width.

d. In moderate foundations, where a gravity dam requiring sound foundation rock may not be suitable, an arch dam can be constructed since cantilever action only transfers a tiny portion of the water load to the foundation.

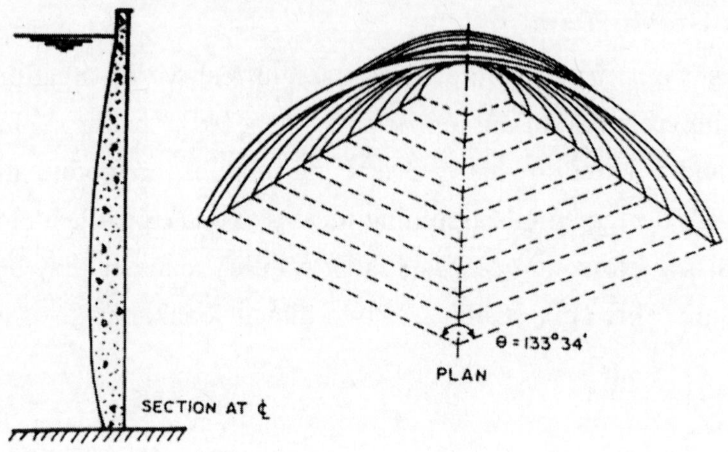

Figure 8.14: Arch Dam

8.4.2 Disadvantages

a. It needs advanced form of work and specialized labour. An arch dam's design is likewise very specialized.
b. Construction of arch dam typically moves slowly.
c. It needs extraordinarily solid rock abutments that can support arch thrust. Because of this, it is inappropriate in areas lacking stable abutments. Regretfully, there are not many locations that work well for this kind of dam.

8.5 BUTTRESS DAMS

A buttress dam (Figure 8.15) is made up of several piers or buttresses that divide the space that needs to be dammed into several spans. Horizontal arches or flat slabs are used in the construction of panels to support and hold the water in between these buttresses. When the panels are made up of arches, the dam is known as a multiple arches type buttress dam. When panels are made of a single, flat slab, the term "deck type buttress dam" is used.

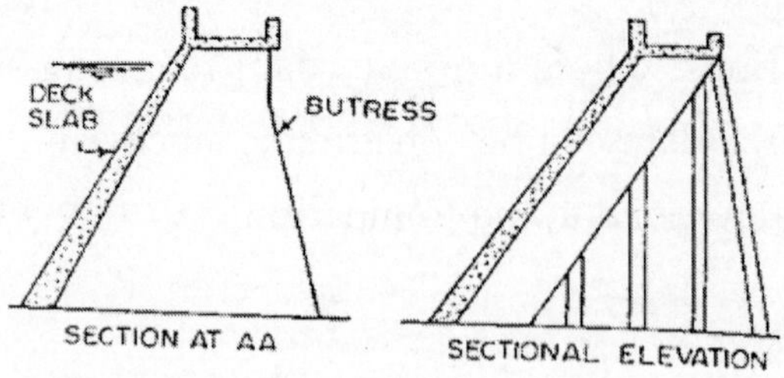

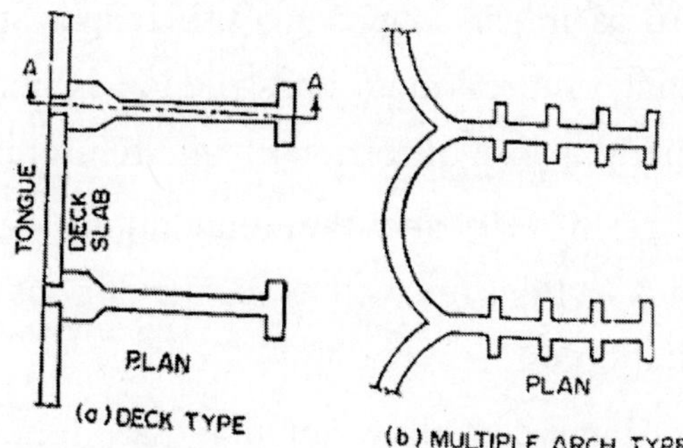

Figure 8.15: Buttress Dams

8.5.1 Advantages of Buttress Dams

a. The mass of a buttress dam is lower than that of a gravity dam. As a result, buttress dams can be built even on unstable foundations that cannot support gravity dams since they have less foundation pressures.

b. The inclined deck experiences normal water load behavior. As a result of this, the vertical component of the water force stabilizes the dam against sliding and overturning, and the buttress dam has a safety factor that is significantly higher than that of a gravity dam.

c. The inclined deck experiences normal water load behavior. As a result of this, the vertical component of the water force stabilizes the dam against sliding and overturning, and the buttress dam has a safety factor that is significantly higher than that of a gravity dam.

d. Since the ice tends to slide over the slanted U/S deck, the ice pressure turns to be irrelevant.

e. For a gravity dam, the only way to increase the dam's height is to install a crest shutter at the overflow section. But in the case of a buttress dam, it is convenient and practical to raise the height even further by extending the slab and buttress as illustrated in Figure 8.16. Therefore, in situations where an increase in reservoir capacity is anticipated in the future, buttress dams are employed.

f. By placing power houses and water treatment facilities between buttresses, building costs can be reduced.

g. About half to one-third as much concrete is utilized in buttress dams as in gravity dams of the same height. Buttress dam construction is not economical in that ratio, though, due to the higher expense of formwork and reinforcing.

h. The back of the upstream face and the foundations between the buttresses are accessible for routine inspections, as well as for the optional sub-sequent grouting and drilling of pressure relief holes.

i. The design of buttress dams can allow for moderate foundation movement without causing significant damage, dependent on the level of articulation or structural isolation offered.

j. Although the financial economy may not always be directly proportionate to the quantity of concrete saved, the mass efficient use of concrete's strength produces an economy in the quantity needed.

k. Because of the greater exposed area and thinner section that do not present a cooling issue, the decreased volume of concrete and increased surface area to volume ratio enables for improved heat dissipation during construction and may even speed up the process.

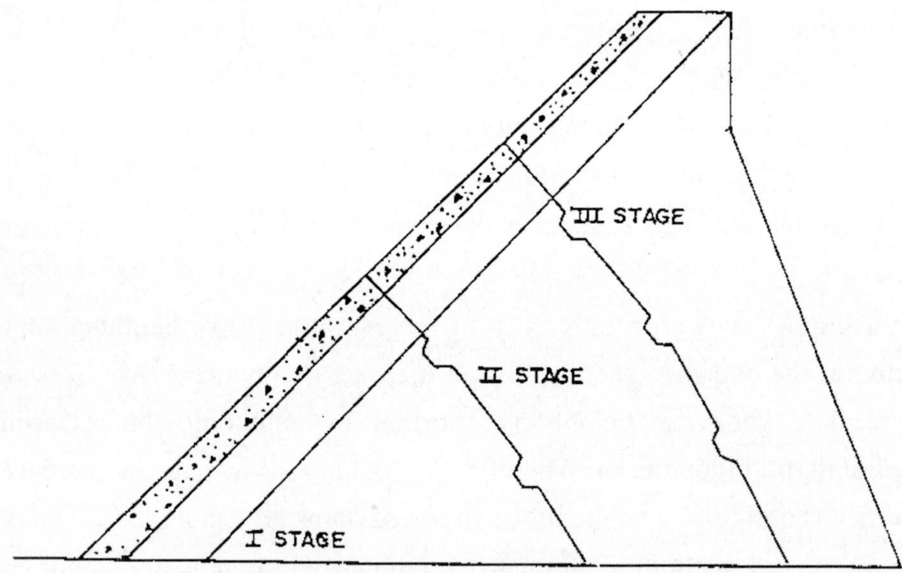

Figure 8.16: Raising Height of Buttress Dam

8.5.2 Disadvantages of Buttress Dams

a. Compared to solid dams, skilled labour is needed more frequently, and the shuttering concrete ratio is greater here. This could result in higher limit rates, compensating some of the savings from using less concrete.

b. Buttress dams with extremely thin concrete faces are severely impacted by the deterioration of the upstream concrete surface.

c. Buttress dams are more prone to intentional harm. The thickness of the upstream face and the difficulty of access from the downstream side will likely to determine the degree of hazards.

8.6 SPILLWAYS

When a flood occurs in an open natural stream, it is considered as a natural phenomenon for which no person or organization is responsible. However, the organization in charge of building the blockage is held responsible if a flood results from the failure of an artificial obstacle (such a dam) built across a natural stream. If overtopped, embankment dams made of rockfill or earth are almost certainly to be destroyed. However, moderate overtopping may be tolerated by concrete dams. A dam failure would result in catastrophic loss of life and property destruction.

As such, there must always be a provision to release excess water safely when the reservoir has been filled to its capacity so that the dam itself is not overtopped. This is achieved by constructing a spillway. Spillways release safely the surplus water which cannot be contained in the reservoir created by the dam. The surplus water is usually drawn from the top of the reservoir and conveyed through an artificial waterway back to the river downstream of the dam or to some other natural drainage channel. Spillway can be constructed either as part of the main dam, such as in overflow section of a concrete dam or as a separate structure altogether. Besides being capable of releasing surplus water, a spillway must be able to meet hydraulic and structural requirements and must be located such that spillway discharges do not damage the toe of the dam. Insufficient spillway capacity and/or the failure of a spillway will cause widespread damage and loss of life. As such, the design criteria for a spillway are usually conservative. The inflow design flood, used to determine the spillway capacity, is also estimated conservatively. In addition to having the capacity to discharge excess water, a spillway must meet the structural and hydraulic specifications and be positioned in such way that spillway discharges do not harm the dam's toe. There will be casualties and extensive damage if a spillway fails or has insufficient capacity. As a result, spillway design specifications are typically conservative. To estimate the spillway capacity, the input design flood is also conservatively estimated.

8.6.1 Components of Spillway

The main components of a spillway are:

- control structure,
- conveyance structure,
- terminal structure, and
- entrance and exit channels.

I. Control Structure

A spillway's control structure controls and directs the reservoir's outflow. It often consists of an overflow crest or a particular kind of opening and is situated at the upstream end of the spillway. On the other hand, the downstream end may have control in some circumstances. For instance, in a "morning glory" spillway, the flow at greater heads is controlled by the downstream tunnel rather than the orifice's crest. The outflow crest in a plan might be circular, semicircular, U-shaped, straight, or curved. The crest may have a different cross-sectional shape, be broad, ogee-shaped, or pointed. Additionally, an orifice can be placed horizontally, vertically, or inclinedly, and it can take on a variety of geometries. Orifices can also have a bell-mouth form, a round edge, or a sharp edge.

II. Conveyance Structure

A discharge channel or waterway typically transports the outflow that is released through the control structure to the river channel downstream. However, no such conveyance system is necessary for free fall spillways. If the spillway has been built into the main body of the dam, the conveyance structure may consist of the downstream face of the dam, or an open channel constructed along the abutment's ground surface, or an underground tunnel constructed through an abutment. Depending on the hydraulic needs and the site's geology and topography, the conveyance structure may also have a range of cross-sections.

III. Terminal Structure

Water moves from the reservoir level to the river level downstream, converting static energy into kinetic energy. If this energy is not adequately released, it could generate enough scour near the dam's toe to cause damage to the spillway, the dam, and other nearby buildings. As a result, appropriate stilling basins are typically installed at the spillway's downstream end to disperse any extra kinetic energy and prevent undesirable scour from being discharged into the river. An appropriate energy dissipator or absorber, such as a roller bucket or hydraulic jump basin, can be used to release the surplus kinetic energy. On the other hand, if the stream bed is made of bed rock that is resistant to erosion, the overflowing water might occasionally be sent straight to the stream. By using structures like flip buckets or cantilevered extensions, the incoming jet should always be expanded downstream from the end of the structure.

IV. Entrance and Exit Channel

While the exit channel transports flow from the terminal structure to the stream channel downstream of the dam, the entrance channel transports water from the reservoir to the control structure. However, as these spillways pull water straight from the reservoir and discharge it directly into the stream channel, as is the situation, for example, with the overflow spillway of a concrete dam, entrance and exit channels are not necessary. Entry and departure channels are required in spillways that are situated next to saddles, ridges, or along abutments.

8.7 TYPES OF GATES FOR SPILLWAY CRESTS

The most basic type of control for a spillway is the free or uncontrolled overflow crest, which releases water automatically whenever the reservoir water level rises above the crest level. Such crests do not require an operator to monitor and regulate the control devices on a continuous basis. Additionally, there are no issues with the controlling device's maintenance or repairs. On the other hand, a regulation gate might be required if an uncontrolled crest or adequate surcharge head cannot be obtained for the required spillway capacity. When the reservoir's water level falls below its typical water surface, these regulating mechanisms allow the spillway to discharge storage. All types of spillways, except for the siphon spillway, can have gates. Installing gates requires extra money for both the initial expense and periodic maintenance and repairs. Numerous factors influence one's decision to control device type and size, such as:

- discharge characteristics of the device,
- climate,
- frequency and nature of floods,
- winter storage requirements,
- need for handling ice and debris, and
- special operating requirements, such as the presence of an operator during flood duration,
- the availability of electricity, operating mechanism,
- economy, reliability, efficiency, and adaptability of the regulating device etc

Typically, the following kinds of regulating devices are employed:

- Flashboards and stoplogs,
- Rectangular lift gates,
- Radial gates, and
- Drum gates.

These can be operated mechanically or hydraulically and can be controlled manually or automatically.

8.7.1 Flashboards and Stoplogs

When the spillway is not needed to release floodwaters, flashboards and stop logs increase the reservoir storage level above a fixed spillway crest level. Typically, flashboards are made up of individual 1.0–1.25 m-tall boards or panels. These have a bottom hinge and struts to support them against the pressure of the water. A bulkhead is formed by stacking individual beams or girders called stoplogs, which are held in place by grooves at both ends of the span. Before the flood, the flashboards or stoplogs are removed to boost the spillway capacity. As an alternative, they are made and placed in a way that allows them to be taken out while being topped. When removed, flashboards and stoplogs, a low-cost, basic kind of controlling device, provide an open crest. Nevertheless, they have the following drawbacks:

a. They pose a hazard if not removed in time to prevent flooding, particularly in areas with small reservoirs and flash floods;

b. Unless they are made to collapse automatically, they require the presence of an operator or crew to remove;

c. They are typically not able to be put back in place while water is flowing over the crest;

d. If they are made to fail when the water reaches a certain point, their operation is unpredictable, and when it performs, it releases large, unexpected outflows and

e. If the spillway operates frequently, replacing flashboards that could be expensive.

8.7.2 Vertical Lift Gates

Normally, rectangular in shape, these are formed by steel and move vertically in their own plane while extending horizontally between guide grooves in supporting piers. An overhead hoist raises and lowers the gates, and an exceeded orifice flow discharges water at each gate opening. Due to the significant sliding friction produced by water pressure, sliding gates necessitate a high lifting capacity. By adding wheels to both sides of the gate, the sliding friction would be decreased allowing a smaller hoist to be used. It has proven possible to employ vertical lift gates with spans and heights of up to 20 and 15 meters, respectively. However, the issue of an elevated operating platform becomes significant at higher altitudes.

8.7.3 Radial (or Tainter) Gates

These consist of steel plates that are joined to a supporting bearing by radial arms to form a section of a cylinder. For the entire thrust of the water load to flow through the supporting pins and for only a small amount of time to be overcome to raise or lower the gate, the cylindrical plate must be kept concentric to the pins. The weight of the gate, the sliding friction, and the frictional resistance at the pins are the only components that make up the hoisting loads. Hand operations at small installations are feasible since the radial gates require a minimal hoisting effort to operate. Additionally, they need less headroom than vertical lift gates. The radial gates are more flexible due to all these benefits.

8.7.4 Drum Gates

The drum gates are composed of steel plates, have a triangular form, and are hollow, making them buoyant. The hydraulic chamber of the weir construction, where the drum gate floats, has a hinge at its upstream lip. The hydraulic chamber's intake of water or withdrawals causes the gate to swing up or down. Controls in the piers adjacent to the chambers regulate the amount of water that flows in and out of the chamber.

8.8 RESERVOIRS

A man-made lake or reservoir is created behind a dam that has been constructed over a river or stream. The most significant and costly components of developing a multipurpose river basin are dams and reservoirs. They need to be designed, planned, and operated with extreme caution. The design, building, and operation of dams and reservoirs provide several challenges, such as site selection, the relative benefits of various dam designs, storage capacity and optimal yield, and the coordinated use of storage for various objectives. Storage works are built with a variety of applications in consideration, such as:

a. Storage and control of water for irrigation
b. Storage and diversion of water for domestic uses
c. Water supplies for industrial uses
d. Development of hydroelectric power
e. Increasing water depths for navigation
f. Storage space for flood control
g. Reclamation of low-lying lands
h. Debris control
i. Preservation and cultivation of useful aquatic life
j. Recreation.

8.8.1 Types of Reservoirs

Based on the purposes served, reservoirs may be classified under the following categories:

- Storage or conservation reservoirs
- Flood protection reservoirs
- Distribution reservoirs
- Multipurpose reservoirs.

I. Storage or Conservation Reservoir

Water supplies for irrigation, hydroelectric development, household, and industrial supplies are the main uses of storage reservoirs. A river may carry significant amounts of water during certain seasons of the year; however, its annual flow varies. When there are plenty of water available, excess water is stored in a storage reservoir, which is subsequently incrementally released when needed.

II. Flood Control Reservoirs

Reservoirs designed to defend against flooding are ones that hold water during a flood and release it gradually and safely as the flood subsides. Flood damage downstream is decreased by providing artificial storage during the floods. Planning for flood avoidance is based on two main principles: flood disposition, which directs floodwaters to travel downstream without causing harm, and riverbed stabilization, which stabilizes the river channel from the upper stream to the estuary. Planning for the entire river system's flood disposition includes flood control via dams, and occasionally it also includes riverbank stabilization.

III. Distribution Reservoir

A distribution reservoir is a low storage tank that utilized by the cities to supply water. The different rates of water throughout the day are defined by a distribution reservoir. Such a distribution reservoir enables the continuous operation of water treatment facilities, pumping plants, etc. The distribution reservoir supplies the variable demand rate that exceeds the constant pumping rate.

IV. Multipurpose Reservoir

A reservoir that fulfils more than one purposes is referred to as multipurpose reservoir. A multipurpose reservoir would be one that is intended to store water for domestic uses, hydroelectric power generation, irrigation, and industrial uses. The multipurpose reservoir can also be used for safeguarding the downstream region from flooding.

8.9 RESERVOIR CAPACITY AND YIELD OF RESERVOIRS

8.9.1 Yield

The volume of water that can be drawn from the reservoir in a predetermined amount of time is known as the yield. For small distribution reservoirs, the design interval is one day; for large conservation reservoirs, it is one year. For example, if 25,000 cubic meters of water is supplied from a reservoir in one year, its yield is 25,000 cubic meters/year or 2.5 hectare-meters/year.

8.9.2 Safe yield or firm yield

The safe yield, sometimes referred to as the firm yield, is the largest quantity of water that can be assured during a severe dry spell.

8.9.3 Secondary yield

Secondary yield is the amount of water that is accessible during heavy flood events in excess of the safe yield.

8.9.4 Average yield

Average yield is defined as the long-term arithmetic mean of the secondary yield and the firm yield.

8.9.5 Mass inflow curve

The demand curve and the mass inflow curve are used to calculate the reservoir capacity that corresponds to a specified yield. A plot of the reservoir's cumulative inflow over a period is referred to as a mass inflow curve.

A flood hydrograph of inflow for multiple years is shown in Figure 8.17. A mass inflow curve generated from the flood hydrograph in Figure 8.17 is displayed in Figure 8.18. Using 1957 as the base year, the hatched area represents the entire amount of water that has passed through the river from 1957 to a period t_1 (let us say 1960). The amount of water represented by the hatched region of Fig. 8.18 will thus be equal to the corresponding ordinate at time t_1 (ordinate AB) in the mass curve (Fig. 8.19). In a similar manner, it is possible to calculate and represent the ordinates of the mass curve for subsequent years using Fig. 8.18. A mass curve that displays the cumulative inflow rises steadily. The mass curve will be horizontal for a period when there is no inflow. The high flood period will cause the mass curve to climb extremely sharply. The mass curve's steepness, therefore, indicates the flow rate for that particular time period. The mass curve's gaps indicate comparatively dry times.

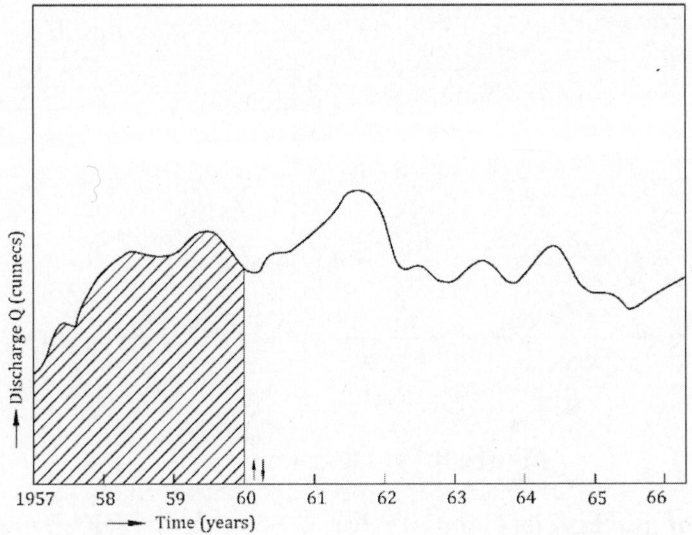

Figure 8.17: Flood Hydrograph of Inflow

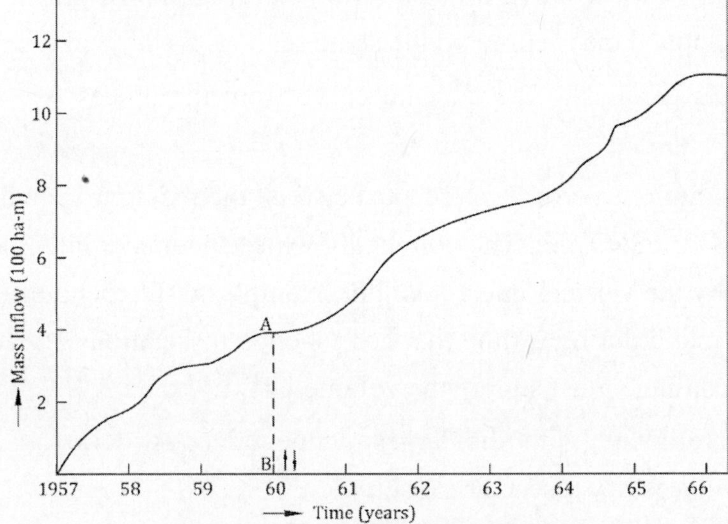

Figure 8.18: Mass Inflow Curve

8.9.6 Demand curve

A graph of the accumulated demand over a period is called a demand curve (Figure 8.19). A straight line having a slope equal to the demand rate represents the demand curve for a uniform rate of demand. Additionally, a demand curve can indicate a fluctuating rate of demand.

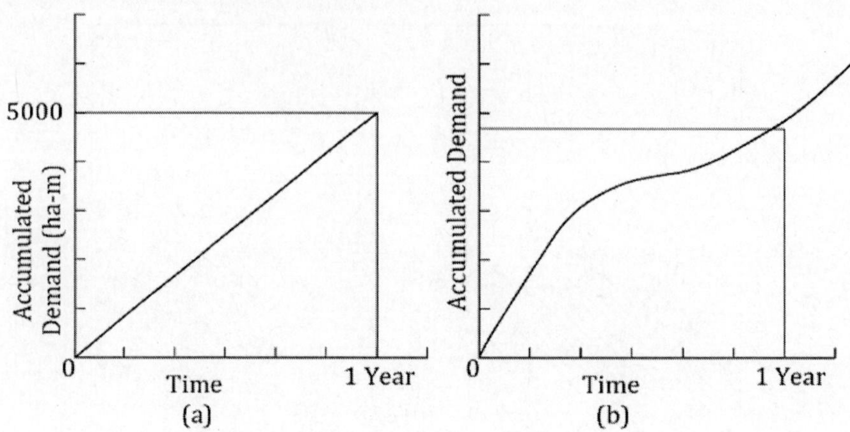

Figure 8.19: Demand Curves

8.9.7 Calculation of Reservoir Capacity for a Specified Yield from the Mass Inflow Curve Procedure (Figure 8.20):

a. Create the mass inflow curve using the flood hydrograph of inflow for several years. Prepare the demand mass curve on the same scale as well.

b. Draw tangents parallel to the demand curve from the apices A1, A2, A3, ... of the mass curve.

c. Calculate the highest vertical intercepts between the mass curve and the tangent, such as E1D1, E2D2, E3D3, etc. The volume by which the intake falls short of the demand is indicated by the vertical intercepts. For example, C1D1 denotes the net inflow and C1E1 the demand during a time that corresponds to locations A1 and C1. Therefore, the reservoir storage must supply the volume E1D1.

d. The vertical ordinate with the largest value, amongst E1D1, E2D2, E3D3, etc., indicates the necessary reservoir capacity.

It should be remembered that the water lost over the spillway is represented by the vertical distance between the subsequent tangents. The spillway's capacity must be adequate to release this amount of floodwater.

In accordance with the numerical figures displayed in Figure 8.21, we observe that:

1) A reservoir capacity of 2100 ha-m is required.

2) Assuming the reservoir to be full at A1, it is depleted to (2100 — 800) =1300 ha-m at D1 and is again full at B1.
3) Assuming the reservoir is full at A2, it is empty at D2 and is full again at B2.
4) The reservoir is full between B1 and A2, and the spill volume is 800 ha-m.

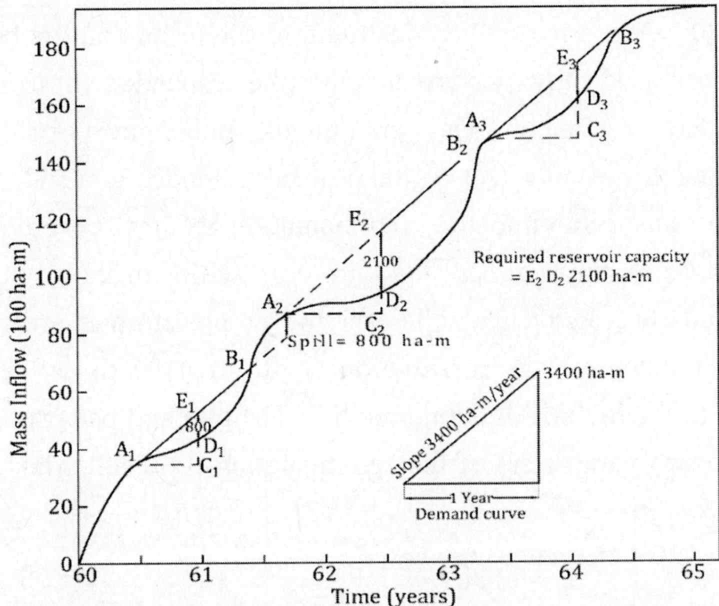

Figure 8.20: Determination of Reservoir Capacity

8.9.8 Determination of Safe Yield from a Reservoir of a Given Capacity

The process for calculating the safe yield from a reservoir with a particular storage capacity using a mass inflow curve is as follows:

1. Get the mass inflow curve ready. Draw straight lines from a similar origin on the same picture to reflect needs at different rates, say ranging from 0 to 5000 ha-m annually.
2. Draw tangents from the mass curve's apices, Al, A2, A3, etc., making sure that their greatest deviance from the mass curve stays within the designated reservoir capacity. As a result, in Figure 8.21, the reservoir capacity (let us say 1500 ha-m) is equivalent to the ordinates E1D1, E2D2, E3D3, etc.

3. Calculate each of these tangents' slopes. The yield that may be obtained annually from a reservoir with a particular capacity is indicated by the slopes. The firm yield is the slope of the flattest demand line.

8.10 SEDIMENTATION

A reservoir is created, when a hydraulic construction, such as a dam, is built across a river. As a result, due to decreased velocity of river water, the suspended silt load settles down and begins to accumulate sediment. The continuous phenomena of reservoir storage encroachment have a detrimental effect on the project's intended goal. Therefore, to evaluate a reservoir's viability and economic life, sedimentation studies are crucial. Depleting the reservoir's design capacity, the sediment load not only settles in the dead storage region as previously considered, but also encroaches on the active storage area. Determining the amount of silt accumulated and its distribution at different levels is therefore crucial for estimating the harm to the reservoir's economic life. The rate and pattern of silt deposition in the reservoir are the two parameters of the sedimentation problem, whose understanding is crucial.

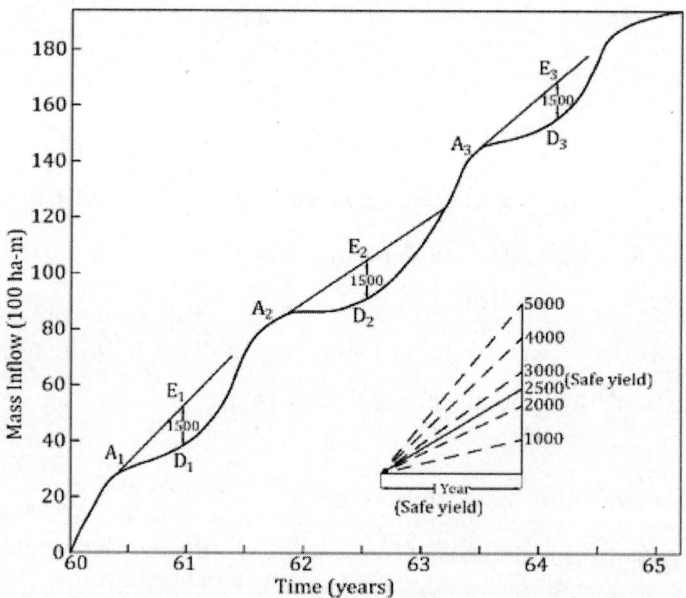

Figure 8.21: Determination of Yield from Reservoir of Specified Capacity

8.10.1 Silting Rate

The river catchment above the dam is the source of silt to the reservoir. The catchment's rate of erosion and the river's transport capacity are the factors which determine the amount of silt that enters the reservoir. When measuring or estimating the sediment yield from the watershed is not practical, similar reservoirs' observed data may be utilized to plan a new reservoir.

In any storage reservoir, the two main factors affecting the rate of silting are:

a. sediment from the river flow and
b. the capacity inflow ratio.

The other factors which affect the loss of storage capacity in the long run are:

a. The trap efficiency
b. Sediment characteristics, and
c. The reservoir operation.

These factors are inter-related. If the capacity is small the reservoir can be filled up by sediment much earlier than big reservoir having large capacity-inflow ratio. On the other hand, reservoirs having large capacity-inflow ratio will trap large amount of sediment entering the reservoir i.e., these will have high trap efficiency.

Brune (1953) established an empirical relationship between the capacity-inflow ratio (c/i) and the observed trap efficiency of reservoirs in the United States to estimate long-term trap efficiency, which is defined as the ratio of deposited sediment to the total sediment inflow in typically impounded reservoirs. In Figure 8.22, the relationship is displayed. A reservoir's trap efficiency varies over the time as sedimentation diminishes capacity. Table 8.2 lists the trap efficiencies found in a few Indian reservoirs.

The observed data of Table 8.2 supports the trend of Brune's curve for trap efficiency. It demonstrates that when the c/i ratio is one or higher, the trap efficiency is almost 100%, and when it is less than 0.2, the trap efficiency drops significantly.

M.A. Churchill created a link between the reservoir's sedimentation index and the proportion of incoming silt using data from the Tennessee Valley Authority. She then provided a curve, which is depicted in Figure 8.23.

Table 8.2: Observed trap efficiency of Indian reservoirs

Name of Reservoirs	Period of observation	Trap efficiency in%	c/i ratio
Matatila	1962-72	66.5 to 89.2	0.19
Hirakund	1957-73	64.3 to 91.1	0.20
Bhakra	1962-73	98.4 to 99.8	1.0
Maithon	1963-67	92 to 95	0.58
Lower Bhawani	1953-65	79 to 89	0.49

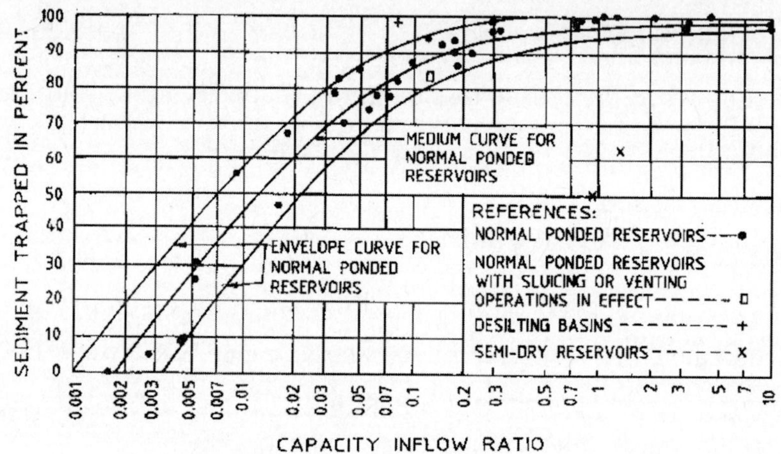

Capacity — Reservoir capacity at FRL; Inflow - Average annual inflow in volumetric units.

Figure 8.22: Brune's curve for trap efficiency

8.10.2 Description of Terms in Churchill's Curve

Capacity: Capacity of the reservoir during the considered time at the mean operational pool elevation.

Inflow Average monthly inflow during the duration of the study

Period of Retention The ratio of capacity to inflow rate

Length length of reservoir at the level of mean operation pool.

Velocity Mean velocity is obtained by dividing the inflow by the average cross sectional area. The average cross-sectional area is computed by dividing capacity by length.

Sedimentation Index Periods of retention divided by velocity.

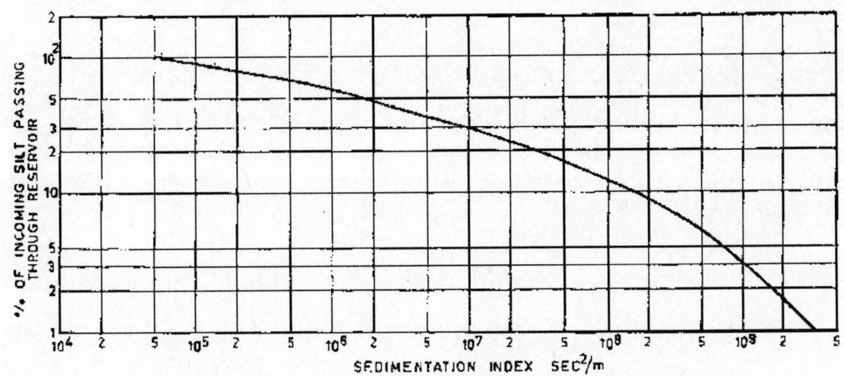

Figure 8.23: Churchill curve for trap efficiency

Example 8.1: Reservoir Data for a reservoir is given as below. Obtain trap efficiency using Brune's method and Churchill's Method.

- Full reservoir level (FRL)= 400 m
- Mean operating pool elevation = 399 m
- Capacity at FRL, C_1= 55.1 x 10^6 m^3
- Capacity at mean operating pool elevation, C_2 = 51.5 x 10^6 m^3
- Average inflow, I, over the study period of 10 years = 1380.06 x 10^6 m^3/year
- Length of the reservoir, L, at the mean operating level = 19312.13 m

Solution

Brune's Method

Capacity inflow ratio (C_1/I) = 55.1/1380.06 = 0.0399 year.

Trap efficiency corresponding to the above ratio as read from the median curve of Fig. 8.22 for normally ponded reservoir = 75 per cent

Dendy added more data to Brune's curve and developed a prediction Equation 8.42 for the median curve as:

E = 100 * 0.97 ** 0.19 ** log (C/I) ....(8.42)

Churchill's Method

Average inflow = 1380.06 m^3/year = 43.76 m^3/sec

Retention period (C_2/I) = 51.5 x 10^6 / 43.76 = 1.1769 x 10^6 sec

Average cross-sectional area (A) = C_2/ L = 515.5 x 10^6 / 1.9312.13 = 2666.7 m^2

Velocity = I/ A = 0.01641 m/sec

Sedimentation Index = period of retention/ velocity = 7.1718 x 10^7 sec^2 / m

Percentage of incoming sediment passing through as read from Fig. 8.24 corresponding to above sedimentation index = 15 percent.

Trap efficiency = 100 - 15 = 85 percent.

Absence of definite trend between C/I ratio and annual rate of silting indicates that there are factors other than C/I ratio, which are responsible for annual rate of silting.

The other factor influencing the rate of silting is sediment characteristic i.e., the size of sediment and its concentration in the inflow. If C/I ratio is less but size of sediment is large the trap efficiency will be more. Similarly, if concentration is large and C/I is less more silting will occur and the converse will be true.

8.10.3 Prediction of Rate of Reservoir Sedimentation

Numerous researchers have attempted to build empirical relations to forecast the rate of sediment deposition in a reservoir based on observations, field measurements, and surveys conducted on numerous reservoirs in India and overseas. These relations relate to sediment volume and catchment area. These are specific to a given area and cannot be applied globally. However, when no other data is available, these should be used carefully when planning a reservoir. Below are a few relationships that were created with Indian conditions in consideration:

a. **CBIP Research Committee Method**: In the lack of long-term records, the Central Board of Irrigation & Power Research Committee has recommended Equation 8.43 for calculating silt deposition in reservoirs. Maximum silt deposition rate suggested for catchments larger than 2600 km^2 = 3.57 ha m/100 km^2/year

$$S = KA^{3/4} \qquad \text{.............(8.43)}$$

where,

S = Sediment volume (in acre-ft/100 sq. mile/year)

A = Catchment area (in sq. mile)

K = Coefficient of proportionality (depending on type of catchment Table 8.3).

Table 8.3: Coefficient of proportionality (K)

Type of catchment	**Coefficient of proportionality (K) (Based on reservoirs of USA, India, and Burma)**
Rocky catchment	0.5
Normal catchment	1.7
Alluvial catchment	5.5

(The Maximum silt deposition rate suggested for catchments larger than 2600 km^2 = 3.57 ha m/100 km^2/year)

b. **Khosla's Method**: Khosla proposes the following empirical relation (Equation 8.44) for catchments with an area of less than 2600 km^2 (1000 sq miles).

$$Y = 5.19/A^{0.28} \qquad \textbf{............(8.44)}$$

where,

Y = annual sediment deposition (in acre ft per 100 sq. miles of catchment)

A = catchment area (in sq. miles);

(The recommended range of annual sedimentation rate for catchments larger than 2600 km^2 is 75 to 90 acre ft/100 km^2 (3.57 to 4.3 ha m/100 km^2).

Alternatively,

$$Q_s = 0.323\,(A)^{-0.28} \qquad \textbf{............(8.45)}$$

where,

Q_s = Annual siltation rate (in M. m^3/100 sq. km/ year)

A = Catchment area (in sq. km).

For example, for the proposed Jamrani project, the catchment area = 450 sq. km. Thus, the Annual sedimentation rate = 0.058 M m^3/ 100 sq. km (**5.8 ha. m/ 100 sq. km / year**).

c. **CWPRS, Pune Method**: The CWPRS, Pune has proposed the following relationship (Equation 8.46), which is comparable to Khosla equation

$$S = 10/A^{0.24} \qquad \textbf{............(8.46)}$$

where,

S = Sedimentation rate (in acre-ft/sq. mile/year);

A = Catchment area (in sq. miles).

For catchments up to 10 sq. miles; $S = 0.2743$ ha-m/km^2 (5.7 acre ft/sq. mile)

and for catchments up to 1000 sq. Miles; $S = 0.03$ ha-m/km^2 (0.63 acre ft/sq. mile).

d. **Raichur Method:** After dividing the data from Indian reservoirs into Himalayan and non-Himalayan regions according to catchment area, Raichur examined the data and proposed the following relationships (Equation 8.47 to 8.51):

- **Catchment area up to 130 km^2**

$Y = 0.395/A^{0.311}$ for Himalayan rivers in mountains ....(8.47)

$Y = 0.392/A^{0.202}$ for Himalayan rivers in trough and plains ....(8.48)

$Y = 0.460/A^{04.468}$ for non-Himalayan rivers ....(8.49)

- **Catchment area larger than 130 km^2**

$Y = 1.534/A^{0.311}$ for Himalayan rivers ....(8.50)

$Y = 0.159/A^{0.01}$ for non-Himalayan rivers ....(8.51)

where,

Y = Annual silting rate (in Mm^3/100 km.^{2});

A = Catchment area (in km.^{2}).

e. **Varshney's Method**: Varshney examined Indian reservoir data and proposed the relationship (Equation 8.52 and 8.53) for Himalayan catchments larger than 5000 km^2 catchment area,

$S = 141/A^{0.264}$ (for) ....(8.52)

Or

$Q_s = 1.534\,(A)^{0.264}$ ....(8.53)

where,

Q_s = Annual siltation rate (in M. m^3/100 sq. km./ year)

A = Catchment area (in sq. km.).

For example, for the proposed Jamrani project, the catchment area = 450 sq. km. Thus, the Annual sedimentation rate = 0.306 M m^3/ 100 sq. km (**30.6 ha. m/ 100 sq. km / year**).

f. **Lagwankar et al.:** Lagwankar proposed the following relationship (Equation 8.54 for catchment area between 100-2500 km^2 based on data of selected Indian reservoirs

$Q_s = 0.278\ (A)^{0.815}$ ....**(8.54)**

where, Qs = Annual siltation rate (in ha. m/ year)

A = Catchment area (in sq. km.)

For example, for the proposed Jamrani project, the catchment area = 450 sq. km. Thus, the annual sedimentation rate =**8.98 ha. m / 100 sq. km.**/ year (0.09 M m^3/100 sq. km./ year)

g. **Jogelkar's Equation:** The following relationship (Equation 8,55) has been proposed by Jogelkar.

$Q_s = 0.59\ (A)^{0.24}$ ....**(8.55)**

where,

Q_s = Annual siltation rate (in M. m^3/100 sq. km./ year)

A = Catchment area (in sq. km.).

For example, for the proposed Jamrani project, the catchment area = 450 sq. km. Thus, the annual sedimentation rate = 0.1362 M m^3/ 100 sq. km. (**13.62 ha. m/ 100 sq. km. / year**).

8.10.4 Prediction of Sediment Distribution

I. Pattern of Sediment Deposition in Reservoirs

Before the capacity surveys of reservoirs were carried out, it was the assumption that sediment gets deposited in the dead storage and so sufficient space should be provided in the reservoir as dead storage to accommodate the sediment entering the reservoir during its lifetime. However, the hydrographic surveys have proved this assumption is incorrect, and it has been revealed that sediment deposits all over the reservoir and starts reducing the live storage from the first year of its operation.

The flow velocity of a river lowers and silt load begins to deposit when it enters a reservoir. Delta deposits are formed in the reservoir's head reach by the bed load and sediment made up of coarse fractions. Lower settling velocity carries fine fractions of the sediment load into the reservoir in the direction of the dam. They either move in a non-stratified or stratified flow.

In this process of sediment deposition, a variety of depositional patterns can occur depending on factors like hydrologic conditions, sediment particle size, reservoir geometry, reservoir operation. Figure 8.24 displays a generalized longitudinal deposition profile.

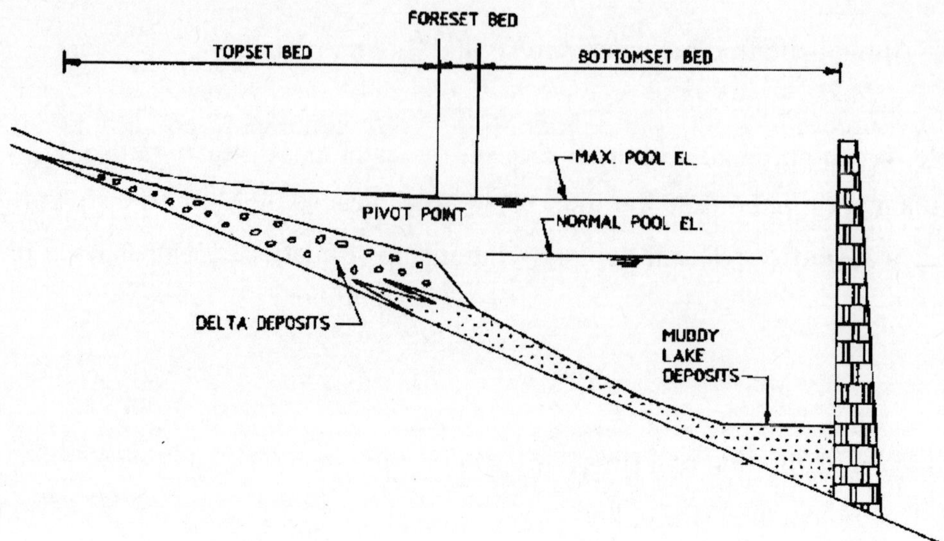

Figure 8.24: Generalized Sediment Deposition Zones in a Reservoir.

It shows the main three zones of deposition in a reservoir. Top set beds match coarse sediment delta deposits. Fine sediment deposits found in the reservoir's deepest sections and close to the dam compose the bottom set beds. The fore set beds, which differ from top set beds in that they have a steeper incline and smaller grain size, are the zones that lie between top set and bottom set zones. Although most of the sediment in bottom set beds is fine-grained, there are occasionally layers or isolated zones of coarser silt due to streams meeting the reservoir, landslides, reservoir decline, severe floods, etc.

Surveys have revealed different patterns of sediment deposition in reservoirs, and the research has demonstrated that a multitude of elements, including the following, control deposition:

a. Reservoir shape.
b. Reservoir operation
c. Sediment characteristics such as size, bed load, total load
d. River valley slope and fluctuations in bed of river
e. Location and size of outlets in dams.
f. Growth of vegetation at the reservoir head.

However, the basic four types of deposition patterns based on sediment characteristics, reservoir shape and reservoir operation are as follows. These are demonstrated in Figure 8.25.

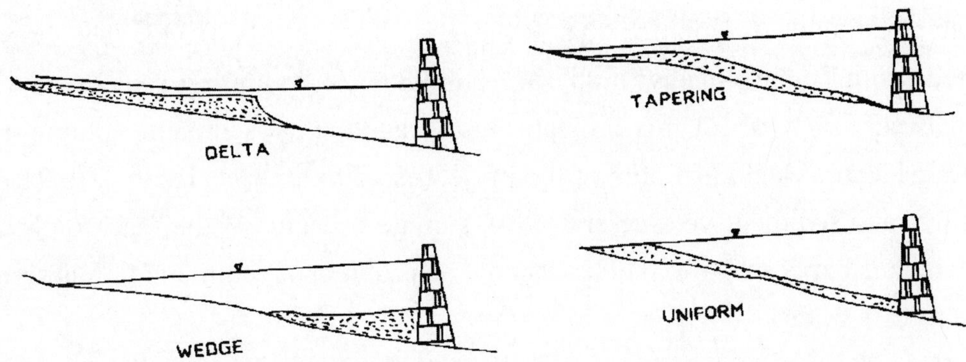

Figure 8.25: Longitudinal Patterns of Sediment Deposition in Reservoirs.

a. **Delta deposits**: These are formed at the head of reservoir and consist of coarse sediment (d >.062mm).

b. **Wedge-shaped deposits**: Most of these deposits are formed near the dam and are made of fine sediments. These are formed in small reservoirs or in large reservoirs operated at low water levels during floods,

c. **Tapering deposits**: In this kind, the deposits gradually get thinner in the direction of the dam. These happen when lengthy reservoirs with fine silt migrating in the direction of the dam are regularly operated at high pool elevation.

d. **Uniform deposits**: These could happen in small reservoirs with a modest percentage of finer silt and a changing operating level. These formations are unique.

The techniques for estimating the amount of sediment present and estimating the likely sediment distribution in the reservoir below the normal (full) reservoir level (F.R.L.) are described in IS 5477 (part 2) 1994. The amount of dead and live storage capacity in a storage reservoir is decreased as sediment enters and deposits gradually over time. As a result, the bed level adjacent to the dam rises; this elevated bed level is known as the "new zero elevation." Therefore, to verify the reservoir performance and the new zero elevation in simulation studies, it is important to evaluate the revised areas and capacities at various reservoir elevations that will become available in the future.

Example 8.2: A case study of a multipurpose project in Uttarakhand is conducted in order to determine the trap efficiency and classify the sediment problem. By calculating the capacity inflow ratio (C/I), the severity of the sedimentation problem can be ascertained.

For example, the C /I ratio for this Project is 208.6/605 = 0.345. The trap efficiency, calculated using Brune's median trap curve, is 93.65%. Assuming sediment rate of 0.1429 Mm^3 / 100 sq. km/ year for this project, the average annual sediment volume works out as 0.643 MCM as the catchment area of the project is 450 km^2 (See Table 8.4). The percentage ratio of average sediment volume and gross storage capacity of the reservoir is found to be 0.31% which is between 0.1 to 0.5 %. Hence the problem of sedimentation is **significant** and requires further studies.

The depth of the proposed reservoir is 127 m (762-635), and the capacity of the reservoir is 20860 ha-m. The depth-capacity curve's reciprocal slope, measured on a log-log scale, is 2.74. Therefore, it is a **Type II** (Floodplain – Foothill) reservoir. The Equation for the design curve used is $A_p = C_p^m (1 - p)^n$

where,

A_p= a non-dimensional relative area situated above the stream bed at a relative distance of p.

C, m, and n = non-dimensional constants that are fixed according to the reservoir type.

For type- II reservoir, C= 2.487, m = 0.57, and n = 0.41.

Table 8.4: Calculation of Trap Efficiency and Sediment Volume

Volume at FRL 762 m				**208.6 Mm^3**
Average Annual Sediment flow				**0.643 Mm^3**
Average Annual Inflow				**605 Mm^3**
Time (years)	**Capacity/Inflow (Mm^3)**	**Trap Efficiency (in %)**	**Sediment Volume (Mm^3)**	**End period capacity (Mm^3)**
				208.600
0-5	0.345	93.65	3.011	205.589
6-10	0.340	93.58	3.009	202.581
11-15	0.335	93.51	3.006	199.574

16-20	0.330	93.45	3.004	196.570
21-25	0.325	93.38	3.002	193.568
26-30	0.320	93.31	3.000	190.568
31-35	0.315	93.23	2.997	187.571
36-40	0.310	93.16	2.995	184.576
41-45	0.305	93.08	2.993	181.583
46-50	0.300	93.00	2.990	178.593
51-55	0.295	92.92	2.987	175.606
56-60	0.290	92.84	2.985	172.621
61-65	0.285	92.75	2.982	169.639
66-70	0.280	92.66	2.979	166.660
71-75	0.275	92.57	2.976	163.684
76-80	0.271	92.48	2.973	160.711
81-85	0.266	92.38	2.970	157.741
86-90	0.261	92.28	2.967	154.774
91-95	0.256	92.18	2.964	151.810
96-100	0.251	92.08	2.960	148.850
	Total sediment Volume = 59.750 Mm³			

8.10.5 Elevation Area Capacity

The elevation area capacity for the proposed project is given in Table 8.5 below. The annual sediment load is 64.3 ha.m. Bed Elevation is 635 m. Normal reservoir level is 762 m. Period of simulation 100 years (total sediment for the period = 6430 ha.m. The gross storage capacity of the dam at F R L (762 m) is 208.6 MCM and dead storage at RL 716.63 m is 64.3 MCM.

8.10.6 Computation of Revised Area and Capacity after 100 Years

Using the empirical area reduction approach, the revised area and capacity are determined by assuming that the reservoir has been silted up to the New Zero Elevation. The resulting silt volume, 64.58 MCM, is found to be just near 64.3 MCM. The Revised Areas and Capacities after 100 years of sedimentation is shown in Table 8.5.

8.11 SELECTION OF SUITABLE SITE

The following elements should be considered to determine the reservoir's final site selection:

a. The catchment area's geology should be such that the area experiences minimum percolation losses as possible and maximum run-off is obtained.
b. The location of the reservoir ought to decrease the rate of spilling through it. The presence of highly permeable rocks at the reservoir site increases the water tightness of the reservoir. Rocks such as gneisses, schists, slates, and crystalline igneous rocks like granite are unlikely to permit much water to flow through them.
c. A suitable location for a dam must exist. There should be very little percolation beneath the dam and it should be constructed on a solid, waterproof rock basis. When choosing a location, the cost of the dam is frequently a deciding challenge.
d. The cost of real estate for the reservoir, including the relocation of homes, railroads, and roads, must be kept to a minimum.

Table 8.5: Elevation-Area-Capacity and Sediment Deposition Computation by Empirical Area Reduction Method

Elevation (m)	**Original**		**Relative Depth (m)**	**A_p (Type II)**	**Sediment**		**Accumulated Sediment Volume (ha-m)**	**Revised Area (ha)**	**Revised Capacity (ha-m_**
	Area (ha)	**Capacity (ha-m)**			**Area (ha)**	**Volume (ha-m)**			
762.00	452.0	20860	1.00	0.00	0.00	0	**6457.88**	452.00	14402.12
760.00	432.5	19750	0.98	0.45	23.10	23.10	6434.77	409.40	13315.23
750.00	365.0	15750	0.91	0.89	45.93	345.18	6089.60	319.07	9660.40
740.00	315.0	11750	0.83	1.09	55.91	509.21	5580.38	259.09	6169.62
730.00	266.5	9800	0.75	1.20	61.58	587.47	4992.91	204.92	4807.09
720.00	222.5	7000	0.67	1.26	64.62	630.99	4361.92	157.88	2638.08
716.63	205.0	6430	0.64	1.27	65.17	218.69	4143.23	139.83	2286.77
710.00	172.5	4750	0.59	1.28	65.67	433.75	3709.48	106.83	1040.52

Elevation (m)	**Original**		**Relative Depth (m)**	**A_p (Type II)**	**Sediment**		**Accumulated Sediment Volume (ha-m)**	**Revised Area (ha)**	**Revised Capacity (ha-m_**
	Area (ha)	**Capacity (ha-m)**			**Area (ha)**	**Volume (ha-m)**			
700.00	132.5	3250	0.51	1.27	65.05	653.64	3055.84	67.45	194.16
690.00	99.9	2260	0.43	1.22	62.89	639.70	2416.14	36.98	0.00
680.00	72.5	1470	0.35	1.15	59.16	610.23	1805.91	13.34	0.00
675.00	62.5	1360	0.31	1.10	56.68	289.60	1516.31	5.82	0.00
671.00	**54.5**	1261	0.28	**1.06**	54.37	222.09	1294.22	0.00	0.00
670.00	52.5	1250	0.28	1.05	53.74	54.05	1240.16	0.00	0.00
660.00	32.5	800	0.20	0.90	46.28	500.11	740.06	0.00	0.00
650.00	15.0	0	0.12	0.70	35.94	411.10	328.96	0.00	0.00
640.00	5.0	0	0.04	0.39	19.90	279.21	49.75	0.00	0.00
635.00	0.0	0	0.00	0.00	0.00	49.75	0.00	0.00	0.00
						6457.88			

e. The reservoir site's topography should allow for a sufficient capacity without overflowing too much land or other resources.

f. The location ought to allow for the formation of a deep reservoir. Because a deep reservoir will cost less to submerge per unit of capacity, it will reduce evaporation losses due to a smaller water spread area, and it will probably be less likely to support vegetation development than a shallow reservoir.

g. The location of the reservoir should be such that water from streams that include a large proportion of silt in water is avoided or prohibited.

h. The location of the reservoir should ensure that the water it holds is appropriate for the project's intended use. At the reservoir site, there must be no undesirable minerals or salts in the soil or rock mass.

SUMMARY

A crucial component of water management systems are spillways and dams. To control water flow, build reservoirs, and store water, dams are structures that are built across rivers or streams. They facilitate recreational activities, flood management, hydroelectric power generation, and water supply. There are several numerous types of dams, such as earthen dams, arch dams, and gravity dams. To properly release surplus water from a dam and avoid overflow and possible damage, spillways are outlets or structures. They are crucial for safeguarding the dam and controlling the reservoir's water levels. This unit covers the classification, design considerations, failure reasons, seepage estimation, and control of different dam types, including embankment dams and arch dams. Furthermore, this unit also covers spillway components and spillway crest gate types, reservoir types, yield, regulation, and sedimentation.

EXERCISE

Revision Questions

1. What are embankment dams, and why are they commonly used for water storage, flood control, and hydropower generation?
2. What are the critical considerations in the design of an embankment dam?
3. How does earthquake activity influence the design features of an embankment dam in seismically active areas?
4. Explain the general design criteria for constructing embankment dams.
5. How does a homogeneous earth dam differ from a zoned earth dam, and what are the advantages of each?
6. What principle governs the estimation of seepage through an embankment dam and its foundation?
7. Explain the various methods used to control seepage in an embankment dam.
8. How are the upstream and downstream faces of an embankment dam protected from wave and rainwater action?

9. Describe the issues caused by surface water runoff and wind erosion on the downstream slope of an embankment dam.
10. How does vegetation contribute to slope protection on the downstream slope of an embankment dam, and what are the challenges in maintaining it in arid regions?
11. Describe the primary forces that act on a gravity dam.
12. Explain the stability conditions that is essential for the design of a gravity dam and how the elementary profile of a right-angled triangular section satisfies these conditions.
13. Identify the primary causes of a gravity dam failure.
14. Write notes on
    a) Gravity, trial-load, and finite element method
    b) Arch and Buttress Dam
    c) Mass inflow and demand curve
15. Explain the importance of spillways in dam construction and describe the consequences of inadequate spillway capacity.
16. What are the primary components of a spillway, and how does each component function to ensure the safe release of excess water?
17. Compare and contrast the flashboards and stoplogs as types of spillway gates, including their operational challenges and advantages.
18. Describe the functioning and applications of vertical lift gates in spillways. What are the limitations of these gates in large installations?
19. Explain how safe yield is determined using a mass inflow curve. Discuss its significance for ensuring water availability during periods of drought.
20. Discuss the factors influencing sediment deposition and its impact on reservoir capacity over time.
21. Discuss the key factors and considerations involved in selecting a suitable site for a reservoir and dam construction.
22. Compare and contrast the different types of reservoirs based on their purposes and functionalities.

Numerical Problems

1. For an earth dam with a homogeneous section and a horizontal drain, as illustrated in Figure8.26, construct the top flow line and the flow net. Additionally, calculate the discharge per meter length through the dam's body, given K = 3 × 10^{-4} cm/s.

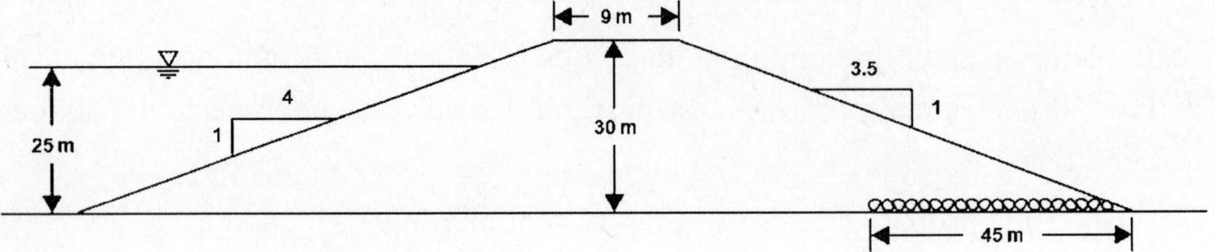

Figure 8.26: Sketch for Problems 1 and 2.

2. For the embankment section shown in Fig. 8.26, with K_x=7×10^{-4} cm/s and K_y=4×10^{-4} cm/s, calculate the discharge through the dam section per meter length.
3. A gravity dam has a profile, as shown in Figure 8.27. If the water on the downstream side is completely removed, what will the uplift force be exerted on the dam? (Assume the unit weight of water = 10 kN/m^3)

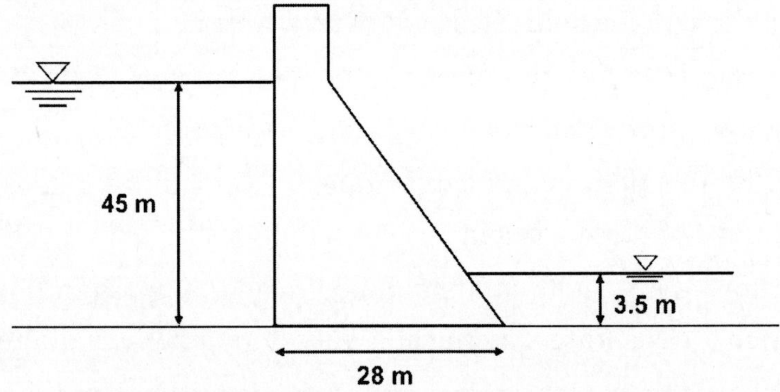

Figure 8.27: Sketch for Problem 3.

4. For a gravity dam with the following parameters: height h= 30 meters, specific gravity of concrete s= 2.4, density of water ρ= 1000 kg/m^3, and the correction factor for uplift force c'=1. Calculate the base width b of the dam to satisfy the no tension condition.
5. Figure 8.28 shows the cross-section of a gravity dam with the following data:

The angle of internal friction of silt: 35°

Submerged unit weight of silt: 16 kN/m³

Horizontal earthquake acceleration: 0.20 g

Vertical earthquake acceleration: 0.10 g

Shear strength between dam and foundation: 1600 kN/m²

Coefficient of friction: 0.80

Determine, for extreme load combinations (excluding ice pressure), the following:

(a) The various relevant stresses at the toe and heel and

(b) The factors of safety against overturning and sliding, as well as the shear friction factor.

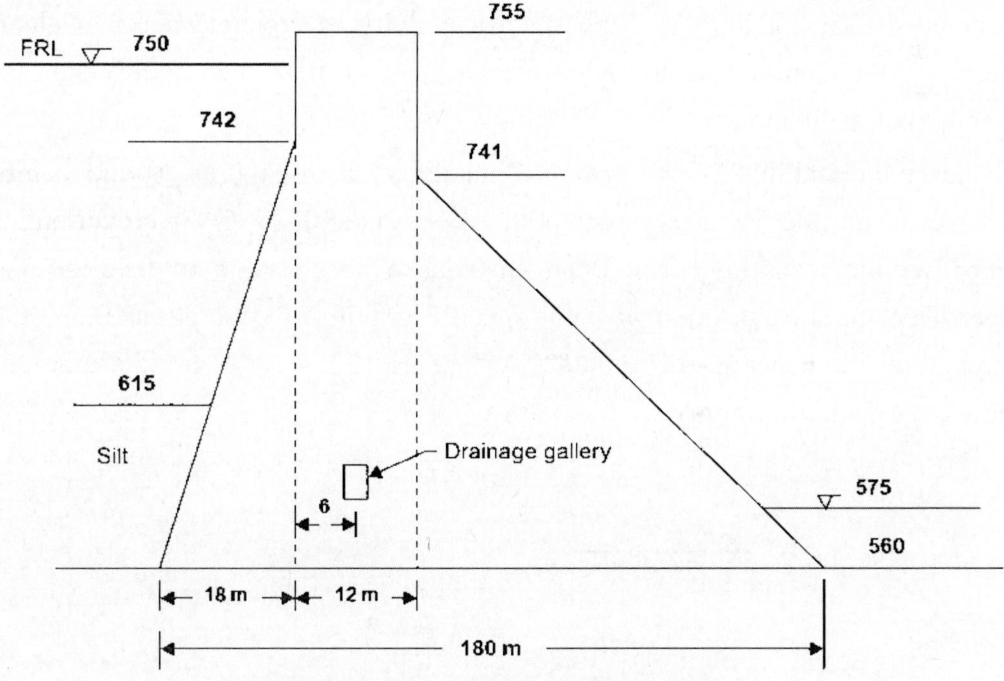

Figure 8.28: Sketch for Problem 5.

6. Given the following data for a reservoir:

   Full Reservoir Level (FRL) = 400 m

   Mean Operating Pool Elevation = 399 m

   Capacity at FRL, C_1= 55.1 x 10^6 m³

Capacity at mean operating pool elevation, C_2= 51.5 x 10^6 m³

Average annual inflow, *I*over 10 years = 1380.06 x 10^6 m³/year

Length of reservoir, *L*, at the mean operating level = 19,312.13 m

Calculate the following:

a) Capacity Inflow Ratio (C/I)for the reservoir

b) Trap Efficiency using Brune's method

c) Sedimentation Index using Churchill's method

d) Trap Efficiency from Churchill's curve

7. A concrete gravity dam has a maximum reservoir level at 240 meters and the foundation level at 120 meters. The maximum permissible compressive stress in concrete is 3500 kN/m², and the specific gravity of concrete is 2.4. Calculate the height of the dam and determine if it qualifies as a high dam or low dam based on standard classifications.

8. Evaluate the stability of the gravity dam depicted in Figure8.29 and compute the stresses at the toe and heel under both empty and full-reservoir conditions. For the full-reservoir scenario, consider an earthquake acceleration of 0.15 g. Assume a coefficient of shear friction of 0.65, specific gravity of concrete as 2.50, and shear strength at the concrete-rock contact surface as 120 × 10⁴ N/m². Assume any other necessary data accordingly.

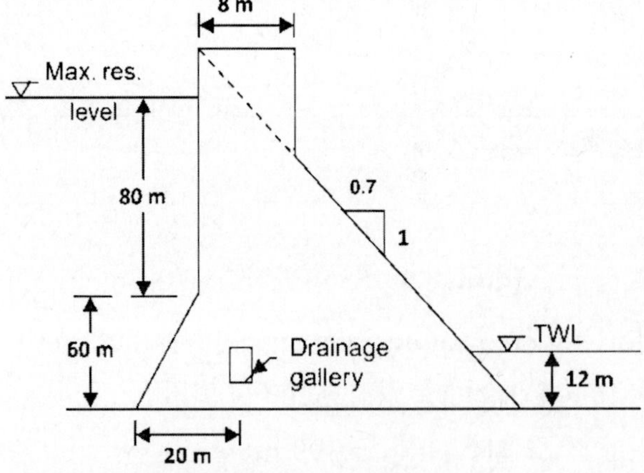

Figure 8.29: Sketch for Problem 8.

Multiple Choice Questions

1. What is the primary purpose of a dam?

A) To provide recreational space
B) To control water flow and create a reservoir
C) To generate electricity
D) All of the above

2. What is a reservoir?

A) A natural body of water
B) An artificial lake created by a dam
C) A river that flows into the ocean
D) A storage tank for drinking water

3. Which type of dam is designed to hold back water using its weight?

A) Arch dam
B) Gravity dam
C) Embankment dam
D) Roller compacted concrete dam

4. What is one major environmental concern associated with dam construction?

A) Increased biodiversity
B) Displacement of communities and wildlife
C) Improved water quality
D) Enhanced recreational opportunities

5. What is the term for the area behind a dam where water accumulates?

A) Spillway
B) Waterway
C) Reservoir
D) Floodplain

6. Which of the following is a benefit of reservoirs?

A) Flood control
B) Water supply for irrigation and drinking
C) Hydroelectric power generation
D) All of the above

7. What is a spillway?

A) A channel for diverting water
B) A structure that allows excess water to flow out of a reservoir
C) A type of dam
D) An area where water is stored

8. Which of the following can happen if a dam fails?

A) Increased water supply
B) Flooding downstream
C) Improved fish migration
D) Enhanced water quality

9. What is the primary design characteristic of an arch dam?

A) It is built on a flat surface

B) It is curved upstream to withstand water pressure

C) It is made entirely of earth materials

D) It has a wide base and narrow top

10. Which of the following materials is commonly used to construct arch dams?

A) Wood B) Steel

C) Concrete D) Earth and rock

11. What is a key advantage of an arch dam compared to a gravity dam?

A) Lower construction cost

B) Greater ability to store water

C) Reduced amount of material needed due to its design

D) Easier to construct in flat terrain

12. What type of topography is most suitable for constructing an arch dam?

A) Flat plains B) Rugged, mountainous areas

C) Coastal regions D) Low-lying areas

13. How does an arch dam transfer the pressure of the water it holds?

A) By pushing water straight down B) By arching the force into the canyon walls

C) By using mechanical pumps D) By creating a vacuum

14. What is a common use for an arch dam?

A) Irrigation B) Flood control

C) Hydroelectric power generation D) All of the above

15. Which of the following is a potential risk associated with arch dams?

A) Erosion of the dam structure B) Failure due to extreme seismic activity

C) Difficulty in controlling water levels D) High evaporation losses

16. What is a "spillway" in the context of an arch dam?

A) A structure to divert water away from the dam

B) An area where excess water can safely flow out of the reservoir

C) A type of water treatment facility

D) A mechanism to generate electricity

17. What is an embankment dam primarily made of?

A) Concrete B) Steel

C) Earth and rock materials D) Wood

18. What is the main purpose of an embankment dam?

A) To generate hydroelectric power

B) To provide recreational areas

C) To store water for irrigation and flood control

D) To facilitate navigation

19. Which of the following is a characteristic feature of embankment dams?

A) Curved design to resist water pressure B) Wide base tapering to a narrower top

C) Built on steep slopes D) Made entirely of reinforced concrete

20. What is a key advantage of embankment dams?

A) Lower risk of failure B) Flexibility in design

C) Reduced construction time D) Minimal environmental impact

21. What is the term for the materials used to construct an embankment dam?

A) Fill B) Aggregate

C) Reinforcement D) Structural material

22. Which type of monitoring is essential for the safety of embankment dams?

A) Temperature monitoring B) Structural integrity assessment

C) Vegetation health assessment D) Air quality testing

23. What is a potential risk associated with embankment dams?

A) High costs of construction B) Erosion and slope failure

C) Limited storage capacity D) Inability to generate electricity

24. What is the purpose of a spillway in an embankment dam?

A) To increase water pressure

B) To divert water for irrigation

C) To safely release excess water from the reservoir

D) To generate hydroelectric power

25. What is the primary cause of sedimentation in reservoirs?

A) Water temperature variations
B) Erosion of surrounding land
C) Biological activity
D) Chemical reactions in water

26. Which of the following factors can influence the rate of sedimentation in a reservoir?

A) Flow velocity of inflowing water
B) Reservoir depth
C) Vegetation cover in the watershed
D) All of the above

27. What is the primary impact of sedimentation on reservoir capacity?

A) Increases water quality
B) Decreases water storage capacity
C) Enhances aquatic habitat
D) Improves recreational opportunities

28. What is a common method for managing sedimentation in reservoirs?

A) Dredging
B) Adding chemical flocculants
C) Constructing sediment traps
D) All of the above

29. Which of the following best describes the term "reservoir sedimentation"?

A) The accumulation of nutrients in water
B) The deposition of sediment particles in the reservoir
C) The evaporation of water from the reservoir
D) The filtering of pollutants by sediment

30. What is the main consequence of excessive sedimentation in a reservoir?

A) Increased fish population
B) Decreased water quality
C) Enhanced water supply
D) Improved flood control

31. Which type of sediment is most found in reservoirs?

A) Organic matter
B) Silt and clay
C) Gravel
D) Sand

32. What role do vegetation and land management practices play in sedimentation?

A) They have no effect on sedimentation
B) They can reduce sediment runoff
C) They increase sedimentation
D) They solely improve aesthetics

33. What is the primary structural feature of a buttress dam?

A) A curved arch
B) Vertical walls
C) Series of buttresses
D) Gravitational foundation

34. Which of the following is a key advantage of buttress dams over gravity dams?

A) Lower construction costs
B) Better resistance to seismic forces
C) Greater water storage capacity
D) Increased height

35. What materials are commonly used in the construction of buttress dams?

A) Steel and plastic
B) Concrete and stone
C) Wood and clay
D) Aluminium and glass

36. In which type of terrain are buttress dams most effectively constructed?

A) Flat plains
B) Rocky or mountainous regions
C) Coastal areas
D) Swampy lowlands

37. What is the primary purpose of the buttresses in a buttress dam?

A) To provide aesthetic appeal
B) To support the dam and transfer water pressure to the foundation
C) To facilitate fish passage
D) To reduce evaporation

38. What type of load does a buttress dam primarily resist?

A) Tensile load
B) Shear load
C) Hydrostatic load
D) Thermal load

39. Which of the following is a disadvantage of buttress dams?

A) Limited height
B) Complex construction
C) Vulnerability to overtopping
D) High maintenance costs

40. What is a common application for buttress dams?

A) Flood control
B) Water supply
C) Hydroelectric power generation
D) All of the above

41. Which of the following is a primary type of reservoir?

A) Natural reservoir
B) Artificial reservoir
C) Both A and B
D) Neither A nor B

42. What is the main purpose of a spillway in a reservoir?

A) To increase water storage
B) To control water level and prevent overflow
C) To enhance recreational activities
D) To generate hydroelectric power

43. Which type of spillway is designed to handle large amounts of water flow during floods?

A) Chute spillway
B) Side channel spillway
C) Controlled spillway
D) Weir spillway

44. What is a key characteristic of a rockfill reservoir?

A) It is entirely man-made
B) It utilizes natural terrain for support
C) It is primarily used for irrigation
D) It has no spillway

45. Which type of spillway uses a gate or valve to control water flow?

A) Ogee spillway
B) Controlled spillway
C) Labyrinth spillway
D) Free-flowing spillway

46. What is an example of a natural reservoir?

A) A dammed river
B) A large lake
C) An artificial pond
D) A water tank

47. Which type of reservoir is primarily constructed for the purpose of hydropower generation?

A) Retention reservoir
B) Run-of-the-river reservoir
C) Flood control reservoir
D) Recreational reservoir

48. What is a common concern with spillways during heavy rainfall?

A) Sedimentation build-up
B) Structural failure
C) Erosion of downstream banks
D) All of the above

Answer: 1-D; 2-B; 3-B; 4-B; 5-C; 6-D; 7-B; 8-B; 9-B; 10-C; 11-C; 12-; 13-B; 14-D; 15-B; 16-B; 17-C; 18-C; 19-B;20-B; 21-A; 22-B; 23-B; 24-C; 25-B; 26-D; 27-B; 28-D; 29-B; 30-B; 31-B; 32-B; 33-C; 34-A; 35-B; 36-B; 37-B; 38-C; 39-B; 40-D; 41-C; 42-A; 43-A; 44-B; 45-B; 46-B; 47-B; 48-D.

KNOW MORE

For more information related to this topic scan the QR code.

OR

Type this link in your browser

https://drive.google.com/file/d/11b65sZmz5_Teki7wXV9oYt7jfiK0mp9V/view?usp=sharing

BIBLIOGRAPHY

1. Asawa, G.L. (1993), "Irrigation Engineering", Published by Willey Eastern Limited, ISBN 81-224-0457-X.
2. Asawa, G.L. (2005), "Irrigation and Water Resources Engineering, New Age International (P) Ltd., ISBN-13: 978-8122416732
3. Chow. V.T, (Ed), (1964), "Handbook of Applied Hydrology", McGraw-Hill, New York, NY, 1964.
4. Jain S. K., Agarwal, P. K. And Singh, V. P. (2007), Hydrology and Water Resources of India, Water Science and Technology Library, Vol 57, Springer, The Netherland, ISBN-10 1-4020-5179-4(HB).
5. Punmia, B.C. and Lal, Pande B.B., (1992), "Irrigation and Water Power Engineering", Twelfth Edition. Laxmi Publications (P) Ltd.
6. Singh, V. P. (1992). Elementary hydrology. Prentice Hall, New Jersy, 973.
7. Subramanya, K. (2015), "Engineering Hydrology", McGraw-Hill Education (India) Private Limited, New Delhi, 2015, ISBN(13): 978-1-25902997-4.
8. Varshney, R.S., Gupta, S.C and Gupta R. L.(2009), "Theory and design of Irrigation structures (Vol-1: Channels and Tubewells)", Published by Nem Chand & Bros, Civil Lines, Roorkee, ISBN 81-85240-73-6.
9. Varshney, R.S., Gupta, S.C and Gupta R. L.(2009), "Theory and design of Irrigation structures (Vol-I1: Canal and Storage works)", Published by Nem Chand & Bros, Civil Lines, Roorkee, ISBN 81-85240-48-5.
10. Working Group (1999), The national Commission for Integrated Water Resources Development, Ministry of Water Resources, Govt. of India.

References for Further Learning

1. Asawa, G.L. (1993), "Irrigation Engineering", Published by Willey Eastern Limited, ISBN 81-224-0457-X.
2. Asawa, G.L. (2005), "Irrigation and Water Resources Engineering, New Age International (P) Ltd., ISBN-13: 978-8122416732
3. Bouwer, H. (1978) Groundwater Hydrology, McGraw-Hill Book, New York, 480.
4. Chow. V.T, (Ed), (1964), "Handbook of Applied Hydrology", McGraw-Hill, New York, NY, 1964.
5. Ghosh, S.N. (2014), "Flood control and drainage engineering", Published by CRC Press/Balkema, The Netherlands, ISBN: 978-1-138-02627-8.
6. Heath, R. C. (1987). Basic ground-water hydrology (Vol. 2220). US Department of the Interior, US Geological Survey.
7. IMD (1971), "Scientific Report No. 136, India Meteorological Department,1971, Government of India.
8. Jain S. K., Agarwal, P. K. And Singh, V. P. (2007), Hydrology and Water Resources of India, Water Science and Technology Library, Vol 57, Springer, The Netherland, ISBN-10 1-4020-5179-4(HB).
9. McQueen, I.S. (1963), Development of a Hand Portable Rainfall-Simulator Infiltrometer, Geological Survey Circular, 482, Washington, 1963.
10. Punmia, B.C. and Lal, Pande B.B., (1992), "Irrigation and Water Power Engineering", Twelfth Edition. Laxmi Publications (P) Ltd.
11. RBA (1980), "Report of National Commission of Flood, Govt. of India.
12. Schendel, V, (1975), "The world's water resources and water balance", Natural Resources and Development, Inst. for Sci. Coop, Hannover, Germany, pp. 8-14, Vol. 1, 1975.

13. Shiklomanov (2000), “Appraisal and assessment of world water resources”, Water Int (25(1):11-32.
14. Singh, V. P. (1992). Elementary hydrology. Prentice Hall, New Jersy, 973.
15. Subramanya, K. (2015), “Engineering Hydrology”, McGraw-Hill Education (India) Private Limited, New Delhi, 2015, ISBN(13): 978-1-25902997-4.
16. Todd, D.K. and L. Mays (2004), Groundwater Hydrology, 3rd Edition, ISBN: 978-0-471-05937-0, 656 Pages, Wiley.
17. UNESCO, (1978), “World Water Balance and Water Resources of the Earth”, Studies and Reports in Hydrology, 25, UNESCO, Paris, France, 1978.
18. Van der Leeden, (1975), “Water Resources of the World”, Water Information Center, Port Washington, N.Y., USA, 1975.
19. Varshney, R.S., Gupta, S.C and Gupta R. L.(2009), “Theory and design of Irrigation structures (Vol-1: Channels and Tubewells)”, Published by Nem Chand & Bros, Civil Lines, Roorkee, ISBN 81-85240-73-6.
20. Varshney, R.S., Gupta, S.C and Gupta R. L.(2009), “Theory and design of Irrigation structures (Vol-I1: Canal and Storage works)”, Published by Nem Chand & Bros, Civil Lines, Roorkee, ISBN 81-85240-48-5.
21. Working Group (1999), The national Commission for Integrated Water Resources Development, Ministry of Water Resources, Govt. of India.
22. https://indiawris.gov.in/wiki/doku.php?id=india_s_water_wealth)

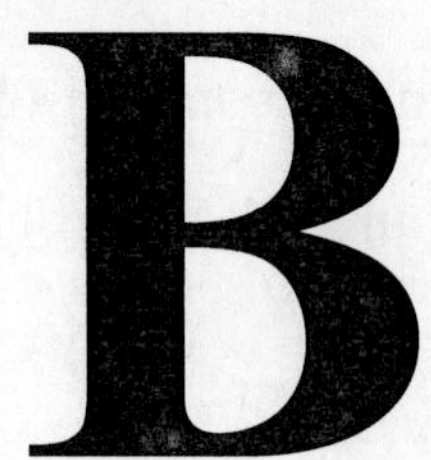

CO and PO Attainable

Course outcomes (COs) for this course can be mapped with the programme outcomes (POs) after the completion of the course, and a correlation can be made for the attainment of POs to analyze the gap. After proper analysis of the gap in attaining POs, necessary measures can be taken to overcome the gaps.

Table for CO and PO attainment

Course Outcomes	**Expected Mapping with Programme Outcomes** *(1- Weak Correlation; 2- Medium correlation; 3- Strong Correlation)*											
	PO-1	**PO -2**	**PO -3**	**PO - 4**	**PO -5**	**PO -6**	**PO -7**	**PO -8**	**PO -9**	**PO -10**	**PO -11**	**PO -12**
CO-1												
CO-2												
CO-3												
CO-4												
CO-5												
CO-6												
CO-7												
CO-8												

C Index

	Page No.
A	
Abstraction loss	1, 25
Abstraction from precipitation	14,92,142
Acoustic rain gauge	51,471
Actual evapotranspiration	92,111,143,146
Adaptive management and monitoring	208
Adequacy of rain gauge stations	52
Advancing monsoon season	43
Afforestation	15,26,168,292
Agricultural land	22,183
Alluvial channels	334,258,383
Analysis of surface water supply	294,323
Annual Precipitation in India	36,45
Annual rainfall	14,26,80,85
Antecedent moisture condition	125,180,190
Applications in engineering	1,21,93,368
Aquiclude	227, 272
Aquifer	171,216,227
Aquifer properties	171,224
Aquifer recharge	19,295
Aquifer test	224,257,261
Aquifuge	228,272
Aquitard	227,229,272

	Page No.
Arch dams	395,354,
Arithmetical-mean method	63
Artesian aquifer	228,233
Atmospheric pressure	33,229,416
Average yield	438
Avulsion and shifts	359
B	
Barren land	183
Base flow	8,207,222
Base flow generation	170,215
Base flow separation	154,172
Basin irrigation	315
Basin lag	212,213
Bed load	449,450
Blaney-Criddle formula	121,147
Braided channel	359
Brune's method	445
Building block method	204
Buttress dams	396,428,465
C	
Calculation of storage volume	198
Canal distribution system	334,386
Canal losses	342,392
Canal outlets	334,385
Capillary action	309
Cash crops	299
Catchment area	3,31,446
Catchment characteristics	154,167
Causes of canal losses	342,384

	Page No.
Causes of floods	288
Channelization	169,292,357
Channel morphology	359
Channel network	168
Characteristics of soil	132
Check dams	382
Chemical and Isotopic Tracers	172
Chemical characteristics	302
Chezy's equation	384
Churchill curve	445
Clark's model	352
Class A evaporation pan	96,99
Classification of canals	336
Clay lining	379,380
Classification of infiltration capacities	139
Climate change adaptation	23,296
Climate variability	32,169,171
Clogging risk	318
Coastal flood	288,327
Coefficient of permeability	235,250,306,406
Coefficient of storage	232,272
Cold winter season	42
Colorado sunken pan	97,99
Command area	339,389
Components of natural hydrograph	164,215
Condensation	15,25,109,151
Confined aquifer	229,234,241,262
Consumptive use	111,311,324
Conventional hydropower	285

	Page No.
Convolution	351,352
Conveyance structure	432
Copula model	354
Crop rotation	298,389
Cropping pattern	298,323
Cultivable command area	344,384
Culverts and bridges	372,382
Curve number (CN)	350,392
Cyclone	43,87
D	
DAD curves	70,71
Dams and reservoirs	206,285,436
Dam safety	296
Darcy's law	272,306
Dead load	414,416
Delay index	234
Delta	281,301,323
Demand curve	438,439,457
Depression storage	125,144,152
Depth-Area-Duration curve	76,78
Depth-Area-Duration frequency curve	71,72
Design discharge	345,391
Design of channels	371
Design storm	62,210,390
Direct runoff	141,158
Discharge in canals	343
Distance drawdown analysis	266
Distributary channel	337
Distribution reservoir	436,437

	Page No.
Domestic water supply	162,282
Double-mass curve	58
Double-ring infiltrometer	133,134
Drainage density	168
Drainage factor	234
Drip irrigation	108,312,317
Drizzle	40,84
Drought analysis	294
Drought stress	110,308
Duty and Delta	281,300,322
E	
Earth dams	397
Earthen lining	379
Earthquake activity	401,456
Ecosystem services	202,356
Effluent stream	166,272
Elevated structures	292
Elevation-Area-Capacity	454
Embankment dams	395,401,431
Emergency preparedness	293
Empirical evaporation equations	95,100
Energy-budget method	104
Entrance and exit channel	432,433
Environmental flow assessment	195,204
Environmental Impact assessment	211,221,296,373
Erosion control	22,134,371
Erosion and Sediment Control	22,158
Estimation of missing data	55
Evaporation losses	107,160,462

	Page No.
Evaporation process	92,126,143
Evaporimeters	92,149
Evapotranspiration	111,112,150
Evapotranspiration equations	115
Excess rainfall	77,141,209
Extreme events	83,169,346
F	
Fertigation	318,322
Field capacity	111,144,307
Field channels	337,338
Field measurements	25,172,446
Flash flood	287,327,434
Flood control canals	338
Flood control reservoirs	401,437
Flood control strategies	290
Flood forecasting	21,90,163,351
Flood hydrograph	154,410,439
Flood management measures	289,323
Floodplains	203,293,359
Floodplain management	293
Flood risk assessment	33,69
Floodwalls	289,292,327
Flow-duration curves	192,196,215
Flow-mass curve	197,215
Flow rates	164,192,220,375
Flow regime design	207
Flow routing models	206
Fluid characteristics	132
Flumes	374

	Page No.
Fluvial processes	356
Food crops	300,477
Forest and woodland	183
Forms of Precipitation	36,78,156
Forms of subsurface water	225
Frequency analysis	72,194,346
Frequency of irrigation	307,312,324
Furrow irrigation	315,326
G	
Gamma model	353
Ganguillet and Kutter equation	360
Geologic formations	226,230
Glaze	40,41
Glaciers	8,33
Global freshwater resources	18
Global Water balance	15,32
Grassland and open spaces	183
Gravity dams	396,408,417
Gross command area	339,384
Groundwater flow	8,21,165,170
Groundwater hydraulics	240
Groundwater seepage	8,170
Grouted lining	380
H	
Hail	36,77,126,156
Hazen’s formula	237
Head discharge	344
Heavy metals	294,303
Horton’s equation	137

	Page No.
Hot summer season	42
Hydraulic conductivity	132,232,268
Hydraulic design	211,358,371
Hydraulic diffusivity	233,272
Hydraulic geometry	69,360
Hydraulic modelling	346,347
Hydraulic resistance	234
Hydroelectric canals	338
Hydrologic budget equation	112
Hydrological connectivity	169,355
Hydrograph analysis	71,133,170
Hydrograph generation	163
Hydrograph separation	171,175
Hydrologic cycle	6,11,26,273
Hydrologic soil groups	179,184,350
Hydrological methods	204,206
Hydrological processes	126,129,355
Hydrological variability	357
Hyetograph	62,210,352
I	
Ice jams	288
Ice pressure	411,429,459
Index of wetness	38,78
Infiltration capacity	93,138,290,309
Infiltration indices	92,140
Infiltration losses	342
Infiltration rate	30,130,308
Initial loss	151,152
Instantaneous unit hydrograph (IUH)	353

	Page No.
Intensity-duration-frequency curves	352
Interception	174,351
Irrigation canals	338,357,387
Irrigation efficiency	313,388
ISI standard pan	96
Isohyet method	66
K	
Kennedy's silt theory	365
Kostiakov model	139
L	
Lacey's theory	384,368
Land use changes	90,158,390
Land use and land cover changes	157,171
Leakage factor	234
Levees	22,288,290,316
Lining of canals	379,385
M	
Main canal	336,344,386
Manning's equation	222,361,384
Mass inflow curve	438,457
Mass outflow	9
Meandering channel	359
Mean precipitation	36,68,82
Measurement of evapotranspiration	113
Measurement of infiltration	133
Measurement of precipitation	47
Methods of irrigation	281,314,324
Meyer's formula	102,145
Minors	338

	Page No.
Model outputs	356
Modular canal outlets	377
Monsoon depressions	44
Monte Carlo simulation	354
Multiple cropping	299,323
Multipurpose reservoir	436,437
N	
Nash model	353
Natural channels	356,357
Natural flow	4,159,199
Natural disasters	34,287,323
Natural events	158,170
Navigation canals	338
Non-food crops	300
Non-linear relationship	300,301
Non-modular canal outlets	376
Non-recording rain gauges	48,78
Non-structural flood control measures	292
Normal ratio method	56,79
Nuclear power	287
O	
Optical rain gauge	50
Orographic precipitation	37,83,85
P	
Pan coefficient	107,144
Peak flow estimation	347
Penman's equation	115
Percolation	115,129,165,276
Permeability	156,159,219

	Page No.
Penman-Monteith equation	119,150
Permafrost	15,19
Permanent wilting point	111,144,324
Perennial stream	159,219
pH value	302
Philip's model	138
Piezometric surface	228,272
Pluvial flooding	288
Point rainfall	62,66,88
Pollution sources	295
Porosity	305,324
Potential evapotranspiration	111,143
Precipitation	36
Precipitation intensity and duration	156,217
Precision irrigation	303,318
Probabilistic methods	353
Probable maximum precipitation	36,79
Pumped storage hydropower	285
Q	
Quality analysis	294
Quality of irrigation water	281,303,383
Quantity analysis	294
Quick flow	164,165
R	
Rabi season	299
Radial gates	434,435
Raichur method	448
Rainfall data in India	76,78
Rainfall - depth - duration frequency	74

	Page No.
curve	
Rainfall distribution	63,89,350
Rainfall excess	147,174,351
Rain gauges	49,50,90
Rainfall intensity	50,51,87,141
Rainfall-runoff models	206,352
Rainfall simulator	333
Rainfall variability	77
Rate of evaporation	95,97,102,48
Rational method	222,348
Recession limb	163,171
Recording Rain gauges	48,49,78
Reference crop evapotranspiration	119,124
Reforestation	292
Regional variations	300
Regulation structures	206,340,344
Reservoir capacity	410,438,440
Reservoir evaporation	106
Reservoir sedimentation	410,446
Retreating monsoon season	43
Retention basin	292
Return period	71,346
Rigid boundary channel	357,383
Riparian zone	217,356
Rising limb	163,167,217
River-aquifer interaction	171
River channels	206,345,368
Riverine floods	287
Rock-fill dam	397,399

	Page No.
Rohwer's formula	100,102
Root depth	151,314
Root zone management	308
Root zone soil water	281,306,322
Rugosity coefficient	361,365
Runoff classification	158
Runoff coefficient	27,144,218
Runoff curve number	183,191,220
Runoff potential	179,182
Runoff volume	14,161,176
S	
Safe yield	438,441
Salinity	7,150,302
Satellite-based Rain Gauge	51
Saturated vapour pressure	105
Saturated zone	129,225
SCS-CN method	177,183
Seasonal cropping patterns	298
Seasonal variation	157,221,294
Seasons in India	78,323
Secondary yield	438
Sedimentation	4,464
Sediment distribution	449,451
Sediment load	170,222,302,358
Seepage analysis	402,405
Seepage losses	144,342,380
Semi-confined aquifers	228
Semi-modular canal outlets	377
Siphons	373

	Page No.
Silting rate	443,448
Simple (tube type) infiltrometer	133
Single well test	268
Sleet	84,156
Slichter's formula	237,238
Slope protection	407,457
Sluice gates	373,385
Snowboard	51
Snow course	52
Snow depth sensor	52
Snow gauges	51
Snowmelt dynamics	169
Snow pillow	52
Snow stake	51
Soil aeration	381
Soil characteristics	24,111,314
Soil salinity	150,381
Soil porosity	305
Soil structure	157,305,309,371
Soil properties	8,89,304
Soil texture	111,305,330
Soil-water relationship	304,324
Solar power	32,287
Specific capacity	234,235,251
Specific capacity index	235
Specific retention	231,273
Specific yield	231,236,273
Spillways	236,243,460
Springs	16,225,227

	Page No.
Sprinkler irrigation	284,315
Storage reservoirs	290,437
Storability	232,272,274
Storm analysis	68
Storm surges	287,288
Stormwater management	22,90,211,354
Streamflow distribution	71
Structural losses	342,343
Structural flood control measures	292
Sublimation	9
Surface drainage	382
Surface irrigation	284,315,321
Surface retention	129,130
Surface runoff	3,131,152,309
Surface water resources of India	199
Synthetic unit hydrograph	211,215,222
Syphon's rain gauge	49
T	
Tennant method	204
Theis equation	259
Thermal power plants	286
Thiem's equation	242
Thiessen polygon method	63,89
Time drawdown analysis	262
Tipping bucket rain gauge	49,86
Trace elements	303
Transboundary cooperation	293
Transmissivity	232,265,272
Transpiration	310,330

	Page No.
Trap efficiency	443,460
Triangular texture diagram	304,324
Trickle irrigation	316
Types of aquifers	229,272
Types of Evaporimeters	95,149
Types of hydropower	285
U	
Unconfined aquifer	224,279
Unit hydrograph approach	209,352
Uplift forces	410,414
Urban flooding	288,309
Urban storm-water canals	338
Urban water logging	379
V	
Vapour pressure	93,120,483
Variation of AET	112,113
Variation of PET	122
Vegetative measures	382
Vertical lift gates	435,457
W	
W- index	142
Water balance equation	1,5,177,214
Water budgeting	112,123,294
Water conveyance system	296,301,345
Water cycle	7,15,193,308
Water demand	281,296,371
Water efficiency	317,318
Water flow control	301
Water for agriculture	284,315,335

	Page No.
Water holding capacity	307
Water logging	379,385
Water management practices	21,206,298,319
Water requirement of crops	297
Water resources at a glance	200
Water resources planning	211
Water resources management	22,134,300,355
Watershed area	67,349
Watershed characteristics	171,345
Water supply canals	338
Water treatment	437,462
Water quality management	33,158,303
Water quality testing	294
Water withdrawals and uses	281,322
Wave pressure	411
Weather conditions	38,293,312,374
Weather stations	25,52,114
Weighing bucket rain gauge	50
Weirs	337,343,372
Well loss	251,270,273
Wilting point	111,144,325
World water balance	15,35
Y	
Yield-drawdown curve	253